WEED SCIENCE:
PRINCIPLES AND PRACTICES

Weed Science:

PRINCIPLES AND PRACTICES

Eli Lilly and Company, Greenfield, Indiana
[Formerly Professor, North Carolina State University]

FLOYD M. ASHTON

Professor of Botany
University of California at Davis
Davis, California

With the editorial assistance of

LYMAN J. NOORDHOFF

Communications Specialist, Extension Service
United States Department of Agriculture
Washington, D. C.

A WILEY-INTERSCIENCE PUBLICATION

JOHN WILEY & SONS, New York • Chichester • Brisbane • Toronto

Library of Congress Cataloging in Publication Data:

Klingman, Glenn C.
  Weed science.

  "A Wiley-Interscience publication."
  "A revision of Weed control: as a science, originally published . . . in 1961."
  Includes bibliographies and index.
  1. Weed control. 2. Herbicides. I. Ashton,
Floyd M., joint author. II. Title.
SB611. K55 1975   632'. 58 75-8908
ISBN O-471-49171-3

Printed in the United States of America

10 9 8 7

Preface

Weeds affect almost everyone. Weed control is one of the most expensive steps in crop production—directly affecting the price of food. Weeds may poison or seriously slow down weight gains of livestock. They cause allergies such as hay fever and poison ivy. They infest home lawns and gardens. Weeds create problems in recreation areas such as golf courses, parks, and fishing and boating areas. They are troublesome along highways, railroads, in industrial areas, and in irrigation and drainage systems.

The science of weed control has advanced more during the last 33 years (since 1942) than in the previous 100 centuries. This book brings the reader up to date in this fast-moving science. Old and so-called reliable methods of weed control are blended with the very newest chemical techniques. In many cases, new chemical methods are far superior to older practices in terms of control as well as reduced costs.

Progressive farmers have rapidly accepted new chemicals for weed control. 2,4-D was first used on a large scale in 1947. Ten years later, 92% of the farmers in certain progressive farming areas depended regularly upon herbicides.

Progress in research in any science can be partly measured by the number of research articles published; *Weed Abstracts* lists 3071 papers for 1973 alone. We have prepared a representative bibliography, but it has been impossible to acknowledge adequately the vast number of workers and the vast number of papers published in the field. This book brings together the modern philosophy and techniques of weed control. It is not intended as a complete review of the literature.

This textbook is designed principally for classroom instruction in the principles and practices of weed science. It will also be helpful to research scientists, extension specialists, county agents, vocational agriculture teachers, herbicide development representatives, and farmers.

This book is intended for worldwide usage but it cannot delve into the varied cultural practices and conditions encountered in different countries. Therefore, nothing in this book is to be construed as recommending or authorizing the use of any particular weed-control practice or chemical in any given area. Authority and instructions for the use of any chemical in a particular locale must be obtained from labels approved by regulatory officials of the appropriate country or political subdivision. Read and follow instructions on the chemical label. No warranties, expressed or implied, are made with respect to any chemical or weed-control practice discussed in this book.

Trade names have been used in some cases along with chemical names and abbreviations. Trade names have been given as a convenience to the reader, not as an endorsement of any one product. Other products of the same composition would be equally effective. Also, mention or omission of any chemical does not constitute either endorsement or criticism of a product.

Although we accept personal responsibility for the information presented, we are indebted to the following authorities for reading different parts of the book and for helping to make it technically correct: E. F. Alder, Indianapolis, Indiana; N. B. Akesson, University of California, Davis; D. E. Bayer, University of California, Davis; J. F. Bell, Indianapolis, Indiana; W. B. Duke, Cornell University, Ithaca, New York; B. J. Eaton, Greenfield, Indiana; C. L. Elmore, University of California, Davis; P. A. Frank, U.S.D.A., A.R.S., University of California, Davis; J. R. Hay, Canada Agriculture Research Station, Regina, Saskatchewan; J. W. Hooks, Indianapolis, Indiana; D. L. Klingman, U.S.D.A., A.R.S., Beltsville, Maryland; R. F. Norris, University of California, Davis; S. J. Parka, Indianapolis, Indiana; G. W. Probst, Indianapolis, Indiana; R. Romanowski, Purdue University, West Lafayette, Indiana; P. Santelman, Oklahoma State University, Stillwater; T. J. Sheets, North Carolina State University, Raleigh; F. W. Slife, University of Illinois, Urbana; T. W. Waldrep, Greenfield, Indiana, and W. L. Wright, Greenfield, Indiana.

We also wish to express appreciation to the chemical manufacturers for providing technical data concerning their products.

This book is a revision of *Weed Control: As A Science* originally published by the senior author in 1961. Responsibility for this revision has been largely assumed by Professor Floyd M. Ashton of the University of California at Davis. Editorial assistance has been provided by Lyman J. Noordhoff. The views expressed are solely those of the authors; such views are not necessarily accepted or endorsed by the authors' employers.

We are grateful to our wives Loree Klingman, Phyllis Ashton, and Ruth Noordhoff for their patience and assistance during the preparation of the manuscript.

GLENN C. KLINGMAN
FLOYD M. ASHTON
LYMAN J. NOORDHOFF

March 1975

Contents

WEED SCIENCE:
PRINCIPLES AND PRACTICES

1 Introduction

Weed control is as old as agriculture. It is one of the most expensive steps in crop production. In a way of life that has learned to control almost everything, it is of scientific interest that man has done so little to control this most persistent problem. Until recently the task surpassed even man's imagination, and the only solution was to battle weeds with brute force.

Slowly, man has learned to mechanize and use power in his fight. The first step was the substitution of a sharpened stick for his fingers; then followed the hoe, which in turn was replaced by the cultivator and plow. The idea of planting crops in rows to permit "horse-hoeing" originates with Jethro Tull (1731), author of *Horse Hoeing Husbandry*. He also was among the first to use the word "weed" in its present spelling and meaning. The horse was gradually replaced by the tractor. Now chemical energy is partially replacing mechanical energy for weed control. Because cultivation requires tremendous amounts of energy in terms of equipment, fuel, and manpower, and there is a need for increased food production, it is important to develop and use improved methods of weed control.

DEFINITION OF A WEED

A weed is a plant *growing where it is not desired*; or a *plant out of place*. Therefore, rye in a wheat field is a weed; so is a cornstalk in a peanut field. Weeds encompass all types of undesirable plants—trees, broadleafed plants, grasses, sedges, rushes, aquatic plants, and parasitic flowering plants (dodder, mistletoe, witchweed).

TYPES OF WEED COSTS

Weeds are everybody's business—including every city resident. Weeds directly affect the cost of food, and the health and comfort of people. Weed

1

Figure 1-1. Weeds usually reduce the yields of crop plants. Left: Alfalfa growing in normal soil. Center: Alfalfa watered from a filtrate drained from ground quackgrass rhizomes. Right: Alfalfa grown in soil containing ground quackgrass rhizomes (T. Kommedahl, University of Minnesota).

losses are encountered nearly everywhere in agriculture and in out-of-door industries. To better appreciate these huge losses, consider the following types of weed costs.

Lower Yields, Less Efficient Use of Land

...Yields may be reduced in terms of crops, pastures, meat, milk, and wool. Weeds not only reduce crop yields by "chemical warfare", as shown in Figure 1-1 and 2-2, but also by competition with the crop for soil water, soil nutrients, carbon dioxide, and light.

...Efficiency is reduced because costs are increased through cultivation, hoeing, mowing, and spraying.

...Costs are increased on rights-of-way for railroads, highways, power and telephone transmission lines.

...Land values may be reduced.

...Crop choice may be limited. Some crops will not compete effectively against heavy weed growth.

...Harvesting costs may be increased—by both hand and machines.

...Root and crop damage may result from cultivation. Soil structure may be destroyed by repeated cultivation, especially when wet.

...Reforestation costs may be increased because of loss of stands as well as a slower rate of growth as a result of weed competition.

Added Protection Costs from Insects and Diseases

Weeds may harbor insect and disease organisms that attack crop plants. For example, the carrot weevil and carrot rust-fly may be harbored by the wild carrot, only to attack later the cultivated carrot. Aphids and cabbage root maggots may live in mustard, and later attack cabbage, cauliflower, radish, and turnips. Onion thrips live in ragweed and mustards and may later prey on the onion crop. The disease of curly top on sugar beets is carried by insect vectors that live on weeds in waste lands. Many insects overwinter in weedy fields and field borders.

Disease organisms such as black stem rust may use the European barberry, quackgrass, or wild oats as hosts prior to attacking wheat, oats, or barley. Some virus diseases are propagated on members of the weedy nightshades. For example, the virus causing "leaf roll" of potatoes lives on black nightshade. It is thought that aphids carry the virus to potatoes.

Poorer Quality Products

...All types of crop products may be reduced in quality. Weed seeds and onion bulblets in grain and seed, weedy trash in hay and cotton, spindly "leaf crops," and scrawny vegetables are a few examples.

...Livestock products may be lower priced or unmarketable from weeds, for example, onion, garlic, or bitterweed flavor in milk, and cocklebur in wool. Poisonous plants may kill animals, slow down their rate of growth, or cause many kinds of abnormalities (see Figure 1-2).

More Problems in Water Management

Weeds are important in irrigation and drainage systems. They affect recreation and fishing on ocean beaches, lakes, and farm ponds (see Figure 1-3). They may give off undesirable flavors and odors in public water supplies, as well as affecting the shipping use of our inland waterways and harbors.

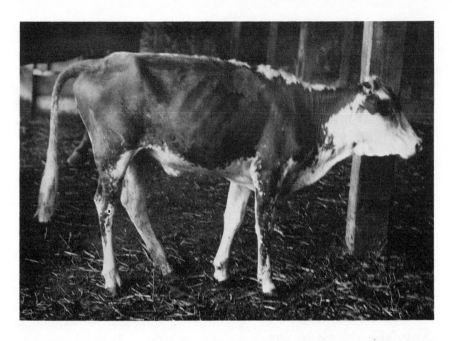

Figure 1-2. Weeds may reduce the yield of livestock products. The animal was fed prostrate spurge. The abnormal stance and digestive disturbances became more pronounced as the spurge feeding continued. (O. E. Sperry, Texas A & M University.)

Figure 1-3. In St. Cloud, Florida, the State Mosquito Control Board sprayed ditches with diuron to eliminate weeds and improve drainage. The weed-control program reduced the expenses of mosquito control enough to permit a one-third savings in the total budget. Left: 1 year after hand cleaning. Right: 1 year after chemical treatment with diuron. (E. I. du Pont de Nemours and Company.)

4

Less Human Efficiency

Weeds reduce human efficiency through allergies and poisoning. Hay fever, caused principally by pollen from weeds, alone accounts for tremendous losses in human efficiency every summer and fall. Poison ivy, poison oak, and poison sumac cause losses in terms of time and human suffering; children occasionally die from eating poisonous plants or fruits.

Weed control comprises a large share of a farmer's work required to produce a crop. This effort directly affects the cost of crop production, and thus the cost of food. In other words, it affects all of us, whether we farm or not.

Farming efficiency has permitted the release of agricultural workers for other means of livelihood. In 1850, about 65% of the work force was employed in agriculture. By 1870, this number had dropped to 52%, 1890 to 42%, 1910 to 31%, 1930 to 11%, and in 1970 to about 5%. Many of the "other 95%" are serving as doctors, teachers, businessmen, servicemen, scientists, and industrial workers. This is possible only because they no longer are needed to grow food. A nation's economy is directly related to agricultural/industrial worker ratio.

Farm output has risen steadily since 1910, the fastest rise occurring after 1940. A farm worker in 1970 produced, on an average, as much in 1 hr as he did in 3 hr in 1950, 5 hr in 1940, and 7 hr in 1910. In 1971, one agricultural worker produced enough agricultural products for 47 persons (see Figure 1-4). It has been said that one worker in a chemical factory producing 2,4-D is equivalent to the work efficiency of 100 hoe-hands in a cornfield.

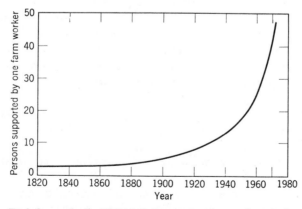

Figure 1-4. Each farmer in the United States produced enough agricultural products to provide for the number of nonfarmers shown in the graph. [Adapted from *U.S.D.A Agricultural Handbook No. 423* (2).]

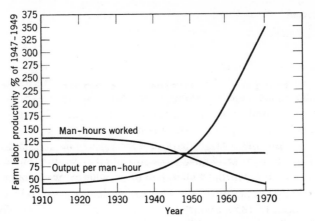

Figure 1-5. Farmers in the United States are working fewer hours and producing more products per hour of work than ever before. [Adapted from *U.S.D.A. Agricultural Handbook No. 423* (2).]

The total number of hours worked in agriculture has continued to decline, despite increased production. This increase in production efficiency is a result of modern agricultural technology involving research, as well as educational programs. Weed control is an important part of this technology (see Figure 1-5).

Weeds have been a plague to man since he gave up a hunter's life. Traveling in developing nations, one may feel that half the world's population work in the fields, stooped, moving slowly, and silently weeding. These people symbolize the great mass of humanity that spends a lifetime simply weeding. Many young people doing such work in Africa, Asia, and Latin America can never attend school; women do not have time to prepare nutritious meals or otherwise care for their families (7). Modern weed-control methods integrated into the economies and cultures of developing nations provide relief from this arduous chore and give opportunity to improve their standard of living through more productive work.

FARM LOSSES FROM WEEDS

Farm losses from weeds are much higher than generally recognized. As such, they represent a major cost item in food production. Because weeds are so common and so widespread, people do not fully appreciate their significance in terms of losses and control costs. Weeds are common on all of our 485 million acres of cropland and almost one billion acres of range and pasture. The whole system of weed control is so intermixed with so-called standard

Table 1-1. Annual Costs of Plant Pests of Crops [Adapted from (3)]

	Losses (× 1,000)	Control (× 1,000)	Total (× 1,000)	Total (%)
Diseases	$3,152,815	$115,000	$3,267,815	27.1
Insects	2,965,344	425,000[1]	3,390,344	28.1
Nematodes	372,335	16,000	388,335	3.2
Weeds	2,459,630	2,551,050	5,010,680	41.6
Total	8,950,124	3,107,050	12,057,174	100.0

[1] Includes control costs for man, animals, and households as well as crops.

practices that separate accounting is difficult. For example, the original reason for planting crops in rows has long since been forgotten. Just how much of plowing and crop cultivation is for weed control, and weed control alone?

The total cost to agriculture as a result of pests (Table 1-1) is slightly over $12 billion/year. Of this amount, the annual cost resulting from plant diseases is about 27%, insects 28%, nematodes 3%, and weeds 42%.

Since the early 1960s there has been a major shift in relative sales among types of pesticides (see Figure 1-6). Dollar value of herbicide sales is now greater than the combined value of insecticides and fungicides.

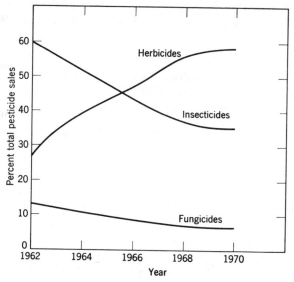

Figure 1-6. Value of pesticide sales. [Adapted from *U.S.D.A. Pesticide Review*, 1971 (3).]

The above weed-control losses are agricultural. If we add the losses to suburban lawns and gardens, public parks and other recreation areas, rights-of-way of highways, railroads, telephone, and electric power transmission lines, industrial plants, lakes, ponds, ditches, and tidewater areas, and public health, the figures become almost unbelievable.

RESEARCH EFFORT

This discussion and Table 1-2 are not meant to depreciate the contributions of entomology and plant pathology to human welfare but to gain a perspective for weed science. Entomology and plant pathology developed as an agricultural science more than 100 years ago. Each has saved billions of dollars annually by reducing damage and can be expected to continue to do so.

Table 1-2 shows that there are far fewer workers and less money for weed science than for insect and disease control. Personnel are categorized into basic research, applied research, extension, graduate teaching, and undergraduate teaching. In the 49 Land-Grant Universities that participated in the October 1969 survey, there were 4.3 times as many professional entomologists and 3.6 times as many plant pathologists as weed scientists.

The future for weed control is extremely promising, judged by losses caused by weeds and the vast potential savings to agriculture and our whole economy. But this can be true only if enough scientists, properly trained, specialize in weed control. Perhaps no other science offers greater potential service to agriculture or to food production efficiency during the next 50 years.

Table 1-2. The number[1] of staff positions, instructor and above, in 49 Land Grant Colleges in the United States in October, 1969 (9)

	Entomology	Plant Pathology	Weed Science
Basic research	232	221	59
Applied research	246	207	75
Extension	130	82	41
Graduate teaching	140	131	19
Undergraduate teaching	100	65	1
Total number	848	706	195

[1] Number given in terms of full-time commitment, using 1 for one person's full-time commitment.

SOCIOLOGICAL BENEFITS OF WEED CONTROL

The sociological benefits resulting from the use of agricultural chemicals are listed (7) as follows:

1. The physical burden of farm work is lessened because the drudgery of hand labour is reduced.
2. Farm children are freed and can attend schools.
3. Fewer people are required to produce food. Sons and daughters of farmers are now surgeons, teachers, researchers, legislators, manufacturers, writers, or labor-unions leaders.
4. The percent of money spent on food is reduced, leaving a larger percentage of the income available for other uses.
5. Improved land use.
6. Health benefits.

HISTORY OF WEED CONTROL

For centuries man fought weeds with his hands, or used sharp sticks, hoes, animal-powered cultivators, and finally mechanical power. As for chemicals, he has used sea salt for ages to kill all plant life. Only about 1900 did he start using purified chemicals for selective weed control. And wide use of selective herbicides began only in the mid-1940s, after the discovery of of 2,4-D [(2,4-dichlorophenoxy) acetic acid].

The science of weed control has advanced more since 1942 (the first use of 2,4-D) than in the previous million years. But chemical weed control is still in its infancy when compared to plant breeding or soil science, for example. Most of us will see the day when a specific chemical will be used for a specific crop. The chemical or chemical combinations will inhibit the growth of all plants but the desired crop.

The following is a listing of the major figures and events that have laid the foundations for our present weed control programs.

Julius Sachs, a German botanist, conducted many experiments (1859–1887) to learn the factors involved in the rooting and flowering of plants. He proposed that minute amounts of root-forming and flower-forming substances were moving either upward or downward in the plant. He wrote that "chemical messengers" were related to the flowering behaviors of begonias and squash. This is one of the earliest articles on translocation of growth-regulating substances.

Charles Darwin, an English biologist better known for his theory of evolution, wrote the book *Power of Movement in Plants* published in 1900. In it he reported plant movement in response to light. Darwin germinated oats and Phalaris in darkness and then exposed the coleoptiles to light coming from one direction. The coleoptiles bent toward the light. By covering with tin-foil or by painting different areas of the coleoptile with india ink, he determined that the *coleoptile tip produced the growth-regulating substance.* This substance moved downward, causing curvature below. He found that the stem tips of Cruciferae (mustards) and Chenopodiaceae (goosefoot) also produced growth-regulating substances. Darwin wrote: "In several respects, light seems to act on plants in nearly the same manner as it does on animals by means of the nervous system ...the effect... is transmitted from one part to another."

1908:   Bolley (United States) reported successful weed control in wheat using common table salt, iron sulfate, copper sulfate, and sodium arsenite. In his words: "When the farming public has accepted this method (selective weed control) of attacking weeds ...the gain to the country at large will be much larger in monetary consideration than that which has been afforded by any other single piece of investigation applied to field work in agriculture."

1941:   R. Pokorny (United States) reported the chemical synthesis for 2,4-D. It was later tried by other research workers as a fungicide and insecticide and found ineffective.

1942:   P. W. Zimmerman and A. E. Hitchcock (United States) first reported 2,4-D to be a growth substance.

1944:   P. C. Marth and J. W. Mitchell (United States) established the selectivity of 2,4-D. Dandelion, plantain, and other broadleaved weeds were removed from a bluegrass lawn. C. L. Hamner and H. B. Tukey (United States) successfully used 2,4-D in field weed control.

1945:   W. G. Templeman (England) established the preemergence principle of soil treatment for selective weed control.

1951:   *Weeds,* the journal of the Association of Regional Weed Control Conferences, was published.

1956:   Weed Science Society of America was organized and assumed publication of *Weeds.* The publication was renamed *Weed Science* in 1968.

1961:   *Weed Research,* official journal of the European Weed Research Council, was published.

1968:   *Weed Control, Principles of Plant and Animal Pest Control,* Vol. 2, Publication 1597, was published by the prestigious National Academy of Sciences—thus confirming the importance of weed science as a discipline.

1970:   The International Conference of Weed Control was organized by the Food and Agricultural Organization (FAO) of the United Nations. This meeting shows the international concern for solutions to the world's weed problems.

PREVENTION, CONTROL, AND ERADICATION

Prevention

Prevention means stopping a given species from contaminating an area. Prevention is often the most practical means of controlling weeds. This is best accomplished by (1) making sure that new weed seeds are not carried onto the farm in contaminated crop seeds, feed, or on machinery; (2) preventing weeds on the farm from going to seed; and (3) preventing the spread of perennial weeds which reproduce vegetatively.

Control

Control is the process of limiting weed infestations. In crops the weeds are limited so that there is a minimum of weed competition. Thus, the amount of control is usually balanced between the costs involved and the amount of possible injury to the crop. Control is the usual aim toward annual weeds competing with farm crops.

Eradication

Eradication is complete elimination of all live plants, plant parts, and seeds from an area. Two main problems are involved: (1) eliminating the living plants and (2) extermination of seeds in the soil. Because seeds may remain dormant in the soil for many years, it is usually easier to eliminate living plants than seeds. For eradication, both must be exterminated.

Usually serious perennial weeds are first located in small areas on the farm. Effective soil sterilants applied at this time may prevent them from spreading.

WEEDS CLASSIFIED

The plant's life span, season of growth, and its methods of reproduction largely determine the methods needed for control or eradication. In temperate climates there are three principal groups: annuals, biennials, and perennials.

Annuals

Annual plants complete their life cycle in less than one year. Normally they are considered easy to control. This is true for any one crop of weeds. However, because of an abundance of dormant seed and fast growth, annuals are very persistent, and they actually cost more to control than perennial weeds. Most common field weeds are annuals. There are two types–summer annuals and winter annuals.

Summer Annuals

Summer annuals germinate in the spring, make most of their growth during the summer, and the plants mature and die in the fall. The seeds lie dormant in the soil until the next spring. Summer annuals include such weeds as cockleburs, morning glories, pigweeds, common lambsquarters, common ragweed, crabgrasses, foxtails, and goosegrass. These weeds are most troublesome in summer crops like corn, sorghum, soybeans, cotton, peanuts, tobacco and many vegetable crops.

Winter Annuals

Winter annuals germinate in the fall and winter and usually mature seed in the spring or early summer before the plants die. The seeds often lie dormant in the soil during the summer months. In this group, high soil temperatures (125°F or above) have a tendency to cause seed dormancy—inhibit seed germination.

This group includes such weeds as chickweed, downy brome, hairy chess, cheat, shepherdspurse, field pennycress, corn cockle, cornflower, and henbit. These are most troublesome in winter-growing crops such as winter wheat, winter oats, or winter barley.

Biennials

A biennial plant lives for more than 1 year but not over 2 years. Only a few troublesome weeds fall in this group. Wild carrot, bull thistle, common mullein, and common burdock are examples.

There is confusion between the biennials and winter annual group, because the winter annual group normally lives during 2 calendar years and during 2 seasons.

Perennials

Perennials live for more than 2 years and may live almost indefinitely. Most reproduce by seed and many are able to spread vegetatively. They are classified according to their method of reproduction as *simple* and *creeping*.

Simple Perennials

Simple perennials spread by seed. They have no natural means of spreading vegetatively. However, if injured or cut, the cut pieces may produce new plants. For example, a dandelion or dock root cut in half longitudinally may produce two plants. The roots are usually fleshy and may grow very large. Examples include common dandelion, dock, buckhorn plantain, broadleaf plantain, and pokeweed.

Creeping Perennials

Creeping perennials reproduce by creeping roots, creeping above-ground stems (stolons), or creeping below-ground stems (rhizomes). In addition, they may reproduce by seed. Examples include red sorrel, perennial sow thistle, field bindweed, wild strawberry, mouseear chickweed, ground ivy, bermudagrass, johnsongrass, quackgrass, and Canada thistle.

Some weeds maintain themselves and propagate by means of tubers, which are modified rhizomes adapted for food storage. Nutsedge (nutgrass) and Jerusalem artichoke are examples.

Once a field is infested, creeping perennials are probably the most difficult group to control. Cultivators and plows often drag pieces about the field. Continuous and repeated cultivations, repeated mowing for 1 or 2 years, or persistent herbicides are often necessary for control. Cultivation in combination with herbicides is proving effective on some weeds. An effective eradication program also requires the killing of seedlings.

METHODS OF WEED CONTROL

There are six principal methods of weed control: mechanical, crop competition, crop rotation, biological, fire, and chemical.

Often the best and most economical way to control weeds is the wise combination of two or more of these methods. For example, you might combine mechanical and chemical, as in Figure 1-7, for 1 year; or you could continue any appropriate combination for several years.

Figure 1-7.　Shielded sweeps killed weeds between the rows and a parallel action-spray applicator applied the herbicide to the weeds in the row. See also Figure 1-12. (O. B. Wooten, Delta Branch Experiment Station, Stoneville, Miss.)

Mechanical

Mechanical control involves two well known weed-control techniques—tillage and mowing.

Tillage

One type of tillage is *burial*. This is effective on most small annual weeds. If all growing points are buried, most annual weeds are killed. Burial is only partly effective on weeds with underground stems and roots that are capable of sprouting; examples are field bindweed, Canada thistle, quackgrass, bermudagrass, johnsongrass, and nutsedge. For eradication, such perennials must be repeatedly cut off or buried until the underground parts are killed by carbohydrate starvation. (See Figure 1-9.)

The second method of tillage is the *disturbed rooting system*. Shallow cultivation equipment such as sweeps, knives, harrows, finger weeders, and rotary hoes, is used. The objective is to loosen or cut the root system so the plant dies from desiccation (drying out) before it can reestablish its roots. Small weeds are easily controlled by tillage, and it is most effective in hot, dry weather with dry soils. In moist soils, or if it rains soon after tillage, the roots may quickly reestablish themselves. In effect the weed is transplanted with little or no injury. Some herbicides reduce the ability of the weed's root system to reestablish itself, greatly increasing the effectiveness of cultivation.

Most serious perennial weeds are easily destroyed by tillage as seedlings, but are difficult to kill after they develop rhizomes, stolons, tubers, or reproductive roots. Combination of tillage with certain chemicals often increases the degree of control.

One of the most important reasons for growing crops in rows, instead of broadcast, is to permit tillage or cultivation. Only one benefit from cultivation stands out as distinct and clear. This is the benefit of weed control.

Other benefits for cultivation are occasionally claimed, such as increased soil aeration, the breaking of surface crusts, and increased rainfall penetration. In some cases these may be enhanced, in other cases hindered.

Cultivation, especially late deep cultivation, may injure crop roots. It may also cause rapid moisture loss and rapid drying of the soil. Cultivation is nearly always associated with a decreasing organic-matter content in the soil.

Experiments with corn have shown that if weeds are controlled without cultivating, at least on certain soils, the yields will equal those of plots cultivated as often as four times during the season. In New York, Sturtevant and Lewis (11) compared different cultivation treatments for corn as early as 1886. Here are their results:

Treatment	Yield per Acre
Free weed growth	18.1 bushels
Weed removal, no cultivation	70.5 bushels
Ordinary cultivation plus hoeing	56.8 bushels

The New York report states, "Strangely enough, we have, during the existence of this station, not been able to obtain decisive evidence in favor of cultivation."

The results of two Illinois studies on cultivation of corn are given in Table 1-3. In the first experiment the weed-scraped plots yielded best for 3 years out of 7, whereas plots given ordinary shallow cultivation yielded highest during the other 4 years; but these differences are probably not significant. Deep cultivation was not better than scraping the weeds off with a hoe.

In the later study (1916–1922), blades were used to stir the soil from 1 to 1.5 in. deep and shovels to stir the soil from 2 to 3 in. deep. The soil in these

Table 1-3. Methods of Cultivating Corn Compared (10, 13)

	Bushels per Acre	
Nature of Cultivation	Average (1888–1893 and 1896)	Average (1916–1922)
Weeds allowed to grow	—	7
None—scraped with hoe	68	53
Shallow—scraped 4 or 5 times	70	—
Deep—scraped 4 or 5 times	67	—
Shallow—scraped 12–14 times	73	—
Deep—scraped 12–14 times	65	—
Cultivation blades	—	53
Cultivation shovel	—	51

studies was a brown silt loam typical of the gently rolling corn belt land in Illinois.

Missouri studies (6) on corn production, summarized in Table 1-4, show more favorable results from surface scraping than from any other cultivation technique tested.

Table 1-4. Cultivation of corn at Various Depths Compared with No Cultivation (6) (bushels per acre)

Kind of Cultivation	Number of Cultivations	Warrensburg 3-Year Average	Maryville 3-Year Average	Springfield 1-Year Average	Cumulative Average
Shallow	4	23.8	45.7	28.4	37.6
Late	6	23.2	37.9	28.3	29.8
Deep	4	21.8	33.7	18.0	23.5
None—surface scraped	0	25.6	58.9	34.9	39.8

Cates and Cox (5) reported corn yields of 124 cooperating farmers in various parts of the United States. They compared ordinary cultivation with no cultivation, where weeds were scraped off with a hoe.

In these 124 trials, when the yield of grain from ordinary cultivation was given a value of 100, the yield on the uncultivated or weed-scraped plots averaged 99.1.

In many trials the soil type was named. The types were grouped into four classes—medium black loam, heavy black loam, clay, and sand—with these comparative yields:

	Cultivated Plots Gave	
	Larger Yields	Smaller Yields
Medium black loam	12 times	5 times
Heavy black loam	11 times	10 times
Clay	15 times	8 times
Sand	14 times	12 times

Kiesselbach et al. (8) state that weed control is the main consideration in corn cultivation. In their experiments, late cultivation tended to reduce the yield.

In all of the above studies the investigators had to disturb the soil to some extent by scraping, hoeing, or by hand-pulling the weeds. These may produce the effect of shallow cultivation.

Today, herbicides make it possible to grow weed-free corn without scraping and hoeing. Research work in North Carolina was conducted by the senior author on three distinct soil types. They were: Tidewater soil high in organic matter, which remains moderately loose and friable when dry; a Coastal Plain sandy loam which remains loose and friable even though dry; and a Piedmont clay loam which becomes hard and cracked when dry.

There were three treatments on each soil type:

1. No cultivation and sprays—gave weed-free corn.
2. One cultivation and sprays to control weeds.
3. Three cultivations.

These tests show there is no benefit from cultivation, beyond weed control, on the Coastal Plain sandy loam soil which remains loose and friable even though dry. Yields of corn were essentially the same, with or without cultivation, provided the weeds were controlled.

Results were similar on the Tidewater soil, with one related finding: there appeared to be a trend toward favoring one early, shallow cultivation, especially in a dry year.

The Piedmont clay loam soil, which becomes very hard and compacted when dry, showed a small advantage, above and beyond weed control, from one early cultivation in a dry season. But in seasons with favorable moisture these differences disappeared. In all cases, one cultivation when the corn was 12–15 in. tall was as effective in favoring high corn yields as three cultivations.

Figure 1-8. Corn averaged 131 bushels/acre for 4 years, without cultivation. The experiment was performed on North Carolina sandy loam soil, and weeds were controlled by the use of herbicides. (North Carolina State University.)

To generalize, porous, well-aerated soils show no benefit from soil tillage beyond weed control. If you control the weeds without corn injury, yields on noncultivated soil will equal those of cultivated corn. Late cultivation, which injures the corn roots, is likely to reduce corn yields.

Increased corn growth and corn yields may be expected from one early cultivation on soils that become hard and compacted when dry—in a dry year. There is no advantage, other than weed control, in giving three cultivations as compared to one.

We have ample proof—from nature—that plants can flourish without cultivation. The proof is simply that many plants grow abundantly under natural conditions. Some of these soils have never been disturbed but still produce large, healthy plants.

Mowing

Mowing is effective on tall-growing plants, but not on short-growing plants. Tall annual weeds are mowed primarily to reduce competition with crop plants and to prevent seed production.

Repeated mowing not only prevents seed production of perennial weeds, but also may starve the underground parts. To control tall perennial weeds, repeated and frequent cutting may be required for 1–3 years. At no time must the plant be permitted to replenish its underground stored food supply. The

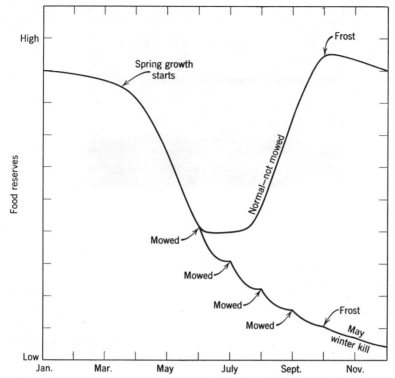

Figure 1-9. Food reserves of a perennial unmowed plant, compared with a repeatedly mowed plant.

best time to start mowing or cultivation is usually when the underground root reserves are at a low ebb. For many species this is between full leaf development and the time when flowers appear during late spring (see Figure 1-9).

Stem tips produce growth-inhibiting substances. These substances hinder growth of buds on the lower stem and on below-ground stems and roots. Thus, the stem apex (tips) produce substances which tend to dominate the growth of lateral buds. This is known as *apical dominance.*

Following mowing or cultivation, apical dominance is no longer present. The dormant buds may start to grow, with more new stems developing than were originally present. This may appear to thicken the stand; however, it is desirable—if you mow this new growth repeatedly. The new stems grow at the expense of the below-ground stored food, and repeated cutting hastens food depletion and death of the plant.

Annual weeds are often mowed to prevent seed production. Mow when the first flowers appear because some weed seeds will germinate even though you cut the plant soon after pollination. (See Chapter 2.)

Some annual weeds sprout new stems below the cut. You can often control this growth by cutting rather high at the first mowing and enough lower with the second mowing to cut off the sprouted stems. By the second mowing the stem is often hard and woody and cannot put out new sprouts below the cut. This procedure is effective on bitter sneezeweed, horseweed, and many others.

Weeds that form rosettes, mats, or a close-growing sod cannot be controlled by mowing. In fact, mowing tends to favor such weeds because it removes competition. Also, weeds that produce seed heads close to the ground are not controlled by mowing. That is why mowing does not kill established plants such as common dandelion, buckhorn plantain, dock, bermudagrass, crabgrass, goosegrass, and the foxtails.

Crop Competition

Crop competition is one of the cheapest and most useful methods available to farmers. It means using the best crop production methods—methods so favorable to the crop that weeds are crowded out. Actually, competition makes full use of one of the oldest laws of nature—survival of the fittest.

Carried one step further a certain "balance of nature" is implied—a sort of "Utopia" of peaceful harmony. Of course, this is not the case, because every living organism in nature carries on the most ruthless kind of competition. "Nature's balance" changes day by day. It is truly a survival of the fittest.

Man is the most disturbing influence of all when his wishes conflict with his environment. He uses his intelligence to bend most factors of the environment to his own advantage. Otherwise we could not feed the 3.5 billion people on this earth.

Herbicides are a relatively new tool in man's fight to control plant competition in his environment. If used properly, chemicals can often do the job better and more efficiently than other methods.

In spite of all this, nature is persistent. Man may control one type of plant, only to find that he has made conditions favorable to another. If man is not careful, he may solve one problem, only to create another equally serious problem.

Weeds are naturally strong competitors. If not, they would fail nature's test of "survival of the fittest." Those weeds that can best compete always tend to dominate. Domination may involve quick germination and very rapid early growth. For example, quick-growing annual weeds tend to dominate in desert areas having a moist spring but a dry summer and fall.

Weeds compete with crop plants for *light, soil moisture, soil nutrients*, and *carbon dioxide*. One mustard plant requires twice as much nitrogen and

Table 1-5.   Chemical Composition of Corn and Corn Weeds, in September (12)

		Mean Percentage Composition, Air-Dry Basis				
Plant	Growth Stage	Nitrogen	Phos-phorus	Potassium	Calcium	Mag-nesium
Corn	Total Plant, early milk stage	1.20	0.21	1.19	0.18	0.15
Pigweed	Green plants, 25–50% seeds ripe	2.61	0.40	3.86	1.63	0.44
Lambs-quarters	Green plants, 25–50% seeds ripe	2.59	0.37	4.34	1.46	0.54
Smart-weed	Green plants, 25–50% seeds ripe	1.81	0.31	2.77	0.88	0.56
Purslane	Lush plants, seeds partly ripe	2.39	0.30	7.31	1.51	0.64
Galinsoga	Lush plants, seeds partly ripe	2.70	0.34	4.81	2.41	0.50
Ragweed	Green plants, seeds partly ripe	2.43	0.32	3.06	1.38	0.29
Crabgrass	After bloom, seeds not ripe	2.00	0.36	3.48	0.27	0.54

phosphorus, 4 times as much potassium, and 4 times as much water as a well-developed oat plant. The average common ragweed requires 3 times as much water as a corn plant.

Weeds, as a group, have much the same requirements for growth as crop plants. For every pound of weed growth, the soil produces about 1 lb less of crops. The competition for nutrients can be easily seen from chemical analysis of weeds growing in corn and competing with corn.

The average content of all the weeds in Table 1-5 shows that the weeds had approximately twice as much nitrogen, 1.6 times as much phosphorus, 3.5 times as much potassium, 7.6 times as much calcium, and 3.3 times as much magnesium as corn (7).

Early weed competition usually reduces crop yields far more than late-season weed growth. Therefore, early weed control is extremely important. Late weed growth may not seriously reduce yields, but it makes harvesting difficult, reduces crop quality, reinfests the land with seeds, and harbors insects and diseases during the fall and winter.

In planning a control program, it is important to know the weed's life cycle. Possibly the cycle can be interrupted to gain easy but very effective control. In crop production this may be a well-timed chemical spray or a shift in planting date; thus, the crop gets the upper hand or competitive advantage.

Smothering with plastic, tar paper, old linoleum, straw, sawdust, or any other similar material is largely a matter of competition for light. Most weed seedlings cannot penetrate the thick coverings and are killed because of lack of light.

Crop Rotations

Certain weeds are more common in some crops than in others. For example, pigweed, lambsquarters, common ragweed, and crabgrass are often found in corn and in other summer-cultivated crops. Wild mustard, wild oats, wild garlic, cornflower, thistles, and corn cockle are a few that occur frequently in the grain crops. Dock, bitter sneezeweed, ironweed, thistles, mullein, and weedy bromegrasses are often found in pastures.

Rotation of crops is an efficient way to reduce weed growth. A good rotation for weed control usually includes strong competitive crops grown in each part of the rotation involving both (1) summer row crops and (2) winter or early spring grain crops, drilled or broadcast.

Biological

In biological weed control, a "natural enemy" of the weed is introduced which must be otherwise harmless. Insects have been the most successful natural enemies to date. Other control agents include disease organisms, parasitic plants, selective grazing by livestock and rodents, highly competitive replacement plants, and fish.

In general, biological control works best on large infestations of a mono-weed species, such as rangeland or water ways. It has been successful where the weed was introduced and in the move was freed of its natural enemies. The predator is then brought from areas of the world where the weed is native and is being controlled by its natural enemies. The method has utility where one aggressive weed devastates thousands of square miles. An excellent summary of procedures and hazards involved has been prepared (4).

Biological control cannot solve every weed problem because there are no effective biological agents for some weeds. It is not likely to solve the weed problems of cultivated fields because the technique must selectively control a complex of many weeds without injury to the crop. Further, it cannot *eradicate* a weed with seeds that remain dormant in the soil. Research must be carefully done to assure that the biological control organism is not likely to adapt to desirable plants, with the control organism itself becoming a pest.

The outstanding example of biological weed control is the prickly pear or cactus (*Opuntia* spp.) in Australia. The prickly pear was originally planted for ornamental purposes, but it spread rapidly from 1839 to 1925, covering 60 million acres and threatening much of the cultivated land. In seeking control measures research scientists found insects that attacked the prickly pear and *no other plants*. A moth borer (*Cactoblastis cactorum*) from Argentina was the most effective. It tunneled through the cladodes, underground bulbs, and roots, and became even more effective after several bacterial root organisms were accidentally introduced into the wounds caused by the insect.

By 1931 the moth borer had multiplied to such numbers that nearly all the prickly pears were destroyed by midseason. With little or no food supply, much of the moth borer population starved. Then followed several years when prickly pear increased, with a later increase in the moth borer. Several "waves" of each type of growth occurred before an equilibrium or "balance of nature" was reached. For this reason, biological methods *control* rather than *eradicate*. This is especially true of weed species reproducing from seeds that may lie dormant in the soil for years.

There are other cases of effective biological control. In the western United States, St. Johnswort (*Hypericum perforatum*), a poisonous range weed, is being controlled by leaf-eating beetles (*Chrysolina* spp.) (see Figure 1-10). In Hawaii, pamakani (*Eupatorium adenophorum*), a range weed, is being controlled by a stem gallfly (*Procecidochares utilis*), and in Fiji, curse (*Clidemia hirta*), another range weed, is being controlled by a shoot-feeding thrip (*Liothrips urichi*).

Fire

You can use fire to remove undesirable plants from ditch banks, roadsides, and other waste areas; to remove undesirable underbrush and broadleaved species in conifer forests; and for annual weed control in some row crops. Burning must be repeated at frequent intervals if it is to control most perennial weeds. In alfalfa, weeds and some insects can also be controlled by burning.

In waste areas, if the vegetation is green, a preliminary searing will usually

Figure 1-10. St. Johnswort or Klamath weed control by *Chrysolina quadrigemina* at Blocksburg, California (*a*) Photograph taken in 1946: the foreground shows weeds in heavy flower, while the rest of the field has just been killed by beetles; (*b*) portion of the same location in 1949 when heavy cover of grass had developed; (*c*) photograph taken in 1966 showing the degree of control that has persisted since 1949. Such results were reported throughout the state. (C. B. Huffaker, University of California, Berkeley.)

24

Figure 1-11. Lightweight, highly maneuverable flame throwers are available. (Damac Corporation, Evans, Colo.)

dry the plants enough so that they will burn by their own heat 10–14 days later. Large trucks with maneuverable booms equipped with oil-burning nozzles are occasionally used for the preliminary sear. Highly maneuverable lightweight burners are also available (see Figure 1-11).

When you want to favor pine trees over hardwoods, proper burning is usually successful in removing many of the hardwoods. Undesirable underbrush can also be removed by controlled burning at regular intervals. Burn often enough so that an excessively hot fire does not develop. Also, burn when there is a slight but steady breeze to sweep the fire along and when the debris close to the ground is slightly moist. "Controlled burning" reduces the waste and hazard of the destructive, uncontrolled forest fires (see Chapter 27).

Fire from special burners is effective in "flaming" or "sizz-weeding" cotton and corn as well as on small annual weeds, but is much less effective on perennial weeds. The flame weeder must be used when the weeds are very small, and thus must be used at regular intervals. Proper adjustment and proper speed are important to avoid injuring the crop.

Burning dried vegetation seldom kills the weed seeds; thus, the practice has little value for this purpose.

Chemical

The use of chemicals for weed control has rapidly increased since 1944, when 2,4-D was first used as a herbicide (weed-killing chemical).

Time of Chemical Treatment

The time of chemical application may largely determine its usefulness in various crops. The time of application may be given *with respect to the crop*, or *with respect to the weed*.

Preplanting

Preplanting treatment is any treatment made before the crop is planted. For example, methyl bromide may be used in seed beds to kill most weed seeds and vegetative reproductive parts before planting. It also controls many plant disease organisms, nematodes, and insect pests.

Preplant soil-incorporation applications are commonly used to place the herbicide in the soil area of the germinating weed seeds. This greatly increases the reliability of weed control regardless of additional rainfall, but the crop seed must have "true tolerance" to the herbicide. Several herbicides, such as EPTC and trifluralin, are soil-incorporated.

Preemergence

Preemergence treatment is any treatment made prior to emergence of a specified crop or weed. The treatment can be applied preemergence to both the crop and weeds, or to just the weeds. Therefore, a statement as to "preemergence to the crop," "preemergence to the weeds," or "preemergence to both crop and weeds" will clearly establish the timing of the treatment. The treatment is usually applied to the soil surface.

Postemergence

Postemergence treatment is any treatment made after emergence of a specified crop or weed. For example, 2,4-D gives effective postemergence control for most broadleaved weeds in corn, sorghum, small grains, and grass pastures.

The chemical may be applied postemergence to the crop, but preemergence to the weeds. For example, corn may be cultivated when it is 24–30 in. tall, leaving the field free of weeds. A herbicide sprayed on the soil surface between the rows at this time may inhibit weed seed germination; thus, the treatment is postemergence to the corn and preemergence to the weeds. Also, the chemical may be applied and incorporated into the soil between the crop rows.

Area of Application

Chemicals are applied *broadcast*, as a *band*, as a *directed spray*, and as *spot* treatment.

Broadcast Treatment

Broadcast treatment or blanket application is uniform application to an entire area.

Band Treatment

Band application usually means treating a narrow strip—usually directly over or in the crop row. The space between the rows is not chemically treated, but is usually cultivated for weed control. Without the chemical, however, cultivation effectiveness may be reduced.

This method reduces the chemical cost, because the treated band is often one-third of the total area with comparable savings in chemical cost. In addition, where the chemical has a long period of residual soil toxicity (remains toxic in the soil for a long time), the smaller total quantity of the chemical reduces the residual danger to the succeeding crop. However, lack of chemical treatment between the rows may result in less effective weed control from cultivation between the rows.

Directed Sprays

Directed sprays are applied to a particular part of the plant, usually to the lower part of the plant stem or trunk. Such sprays are usually directed at or just above the ground line.

Dropped nozzles to spray between row crops give a directed spray treatment. The height of the nozzles is influenced largely by the size of the crop and also the size of the weeds to be controlled (Figure 1-12).

Trees are often basally treated by directing the spray to the base of the trunk. The chemical may kill by chemically girdling the tree and by preventing sprout growth (Figure 1-13).

Spot Treatment

Spot treatment is treatment of a restricted area, usually to control an infestation of a weed species requiring special treatment. Soil sterilant treatments are often used on small areas of serious perennial weeds to prevent their spread.

Figure 1-12.   Nozzles direct spray across the row, killing small weeds. The cotton stem is tolerant of the aromatic oil, but the leaves are easily killed. (Delta Branch Experiment Station, Stoneville, Miss.).

Figure 1-13.   Special shields keep the spray off the onions. (U.S. Department of Agriculture.)

SAFETY OF HERBICIDES TO THE USER AND TO THE PUBLIC

Before herbicides are sold to the public, they are thoroughly tested to assure that they are effective as well as safe. The main Federal agency responsible for this in the United States is the Environmental Protection Agency (EPA). Local government agencies may also add their requirements, more restrictive than those of EPA, for unique local conditions.

EPA's prime concern with pesticide regulations is to see that our food supply is free of harmful pesticide residues and that our environment is not adversely affected. To achieve these goals, federal laws require manufacturers of pesticides to conduct exhaustive research on potential products before EPA will approve their use. These studies include:

1. Efficacy—that the product effectively does the job as claimed on the label.

2. Toxicology—both acute and chronic.

3. Determination of herbicide residues, or absence of residues, in food and feed crops. If there are detectable residues, a tolerance level must be established.

4. Fate in the environment (residues in soil, runoff and ground water, and wild life).

5. Impact on the environment (induced changes in natural populations of animals, plants, and soil microorganisms).

The precise research requirements are being prepared by EPA as the guidelines as this book is being prepared. These guidelines alone are about 400 pages long. They involve research in many phases of toxicology, biology, chemistry, and biochemical degradation of the compound, and the chemical's effect on air, water, soil microorganisms, wild life, birds, fish, and other aquatic organisms. The cost of doing the research to develop a new herbicide in a new area of chemistry is between 6 and 10 million dollars and requires 6–10 years of research, indicating the thoroughness and complexity of the required research.

This research examines not only the applied compound, but also breakdown products. As a result, we can be assured that when applied according to label instructions, herbicides are safe to use.

Toxicity values, determined from experimental animals, are very helpful in assessing the toxic effect of a chemical on humans or other animals. Toxicity is the inherent capacity of a known amount of a substance to produce injury or death. Often, high-dosage levels are used to establish an "effect." There may be no relationship between dosage levels required to produce an effect and exposure levels experienced in normal usage.

"Safety in use" is a function of toxicity and exposure (exposure level and length of time of exposure). It is a total estimate of the hazards of use. Thus, "safety in use" usually signifies safety or hazard better than toxicity data alone.

Risks from toxic compounds may be reduced by using:

...a very dilute formulation (where direct ingestion is possible).

...a formulation not readily absorbed through the skin or readily inhaled.

Table 1-6. Categories of Toxicity for Precautionary Labeling in Federal Insecticide, Fungicide, and Rodenticide Act.[1]
(Adapted from EPA toxicology guidelines)

Toxicity Category	Signal Word	Oral LD$_{50}$ (mg/kg)	Dermal LD$_{50}$ (mg/kg)	Inhalation LC$_{50}$ Dust or mist (mg/liter)	Inhalation LC$_{50}$ Gas or vapor (ppm)	Eye Effects	Skin Irritation
I	DANGER POISON[2] (fatal)	50 or less	200 or less	2 or less	200 or less	Irreversible corneal opacity at 7 days	Severe irritation or damage at 72 hr
II	WARNING (may be fatal)	50 thru 500	200 thru 2000	2 thru 20	200 thru 2000	Corneal opacity reversible within 7 days, or irritation persisting for 7 days	Moderate irritation at 72 hr
III	CAUTION	500 thru 5000	2000 to 20,000	20 to 200	2000 to 20,000	No corneal opacity, irritation reversible within 7 days	Mild or slight irritation at 72 hr
IV	CAUTION	5000 or greater	20,000 or greater	200 or greater	20,000 or greater	No irritation	No irritation at 72 hr

[1] All labels must state: "Keep Out of Reach of Children."
[2] Skull and Crossbones must appear near word poison.

...only occasionally and where humans are not directly exposed.

...only by knowledgeable applicators who are properly equipped to handle the chemical safely.

Terms Used in Expressing Toxicity

LD means lethal dose and LD_{50} means that the lethal dose that will kill 50% of a population of test animals; LC means lethal concentration and LC_{50} means the lethal concentration that will kill 50% of the animals.

"*Acute oral*" refers to a *single dose* taken by mouth or ingested.

"*Acute dermal*" refers to a *single dose* applied directly to the skin (skin absorption).

"*Inhalation*" refers to exposure through breathing or inhaling.

"*Parenteral*" refers to introduction otherwise than by the intestine. More specifically it usually refers to injection beneath the skin, into the veins, into the muscle or as a dermal treatment.

Toxicity values are expressed as *acute oral* LD_{50} in terms of milligrams of the substance per kilogram (mg/kg) of body weight of the test animal; or *acute dermal* LD_{50} in terms of mg/kg; or inhalation data LC_{50} in milligrams of mist or dust per liter of air (mg/liter) or parts per million by volume of gas or vapor (ppm).

Toxicity ratings, groupings, and label requirements are shown in Table 1-6.

LITERATURE CITED

1. *U.S.D.A. Agricultural Handbook No. 291*, Losses in Agriculture, 1965.
2. *U.S.D.A. Agricultural Handbook, No. 423*, Charts, 1971.
3. *U.S.D.A. Pesticide Review*, 1971.
4. *Weed Control*, National Academy of Science, U.S. Publication No. 1597 (1968).
5. Cates, J. S., and H. R. Cox. *U.S.D.A. Bur. Plant Ind. Bull.* 257 (1912).
6. Helm, C. A. *Missouri Agricultural Experiment. Station Bulletin* 185 (1921).
7. Holm, L. *Weed Sci., 19*, 485 (1971).
8. Kiesselbach, T. A., Arthur Anderson, and W. E. Lyness. *Nebraska Agricultural Experimental Station Bulletin* 232 (1928).
9. Klingman, G. C. *Weed Sci., 18*, 541 (1970).
10. Mosier, J. G., and A. F. Gustafson. *Illinois Agricultural Experiment Station Bulletin* 181 (1915).
11. Sturtevant, E. Lewis. *New York (Geneva) Agricultural Experiment Station Annual Report, 5*, 50 (1886).
12. Vengris, J., M. Drake, W. G. Colby, and J. Bart, *Agron. J., 45*, 213 (1953).
13. Wimer, D. C., and M. B. Harland, *Illinois Agricultural Experiment Station Bulletin* 259 (1925).

2 Biology of Weeds and Weed Seeds

The biology of weeds is concerned with the establishment, growth, and reproduction of weeds as well as the influence of environment on these processes. Heredity and environment are the master factors governing life. Heredity determines what an organism becomes—that is, a dog or a corn plant—by controlling the life form, growth potential, method of reproduction, length of life, and so on. The environment largely determines the extent to which these life processes proceed.

ECOLOGY

Ecology is the relationship of organisms to their environment. The ecology of weeds is primarily concerned with the effects of *climatic*, *physiographic*, and *biotic* factors.

Climatic

   ...light (intensity, quality, length of day)
   ...temperature (extremes, range, average, frost-free period)
   ...water (amount, percolation, runoff, evaporation)
   ...wind (velocity, duration)
   ...atmosphere (CO_2, O_2, humidity, toxic substances)

Physiographic

   ...edapic (soil factors including pH, fertility, texture, structure, organic matter content, CO_2, O_2, water drainage)
   ...topographic (altitude, slope, exposure to the sun)

Biotic

...plants (competition, diseases, toxins, stimulants, parasitism, soil flora)
...animals (insects, grazing animals, soil fauna, man)

Many of the most common weeds have a broad tolerance to ecological conditions. For example, common weeds such as lambsquarters, chickweed, and shepherdspurse grow on almost all types of soils. Rarer species such as saltgrass, halogeton, and alkali heath are usually found only on alkali soils.

Similar environmental requirements of certain weeds and selected crop species produce some rather common crop-weed associations. Some examples are: mustard in small grain, barnyardgrass in tomatoes, burning nettle in lettuce, chickweed in celery, and pigweeds in sugarbeets.

REPRODUCTION OF WEEDS

Weeds multiply and reproduce by both sexual and vegetative (asexual) means. Sexual reproduction requires pollination of a flower which in turn produces seed. Vegetative reproduction is carried out by stems, roots, leaves, or modifications of these basic organs such as rhizomes (underground horizontal stem), stolons (aboveground horizontal stem), tubers, corms, bulbs, and bulblets. The roles these organs play in weed dissemination are discussed in the last section of this chapter. Seed dissemination is emphasized and discussed first because weeds spread most widely by that method.

SEED DISSEMINATION

Seeds in general have no method of movement; therefore, they must depend on other forces for dissemination. Regardless of this fact, they are excellent travelers. The spread of seeds, plus their ability to remain viable in the soil for many years (dormant), poses one of the most complex problems of weed control. This fact makes "eradication" nearly impossible for many seed-producing weeds.

Weed seeds are scattered by (1) crop seed, grain feed, hay, and straw; (2) wind; (3) water; (4) animals, including man; (5) machinery; and (6) in weed screenings.

Crop Seed, Grain Feed, Hay, and Straw

Weeds are probably more widely spread through crop seeds, grain feed, hay, and straw than by other means.

Studies of wheat, oats, and barley seed sown by farmers in six North Central States reveal the seriousness of the problem. Researchers took seed directly from drill boxes and analyzed for weed seeds. About 8% of the samples contained primary noxious weeds, and about 45% contained secondary noxious weed seeds. About 80% of the seed had gone through a "recleaning" operation. Much of the recleaning was of limited benefit, because only part of the weed seeds were removed (18).

In the above study, *certified, registered,* and *foundation* seed were free of primary noxious weeds and contained only a small percentage of the common weeds (18).

Remember the difference between the serious, difficult-to-control weeds and those that are common on the farm and easily controlled. If the weed is already common on the farm, the soils are probably already infested; a few more seeds will make little difference. However, if a farm is free of a serious weed, one seed may be enough to start an infestation.

Farmers often feel that a low percentage of weed seeds on the seed label means that the few weeds present are of little importance. This may be a serious mistake (see Table 2-1). One dodder plant may easily spread to occupy one square rod during one season; thus, only 0.001% dodder seed is enough to completely infest a legume crop the first year!

The prevalence of weed seed in legume seed is clearly shown by data collected by official state seed analysts. In 3643 samples, weed seeds averaged from 0.10% to 0.38% by weight (13). These figures emphasize that certain other weed species also can completely infest a field the first year, despite an extremely low percentage of weed seeds present in the crop seed. These percentages are often below the legal tolerances, which make it necessary to state their presence on the seed label.

Table 2-1.   Field Dodder[1] and its Rate of Planting in Contaminated Legume Seed Sown at the Rate of 20 lbs/Acre

Dodder Seed by Weight (%)	No. of Dodder Seeds (/lb of Legume Seed)	No. of Dodder Seeds Sown (/Acre)	(/rod²) (square rod)
0.001	8	160	1
0.010	80	1,600	10
0.025	200	4,000	25
0.050	400	8,000	50
0.100	800	16,000	100
0.250	2,000	40,000	250

[1] There are 550,000 to 800,000 dodder seeds/lb.

Aside from crop seeds, weeds are also commonly spread through grain feed, hay, and straw. Where straw is used for mulching, it is important that the straw be free of viable weed seeds as well as grain seeds. The grain seed in the straw may prove to be a weed under such circumstances. Most of the grain seed will germinate and die if the straw is kept moist for 30 days with temperatures favorable to germination.

As shown later in this chapter, an appreciable number of weed seeds in grain and hay are viable after passing through the alimentary canal of the animal. If the manure is allowed to become "well rotted," the weed seeds will be killed.

Wind

Weed seeds have many special adaptations that help them spread. Some are equipped with parachutelike structures (pappus) or cottonlike coverings which make the seed float in the wind. Common dandelion, sow thistle, Canada thistle, wild lettuce, some asters, and milkweeds are examples (see Figure 2-1).

Water

Weed seed may move with surface water runoff, in natural streams and rivers, in irrigation and drainage canals, and in irrigation water from ponds.

Some seeds have special structures to help them float in water. For example, curly dock has small pontoon arrangements on the winged seed covering. Other seeds are carried in moving water or along the river bottom. Flooded areas from river overflows are nearly always heavily infested with weeds.

Irrigation water is a particularly important means of scattering seed. In 156 weed-seed catches in Colorado, in three irrigation ditches, 81 different weed species were found. In a 24-hr period several million seeds passed in a 12-ft ditch (16).

Scientists have found great variation in the length of time that seeds remain viable in fresh water. Results in Table 2-2 are based on weed seed suspended in bags (luminite screen sewn with nylon thread) at 12- and 48-in. depths in a fresh water canal at Prosser, Washington. Water could circulate freely within the bags. The bags were removed periodically over a 5-year period and germination counts were made.

There was little or no variation in germination between the 12- and 48-in. depths, but there was considerable difference in viability among species. Clearly, some seeds can be stored in fresh water for 3 to 5 years and still

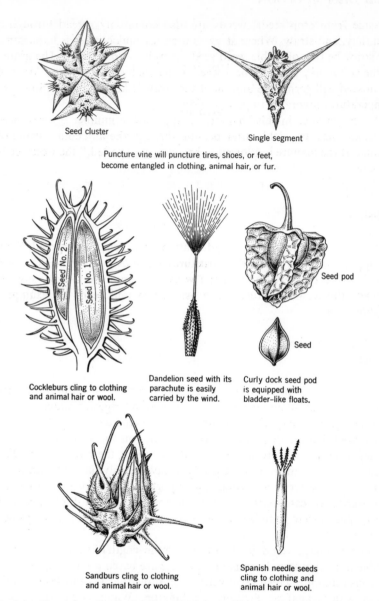

Seed cluster

Single segment

Puncture vine will puncture tires, shoes, or feet, become entangled in clothing, animal hair, or fur.

Seed No. 2

Seed No. 1

Seed pod

Seed

Cockleburs cling to clothing and animal hair or wool.

Dandelion seed with its parachute is easily carried by the wind.

Curly dock seed pod is equipped with bladder–like floats.

Sandburs cling to clothing and animal hair or wool.

Spanish needle seeds cling to clothing and animal hair or wool.

Figure 2-1. Examples of seeds that have special adaptations aiding in their spread.

Table 2-2.   Germination of Weed Seed after Storage in Fresh Water (7, 8, 9)

Field bindweed	After 54 months, 55 % germinated
Canada thistle	After 36 months, about 50 % germinated
	After 54 months, none germinated
Russian knapweed	After 30 months, 14 % of seed still sound; none germinated after 5 years
Redroot pigweed	After 33 months, 9 % still sprouted
Quackgrass	None sprouted after 27 months
Barnyardgrass	After 3 months less than 1 % germinated and none after 12 months
Halogeton	After 3 months less than 1 % germinated and none after 12 months
Hoary cress	After 2 months germination dropped to 5 % or less, and to zero after 19 months

germinate. In some cases storage in water tended to break dormancy and increased the percent germination; this was especially true after 2 to 4 months in the water.

Every effort should be made to free irrigation channels and water storage facilities from weed seed-producing plants that will be troublesome in irrigated fields.

Animals including man

Animals, including man, are responsible for scattering many seeds. They may carry the seed on their feet, clinging to their fur or clothes, or by carrying ingested seed.

Many seeds have specially adapted barbs, hooks, spines, and twisted awns that cling to the fur or fleece of animals or to man's clothing. Sandburs, cockleburs, stick tights, and beggar ticks are examples. Others may imbed themselves in the animal's mouth, causing sores. Examples are wild barley, downy brome, and various needle grasses. Other seeds have cottonlike lint or similar structures that help the seed cling to fur or clothes. Annual bluegrass and bermudagrass seed will stick to the fur of rabbits and dogs. Mistletoe seeds are sticky and become attached to the feet of birds. Birds often carry away fleshy fruits containing seeds for food.

Many weed seeds pass through the digestive tracts of animals and remain viable. Often weed seedlings are found germinating in animal droppings. Table 2-3 shows results of feeding weed seeds to different kinds of livestock and washing the seed free from the feces. The figures are given as a percentage of viable seeds recovered.

Table 2-3. Percentage of Viable Seeds Passed by Animals Based Upon Total Number of Seeds Fed (20)

Kind of Seeds	Percentage of Viable Seeds Passed by					
	Calves	Horse	Sheep	Hogs	Chicken	Average
Field bindweed	22.3	6.2	9.0	21.0	0.0	11.7
Sweetclover	13.7	14.9	5.4	16.1	0.0	10.0
Virginia pepperweed	5.4	19.8	8.4	3.1	0.0	7.3
Velvetleaf	11.3	4.6	5.7	10.3	1.2	6.6
Smooth dock	4.5	6.5	7.4	2.2	0.0	4.1
Pennsylvania smartweed	0.3	0.4	2.3	0.0	0.0	0.6
Average	9.6	8.7	6.4	8.8	0.2	6.7

Machinery

Machinery can easily carry weed seeds, rhizomes, and stolons. Weed seeds are often spread by harvesting equipment, especially combines. Cultivation equipment, tractors, and tractor tires often carry dirt which may include weed seeds. Also, cultivation equipment may drag rhizomes and stolons, dropping them later to start new infestations.

Weed Screenings

Most weed seeds have a reasonably high feed value. Common ragweed seed, when chemically analyzed, had 20.0% crude protein, 15.7% crude fat, 18% nitrogen free extract, and 4.83% ash. Because of their relative cheapness, weed seed screenings are often included in livestock feeds. (See Tables 1-5 and 26-1).

Over the years, considerable attention has been given to the problem of destroying the viability of weed seeds in screenings. Seeds are usually finely ground, or soaked in water and then cooked. Fine grinding with the hammer-mill has reduced to a minumum the hazard of scattering live weed seeds.

Pelleting where high temperatures are used may destroy the seed's viability. Fish solubles with a high protein content may be added to screenings. After heating to about 200°F, the mixture was pressed into pellets. Germination tests indicated that the seeds were no longer viable (25).

NUMBER AND PERSISTENCY OF WEED SEEDS

Number in the Soil

Some species produce a remarkable number of living seeds per acre. Wild poppies seriously infested the Rothamsted, England, Experiment Station in 1930 with an estimated 113 million poppy seeds/acre (5).

In Minnesota, weed-seed counts at four different locations on 24 different plots showed from 98 to 3068 viable weed seeds/ft^2 of soil, 6 in. deep. Converted to an acre basis, this is between 4.3 million and 133 million seeds/acre in the upper 6 in. (29).

Soil samples were taken on ten plantations in Louisiana heavily infested with johnsongrass. The average number of viable johnsongrass seeds per acre was 1,657,195 (28). A sugar cane crop was grown for 3 years without permitting the addition of new johnsongrass seed; after 3 years the johnsongrass seed population in the upper 2.5 inches of soil dropped to 1.3% of the original number (32).

Number Produced by Plants

The persistence of annual and biennial weeds depends mainly on their ability to reinfest the soil. The first infestation of most perennial species depends on seed. Obviously, if we could control the production of seed, we could eventually eliminate many species.

One scientist, reporting on 245 species, found the number of weed seeds produced by one plant ranges from 140 for leafy spurge to nearly a quarter of a million seeds for common mullein (35). Of these, 24 species have been selected and are listed in Table 2-4. Only plump, well-developed seeds were counted from well-developed plants growing with comparatively little competition. Those 24 species averaged 25,688 seeds per plant. A second report listed 263 species (36). Other reports have also listed the number of weed seeds produced (23, 30). One witchweed plant may produce as many as one-half million seeds.

Age of Seed and Viability

The length of time that a seed is capable of producing a seedling varies widely with different kinds of seeds and with different conditions. Certain

Table 2-4. Number of Seeds Produced per Plant, Number of Seeds/lb, and Weight of 1,000 Seeds (35)

Common Name	Number of Seeds (per Plant)	Number of Seeds (/lb[1])	Weight of 1,000 Seeds (g)
Barnyardgrass	7,160[2,3]	324,286	1.40
Buckwheat, wild	11,900	64,857	7.0
Charlock	2,700	238,947	1.9
Dock, curly	29,500	324,286	1.4
Dodder, field	16,000[3]	585,806	0.77
Kochia	14,600	534,118	0.85
Lambsquarters	72,450	648,570	0.70
Medic, black	2,350	378,333	1.2
Mullein	223,200	5,044,444	0.09
Mustard, black	13,400[4]	267,059	1.7
Nutsedge, yellow	2,420[2]	2,389,474	0.19
Oats, wild	250[2]	25,913	17.52
Pigweed, redroot	117,400[2]	1,194,737	0.38
Plantain, broadleaf	36,150	2,270,000	0.20
Primrose, evening	118,500	1,375,757	0.33
Purslane	52,300	3,492,308	0.13
Ragweed, common	3,380[2]	114,937	3.95
Sandbur	1,110[2]	67,259	6.75
Shepherdspurse	38,500[2,3]	4,729,166	0.10
Smartweed, Pennsylvania	3,140	126,111	3.6
Spurge, leafy	140[4]	129,714	3.5
Stinkgrass	82,100[2,3]	6,053,333	0.07
Sunflower, common	7,200[2,3]	69,050	6.57
Thistle, Canada	680[2,3]	288,254	1.57

[1] Calculated from the weight of 1000 seeds.
[2] Many immature seeds also present.
[3] Many seeds shattered.
[4] Yield of one main stem.

seeds maintain their viability for many years, while others die within a few weeks after maturing if they do not find a suitable environment for germination.

A short-lived species, silver maple, has seeds with about 58% moisture when shed and they will germinate immediately. However, when moisture content drops to 30% to 34%, they die. In nature this occurs within a few weeks.

Lotus (*Nelumbo nucifera*) seed found in a lake bed in Manchuria were approximately 1000 years old and still viable. The seeds were 1040 ± 210 years old, as measured by a radioactive carbon dating technique (24). They had been covered in deep mud and very cold water.

To determine the longevity of seed, Duvel placed 107 species in porous clay pots and buried them 8, 22, and 42 in. deep in the soil. He removed samples at various intervals and left others undisturbed. His findings were reported by Toole and Brown (39):

After 1 year, seed of 71 species germinated
After 6 years, seed of 68 species germinated
After 10 years, seed of 68 species germinated
After 20 years, seed of 57 species germinated
After 30 years, seed of 44 species germinated
After 38 years, seed of 36 species germinated

After 38 years
{
91% of jimson weed seed germinated
48% of mullein seed germinated
38% of velvet leaf seed germinated
17% of evening primrose seed germinated
7% of lambsquarters seed germinated
1% of green foxtail seed germinated
1% of curly dock seed germinated
}

The actual percentage of germination is not as important as the fact of survival. Programs to eradicate plants with long seed dormancy are usually doomed to failure.

The three depths did not greatly influence the data; however, seed at the 42-in. depth lived slightly longer. The 8-in. depth is far below the ideal depth for most seeds to germinate. If one sample had been buried 1 in. deep or less, greater differences caused by depth would have been likely.

In another test started in 1879, Beal mixed seeds of 20 different weed species with sand and buried the mixtures in uncorked bottles, with the opening tipped downward (11). After 20 years, 11 of the species were still alive. After 40 years, nine species were still alive. These nine were redroot pigweed, prostrate pigweed, common ragweed, black mustard, Virginia pepperweed, evening primrose, broadleaf plantain, purslane, and curly dock.

After 50 years, moth mullein germinated for the first time. Therefore, the seeds were still alive at 40 years but did not germinate because they were dormant. After 70 years, moth mullein had a germination of 72%, evening primrose 17%, and curly dock a few percent. After 80 years, moth mullein still had a germination of 70 to 80%, but with the other two species only a few seeds germinated. It appeared that these two species were nearing the end of their survival period (12).

Thus, many weed seeds retain their viability for 40 years and longer when buried deep in the soil. Scientists believe that much of the longevity depends upon the seeds being buried deep, with a reduced oxygen supply available to the seed. If brought to the surface, with other conditions favorable, many of the seeds will germinate. It is evident that repeated cultivation for several years without opportunity of reinfestation effectively reduces the weed seed population in a soil.

Depending on climatic conditions, abandoned cultivated land will usually return to a forest or grass-land vegetation. If recultivated after 30–100 years, the original field weeds immediately appear. The weed seeds have remained dormant while buried deep in the soil.

GERMINATION AND DORMANCY OF SEEDS

Germination includes several steps that result in the quiescent embryo changing to a metabolically active embryo as it increases in size and emerges from the seed. It is associated with an uptake of water and oxygen, use of stored food and, normally, release of carbon dioxide. For a seed to germinate, it must have an environment favorable for this process. This includes an adequate, but not excessive, supply of water, a suitable temperature and composition of gases (O_2/CO_2 ratio) in the atmosphere, and light for certain seeds. Specific requirements for seed germination differ for various species. Although these factors are optimal, a seed may not germinate because of some kind of dormancy.

Dormancy is a type of resting stage for the seed. Dormancy may determine the time of year when a seed germinates, or it may delay germination for years and thus guarantee the viability of the seed in subsequent years. Five environmental factors affect seed dormancy: *temperature, moisture, oxygen, light, and the presence of inhibitors.* Other factors directly related to the seed and its dormancy include impermeable seed coats (to water, oxygen, or both), mechanically resistant seed coats, immature embryos, and after-ripening. Mayer and Poljakoff-Mayber (26) have written a monograph on the germination of seeds.

Temperature

The temperature that favors seed germination varies with each species. There is a *minimum* temperature below which germination will not occur, a *maximum* temperature above which germination will not occur, and an *optimum* or ideal temperature when seeds germinate quickest. Thus, some seeds germinate only in rather cool soils while others do so only in warm soils.

The temperature requirements for most crop seeds are well established, and farmers recognize these requirements and plant accordingly. Cotton, for example, requires relatively high temperatures for germination, while the small grains will germinate at relatively cool temperatures.

Russian pigweed seed has germinated in ice and on frozen soil (1). Wild oats may germinate at temperatures of 35°F. Comparatively low temperatures (between 40 and 60°F) are necessary before certain winter annuals will germinate. High temperatures may cause a secondary type of seed dormancy, especially with some winter annual weeds. Wormseed mustard was introduced to secondary dormancy by temperatures of 86°F (33), and many summer annuals require temperatures of 65–95°F to germinate. Alternating temperatures are often better than a constant temperature for seed germination.

When redroot pigweed seed (a summer annual) was placed in germinators at 68°F, some seed remained dormant for more than 6 years. The seeds could be induced to germinate at any time in three ways: (1) by raising the temperature to 95°F, rubbing with the hand, and replacing at 68°F; (2) by partial desiccation; or (3) by alternating the temperature (10).

Temperature alone does not completely explain the periodicity of seed germination. Often the seeds have another form of dormancy which temporarily stops germination. This may be a survival mechanism to keep the plant from germinating immediately upon maturity in a season not suited to the plant.

Seeds may lie dormant for as little as several weeks or as much as several years. For example, cheat and hairy chess have a primary dormancy of 4–5 weeks after maturity. During this period they will germinate only if subjected to low temperatures (59°F or below). However, if the seeds are stored for 4–5 weeks, germination will then readily occur at temperatures of 68–77°F (33).

Moisture

Germination is normally a period of rapid expansion and high rates of metabolism or cell activity. Much of the expansion is simply an increase in water, expanding cell walls. If water content of the seed is reduced, the activity of enzymes—and consequently metabolism—slows down. The amount of moisture contained in seeds may determine their respiratory rate. During germination the seed respires at a very rapid rate. Many seeds cannot maintain this high rate of respiration until they reach a moisture content of 14% or more. Thus, in dry soils the seed remains dormant.

Dry seeds can tolerate severe conditions; some have been kept in boiling water for short intervals without injury and others in liquid air (−310°F).

When moist enough for germination, the same seeds may be killed by cold temperatures of 30°F or warm temperatures of 105°F.

Oxygen

In addition to the right temperature and sufficient moisture, the process of germination depends on oxygen. These conditions are reflected by the respiration rates of the plant. Aerobic respiration requires more free oxygen than anaerobic respiration; thus, some seeds start germination under anerobic conditions, then shift to aerobic respiration when the seed coat ruptures.

The percent oxygen found in the soil will vary widely, depending on soil porosity, depth in the soil, and organisms in the soil which use oxygen and release CO_2 (microorganisms, roots, etc.). The percent oxygen in the soil is usually inversely proportional to the percent CO_2. In swampy rice land there may be less than 1% oxygen in the soil atmosphere, in freshly green manured land 6-8%, and where corn is growing rapidly 8-9%, compared to about 21% in a normal atmosphere. The percent carbon dioxide may range from 5 to 15% under such conditions, compared to 0.03% for normal air.

Different species vary considerably in the amount of oxygen needed for seeds to germinate. Wheat seed germinated well when the replenished oxygen supply of the soil was 3.0 mg or more/m²/hr. It failed to germinate when the rate was below 1.5 mg. Rice seed germinated at 0.5 mg (22). Broadleaf cattail and some other aquatic plants germinate better at low oxygen concentrations than with normal air.

The effect of different oxygen concentrations was determined on the seeds of field bindweed, leafy spurge, hoary cress, and horsenettle. Oxygen concentrations of 5% produced little to no germination of hoary cress, horsenettle, and leafy spurge. At 10% oxygen and below, these three weeds germinated at a rate far below normal, while bindweed was reduced somewhat. The highest percentage germination was found at about 21% oxygen (normal air) for leafy spurge and hoary cress; but the best germination for horsenettle was at 36% oxygen and for field bindweed 53% (6).

Wild oats and charlock germination can be greatly suppressed by reducing the oxygen supply by soil compaction. Cultivation increased the number of wild oats that germinated by six times and charlock twice, compared to the compacted plots (3). Cultivation increased soil aeration, and thus increased the content of oxygen in the soil atmosphere.

Excess water in the soil cuts down seed germination of most plant species. Researchers believe that lower germination is related to a smaller supply of oxygen in water-logged soils, rather than merely excess water.

Many small-seeded weed seeds germinate only in the upper 1–2 in. of soil, mostly in the upper 1 in. A limited number, however, germinate below 2 in. In sandy soils seeds germinate deeper than in clay soils as a result of better aeration or better oxygen supply in the sands. Some seeds buried deep in the soils do not germinate but lie dormant for many years. When brought to the surface they germinate promptly. Aeration, involving increased oxygen supply, is probably responsible.

By using herbicides, you may kill successive crops of weed seedlings without disturbing the soil. Few to no viable weed seeds may remain in the upper soil layer. The soil surface may then remain relatively free of weeds. Repeated treatments to kill annual weeds before they produce seed is especially useful in areas where the soil is not disturbed by cultivation, such as in some perennial crops, permanent sod, or turf areas.

Light

Some kinds of seed germinate best in light, others in darkness, and others germinate readily in either light or darkness. Among several hundred species in which the role of light has been investigated, about half require light for maximum germination (31). In addition to the presence or absence of light affecting seed germination, the length of day and quality (color) of light also have influence. Here is a brief review of the electromagnetic spectrum (color of wavelengths).

	Wave Length (Ångströms)
X-ray	10–150
Ultraviolet	Below 4,000
Visible spectrum	
Violet	4,000–4,240
Blue	4,240–4,912
Green	4,912–5,750
Yellow	5,750–5,850
Orange	5,850–6,470
Red	6,470–7,000
Infrared	Greater than 7,000

Germination of lettuce seed was promoted by radiation at 6600 Å (red light) and inhibited at 7300 Å (Infrared light) (17). A later study with lettuce showed that the inhibitory effect of infrared light could be reversed by red light (4). Regardless of the number of alternating periods of red and infrared light to which the seeds were exposed, the final type of light determined the percent germination. For example, when the final light was infrared, germination was about 50%; but when the final light was red, germination reached almost 100%. Without light, germination fell to about 8%.

Here are a few examples of species and the light requirement of their seeds for germination.

Germination Favored by Light		Germination Favored by Darkness	Germination in Either Light or Darkness	
Bluegrass	Dock	Onion	Salsify	Wheat
Tobacco	Primrose	Lilly	Bean	Rush
Mullein	Buttercup	Jimsonweed	Clover	Toadflax

These effects will vary from species to species. In some seeds, the light requirement can be replaced by after-ripening in dry storage, alternating the daily temperature, higher temperatures, and by treatment in potassium nitrate or gibberellic acid solutions.

Presence of an Inhibitor

Many plants produce and release to the soil substances that inhibit other plants. Quackgrass, bermudagrass, johnsongrass, wild oats, and walnut trees are but a few examples (Fig. 1-1 and 2-2). Herbicides may chemically inhibit the germination of a seed, a bud of a plant or a tuber. For example, trifluralin will inhibit the sprouting of a johnsongrass or bermudagrass bud on the rhizome and EPTC may inhibit the sprouting of a nutsedge or a potato tuber.

Seed Coat Impermeable to Water, Oxygen, or Both

Seed coats may be waterproof and prevent the seed from absorbing water. Such seed will not germinate even if soil moisture is plentiful. Seed that fails to germinate because of a waterproof seed coat is called *hard seed*. Hard seed is common in annual morning glory, lespedeza, clovers, alfalfa, and vetch. Researchers believe that many weed species have hard seeds.

Some seed coats are impermeable to oxygen but not water. Cocklebur has

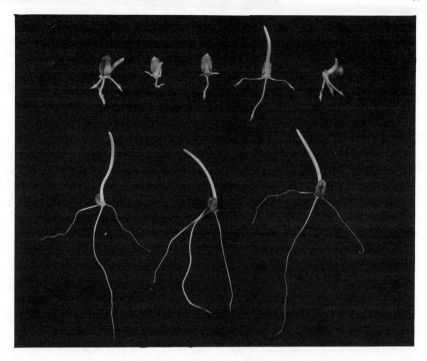

Figure 2-2. Upper row: Stunted wheat seedlings caused by a water leacheate from quack-grass roots and rhizomes. Lower row: Normal wheat seedlings grown in tap water or in a leacheate from cultivated oats. (T. Kommedahl, University of Minnesota.)

two seeds per fruit, one set slightly below the other. The lower seed usually germinates during the first spring and the upper remains dormant until the next year. Both can be made to germinate immediately by breaking the seed coats, or by simply increasing the oxygen supply (see Figure 2-1). Seeds of ragweed, several grasses, and lettuce also show this type of dormancy.

As with waterproof seed coats (hard seeds), anything that breaks the seed coat—scarification, acids, soil microorganisms—will break this type of dormancy. Under laboratory conditions, oxygen dormancy can usually be broken by increasing the oxygen supply. Cultivation of the soil often has a similar effect by increasing the oxygen level in the upper layer of soil.

Mechanically Resistant Seed Coats

A tough seed coat may forcibly enclose the embryo and prevent germination. While the seed absorbs oxygen and water, it builds pressures in excess of 1000 psi. The seed will quickly germinate if the seed coat is removed. Pigweed, wild mustard, shepherdspurse, and pepperweed have this type of dormancy.

As long as the seed coat remains moist, it remains tough and leathery—for as long as 50 years. Any factor that weakens the seed coat will aid in breaking dormancy. Drying at temperatures of 110°F, or mechanical or chemical injury to the seed coat, may break this type of dormancy.

Immature Embryos

The outside of the seed may appear fully developed, but it may have an immature embryo. The embryo needs more growth before the seed can germinate. Therefore, the seed appears dormant, although the embryo is slowly growing and developing. Seeds of orchids, holly, smartweed, and bulrush show this type of dormancy.

After-Ripening

In some species the embryos appear completely developed, but the seed will not germinate although the seed coat has been carefully removed to permit easy absorption of water and oxygen. Light and darkness have no effect. In this case germination occurs normally after a period of *after-ripening*. Occasionally cool temperatures for several months will end this type of dormancy. After-ripening is especially common in the grass, mustard, smartweed, rose, and pink families (34).

After-ripening is a physiological change of a complex physiochemical nature. Although the exact processes are not completely understood, they may be associated with changes in the storage materials present, substances promoting germination may appear, or substances inhibiting germination may disappear (see Figure 2-2).

GERMINATION AS AFFECTED BY CERTAIN CONDITIONS

Burning

Burning fields after weed seeds have matured gives only partial and erratic destruction of seeds. Seeds that lie on the ground may readily escape, whereas those held on the plants often burn completely. The degree of weed-seed destruction depends largely on the intensity of the heat, and this in turn on dryness and the amount of litter and debris to be burned.

Durrel's work (15) showed little control by burning over fields. But Mercer (27) found that Russian knapweed germinated 85% before burning and none afterward; field bindweed germination was 36% before burning and only 7% afterward. Hopkins (21) has shown that many common weed seeds are destroyed when subjected to temperatures ranging from 175 to 212°F for 15 min.

The after-effects of burning are usually pronounced, especially after forest fires. Weed species absent, for the most part, from the area for many years may suddenly appear after the fire and dominate other vegetation.

Data and experience clearly show that moderate heat may end seed dormancy. Five other factors that normally follow a fire may also terminate dormancy; these factors are (1) greater alternation of temperature in the upper soil layers between day and night, (2) more light reaching the surface soil, (3) removal of litter, (4) removal of competition from other plants, and (5) plants previously living in the area having soil-inhibiting substances that prevented seed germination. With removal of those plants, those substances are no longer present.

Cutting (Stage of Maturity)

Cutting weeds to prevent seed production is a common recommendation. The practice is important in agricultural crop lands, in turf, and in hay crops.

To determine the viability of seed cut before maturity, weeds were cut in various stages of maturity and allowed to dry in the sun. Tables 2-5 and 2-6 show the results. Four of 13 weed species showed high seed germination although cut when in flower.

Table 2-5. Germination of Weed Seeds Cut at Various Stages of Maturity (19)

Weed Seeds	Cut When Dead Ripe	Cut When in Flower	Cut in Bud Stage
	(% germinated)		
Groundsel, ragwort	72	80	0
Sowthistle, common	100	100	0
Groundsel, common	90	35	0
Sea aster	90	86	0
Dandelion	91	0	0
Catsear, spotted	90	0	0
Canada thistle	38	0	0

Table 2-6.   Germination of Weed Seeds Cut at Various Stages of Maturity (19)[1]

Weed Seeds	Cut When Dead Ripe	Cut When Medium Ripe	Cut in Flowering Stage
		(% germinated)	
Meadow barley	94	90	0.0
Soft brome	96	81	0.0
Curly dock	84	88	0.0
Shepherdspurse	88	82	0.0
Speedwell, corn	70	69	0.0
Chickweed, common	60	56	0.0

[1] The above table was adapted from the data presented (19).

In South Dakota, Canada-thistle and perennial sow-thistle heads were removed from the plant and dried at daily intervals after the flowers opened. The experiment was continued for 3 years. Perennial sow-thistle heads harvested 3 days after blooming had 0.0% viable seed; 6 days after blooming they had an average of 6% viable seed; and 8 days after blooming, 65% viable seed (15).

Canada thistle harvested 6 days after blooming had an average of 0.03% viable seed; 8 days after blooming, 6.7% viable seed; and 11 days after blooming, 73% viable seed (15).

Another study compared removal of the heads as discussed above with the effect of cutting the entire plant and leaving the heads on the plant during drying. Results of the two methods were similar; thus, little seed development takes place after either type of cutting (14).

In summary, either mowing in or before the bud stage prevents viable seed production. If seeds reach medium ripeness, probably a large percentage of viable seeds will be produced. In the case of many species, cutting after that time does little or no good in preventing viable seed production. With other species, mowing is not effective because part of the heads are short and missed by the mower, and these produce seed.

Storage in Silage

Many seeds lose their viability in silage. Tildesley (38) has shown that many weed seeds lose their germinating power in 10 to 20 days after being placed in silage. Other reports indicate, however, that some seeds will germinate after being in the silo for periods up to 4 years. These variations are possibly a result of differences in the silage as to moisture content, temperature, and amount of organic acids produced.

Storage in Manure

When manure is spread fresh from the stable, viable weed seeds are usually spread with it. But if manure is stored, heating and decomposition will begin and, in time, the weed seeds are destroyed.

In Table 2-7 only three weeds showed any viability after a 1-month storage. All seeds were destroyed at the end of 4 months.

Table 2-7. Effect of Length of Time in Cow Manure upon the Viability of Various Weed Seeds (20)

Kind of Seeds	% Viability before Burial	% Viability after Storage			
		1 month	2 months	3 months	4 months
Velvetleaf	52.5	2.0	0.0	0.0	0.0
Field bindweed	84.0	4.0	22.0	1.0	0.0
Sweetclover	68.0	22.0	4.0	0.0	0.0
Pepperweed	34.5	0.0	0.0	0.0	0.0
Smooth dock	86.0	0.0	0.0	0.0	0.0
Smartweed	0.5	0.0	0.0	0.0	0.0
Cocklebur	60.0	0.0	0.0	0.0	0.0
Puncturevine	52.0	0.0	0.0	0.0	0.0

Stoker, Tingey, and Evans (37) found complete destruction of hoary cress and Russian knapweed seeds in moist, compacted chicken manure at the end of 1 month. However, seed of field bindweed was still viable at the end of 4 months.

These tests were conducted during the summer in cow or chicken manure. Horse and mule manure tends to heat, whereas cow or chicken manure does not. Decomposition of the weed seed would be more rapid at the higher temperature. If the manure is frozen or cold, the seed would live longer.

If the edges or outside of the manure pile dry out, decomposition slows down and viable weed seeds would likely persist. Therefore, turn the manure occasionally to kill all seeds. Well-rotted manure is free of viable seeds.

TWO WAYS TO FAVOR GERMINATION OF DORMANT SEEDS

Scarification

Destruction of the seed coat would presumably affect seed dormancy if the seed coat is impermeable to either water or oxygen, or perhaps mechanically resistant.

Scarification involves scratching or other injury to the seed coat. It may be done with sandpaper or a file. Often the seed is permitted to fall and slide across a sandpaper type of surface, scratching the seed coat. Also the decomposition effect of soil microorganisms may destroy or rupture the coat.

Stratification

Stratification is commonly used to shorten the period of dormancy. Seeds are placed in a moist medium at 33–50°F, and kept there from 2 weeks to 8 months. They are then placed in a warm, moist environment for germination. Seeds overwintering in the soil experience conditions similar to stratification. This is particularly effective in breaking dormancy of immature embryos or embryos needing a period of after-ripening.

STORAGE CONDITIONS FAVORING LONG LIFE OF SEEDS

The conditions during storage are more important than the length of storage-time in maintaining viability. Under farm-storage conditions, many crop seeds remain viable for only 1–3 years. Actually, many crop seeds are killed by warm, moist storage for 1–3 months.

Soybean seed in humid regions usually are not considered good for seed if more than 9 months old. However, when stored at about 9% moisture and about 32°F, they have shown good germination when over 10 years old. Vegetable seeds—carrot, tomato, eggplant, lettuce, onion, and pepper—were stored at about 5% moisture and at 25°F for 17 years with little or no loss in percent germination (10).

For most seeds, three primary factors lengthen the time that seeds can be stored and remain viable. These include storage at a low moisture content (5% for many seeds), low temperature (below 40°F), and a low oxygen supply. To produce a low oxygen supply, you may substitute an inert gas such as nitrogen.

DISSEMINATION BY RHIZOMES, STOLONS, TUBERS, ROOTS, BULBS, AND BULBLETS

Although weeds reproduce and spread most widely by means of seeds, they also multiply by vegetative or asexual methods. Rhizomes, stolons, tubers, roots, bulbs, and bulblets are all vegetative or asexual methods of reproduction. Many perennial weeds classed as "serious" reproduce vegetatively, and most of these also reproduce by means of seeds. Vegetative organs occasionally have a short period of dormancy.

Most plants spread slowly by vegetative means alone. Without help from man and his cultivation equipment, weeds such as quackgrass, field bindweed, johnsongrass, bermudagrass, nutsedge, and Canada thistle would spread vegetatively less than 10 ft/year. However, the rhizomes, stolons, roots, and tubers are dragged about the field with soil-tillage equipment. Wherever these plant pieces drop, a new infestation is likely. Disc-type cultivation equipment is less likely to drag the plant parts than are shovels, sweeps, and plows.

Repeated tillage will kill most plants possessing rhizomes, stolons, roots, and tubers. If cut off and in *dry soils*, the vegetative parts may quickly dry and die, preventing new growth.

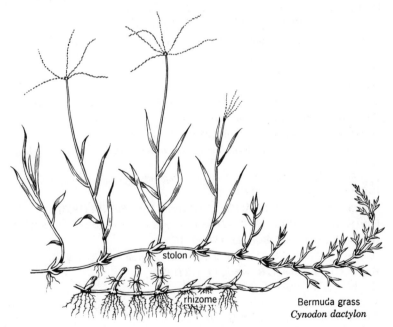

Figure 2-3. Bermudagrass reproduces by seeds, stolons, and rhizomes. (North Carolina State University.)

In moist soils, the cut vegetative parts quickly take root and establish new plants. Under such conditions, repeated tillage may exhaust the underground food reserves. Most such weeds are killed through *carbohydrate starvation* by repeated tillage for 1–2 years. (See figure 1-9.)

Some chemicals mixed into the soil will retard the development of new roots, especially after the plant has been cut off. With a combination treatment of an effective herbicide plus repeated cultivation, many serious perennial weeds may be controlled in a short time. For example, johnsongrass can be controlled by trifluralin plus repeated cultivation.

LITERATURE CITED

1. Aamodt, O. S., *Sci. Agr.*, *15*, 507 (1935).
2. Barton, L. V., *Contribs Boyce Thompson Inst.*, *7*, 405 (1936).
3. Bibbey, R. O., *Sci. Agr.*, *16*, 141 (1935).
4. Borthwick, H. A., S. B. Hendricks, M. W. Parker, E. H. Toole, and V. K. Toole, *Proc. Nat. Acad. Sci. U.S.*, *38*, 662 (1952).
5. Brenchley, W. E., and K. Hanwaring, *J. Ecol.*, *18*, 235 (1930).
6. Brown, E. O., and R. H. Porter, *Iowa Res. Bull. 294*, 1942.
7. Bruns, V. F., L. W. Rasmussen, *Weeds*, *II*, 138 (1953).
8. Bruns, V. F., L. W. Rasmussen, *Weeds*, *V*, 20 (1957).
9. Bruns, V. F., L. W. Rasmussen, *Weeds*, *VI*, 42 (1958).
10. Crocker, W., and L. V. Barton, *Physiology of Seed*, Chronica Botanica, Waltham, Mass., 1953.
11. Darlington, H. T., *Amer. J. Botany*, *38*, 379 (1951).
12. Darlington, H. T., and G. P. Steinbauer, *Plant Physiol. Proc.*, *35*, 31 (1960).
13. Darst, W. H., *Proc. Assoc. Offic. Seed Analysts*, *39*, 11 (1949).
14. Derscheid, L. A., and R. E. Schultz, *Weeds*, 8, 55 (1960).
15. Durrell, L. W., *Colorado Agricultural Experiment Station Report No. 1928*, p. 22 (1929).
16. Eggington, G. E., and W. W. Robbins, *Colorado Agricultural Experiment Station Bulletin No. 253*, pp. 1–25 (1920).
17. Flint, L. H., and E. D. McAlister, *Smithsonian Misc. Pub.*, 94 (1935).
18. Furrer, J. D., *North Central Weed Control Conference Proceedings*, *11*, 26 (1954).
19. Gill, N. T., *Ann. Appl. Biol.*, *25*, 447 (1938).
20. Harmon, G. W., and F. D. Keim, *J. Amer. Soc. Agron.*, *26*, 762 (1934).
21. Hopkins, C. Y., *Can. J. Research*, *14*, 178 (1936).
22. Hutchins, L. M., *Plant Physiol.*, *1*, 95 (1926).
23. Korsmo, E., *Unkräuter in Ackerbau der Neuzeit*, Translated and edited by H. W. Wollenweber, Springer, Berlin (1930).
24. Libby, W. F., *Science*, *114*, 291 (1951).
25. Mather, H. J., *North Central Weed Control Conference*, *9*, 18 (1952).

26. Mayer, A. M., and A. Poljakoff-Mayber, *The Germination of Seeds*, Macmillan, New York (1963).

27. Mercer, W. H., *Annual Report on Weeds: Control, Research and Educational Program on the Uncompahgre Irrigation Project*, Mimeo Dept., Bureau of Reclamation, U. S. Department of the Interior, Washington D.C., 1940.

28. Phillips, R. P., and S. J. P. Chilton, *South Weed Conference Proceedings*, 2, 59 (1949).

29. Robinson, R. G., *Agron. J.*, *41*, 513 (1949).

30. Salisbury, E., *The Reproductive Capacity of Plants*, G. Bell and Sons, London, (1942).

31. Salisbury, F. B., and C. Ross, *Plant Physiol.*, Wadsworth (1969).

32. Stamper, E. R., and S. J. P. Chilton, *Weeds*, *1*, 32 (1951).

33. Steinbauer, G. P., and B. H. Grigsby, *Weeds 5*, 1 (1957).

34. Steinbauer, G. P., and B. H. Grigsby, *Weeds*, *5*, 175 (1957).

35. Stevens, O. A., *Amer. J. Bot.*, *XIX*, 784 (1932).

36. Stevens, O. A., *Weeds*, *V*, 46 (1957).

37. Stoker, G. L., D. C. Tingey, and R. J. Evans, *J. Amer. Soc. Agron.*, *26*, 600 (1934).

38. Tildesley, W. T., *Sci. Agric.*, *17*, 492 (1936–1937).

39. Toole, E. H., and E. Brown, *J. Agr. Res.*, *72*, 201 (1946).

40. Went, F. W., G. Juhren, and M. C. Juhren, *Ecology*, *33*, 351 (1952).

3 Herbicides and the Plant

When a herbicide comes in contact with a plant, its action is influenced by the morphology and anatomy of the plant as well as numerous physiological and biochemical processes that occur within the plant. These processes include: (1) absorption, (2) translocation, (3) molecular fate of the herbicide in the plant, and (4) effect of the herbicide on metabolism. The interaction of these plant factors with the herbicide determines the effect of a specific herbicide on a given plant species. When one plant species is more tolerant to the chemical than another plant species, then the chemical is considered to be *selective*.

The life processes of plants are many and varied; they are complex and delicately balanced. Disturb one, even slightly, and a chain of events may be set off that will cause major changes in the plant's metabolism.

Even a minor change in the plant's environment may result in a major change in its life processes. For example, many perennial plants lie dormant below ground all winter. If the soil temperature is raised but a few degrees, a complex series of reactions is set in motion that may continue through the growing season.

Developing an understanding of the *mode of action* of herbicides is difficult because the problem is complex. However, considerable progress has been made in the last decade because of the thousands of studies made by many hundreds of research workers. Usually these studies have failed to show the mechanism as a single response, but rather as a series of physiological responses. You can find further information in books on the subject by Ashton and Crafts (1) and by Audus (2).

To understand the physiological processes at work in the use of herbicides, we first need to define several terms that will be used repeatedly.

A *herbicide* is a chemical used to kill or inhibit growth of plants.

Mode of action refers to the entire chain of events from the first contact of the herbicide with the plant to its final effect, which could be death of the plant.

Mechanism of action is the primary biochemical or biophysical reaction which brings about the ultimate herbicidal effect.

Symplast (*sym* means together) constitutes the sum total of living protoplasm of a plant. It is continuous throughout the plant; there are no islands of living cells. The phloem is a major component of the symplast. Phloem translocation is via the symplast.

Apoplast (*apo* means separate or detached) constitutes the total nonliving cell-wall continuum of the plant. The xylem is a major component of the apoplast. Xylem translocation is via the apoplast.

ABSORPTION

To be effective, herbicides must enter the plant. Some plant surfaces absorb the herbicide quickly, but other plant surfaces absorb the chemical slowly, if at all. The chemical nature of the herbicide is also involved. Therefore, differential absorption or selective absorption may account for differences in plant responses.

The two most common points of entry in plants are through the leaves and through the roots. In addition, some chemicals are effectively absorbed through the stems, including coleoptiles or young shoots as they grow through treated soil. Also, seeds may absorb the herbicide.

Leaf Absorption

Initial leaf penetration may take place either through the leaf surface or through the stomates. The volatile fumes of some herbicides and some solutions enter through the stomates. Probably of far greater importance, however, is the direct penetration of the leaf surface. Here the herbicide must first penetrate the cuticle. The cuticle is not homogeneous in composition (see Figure 3-1). Externally it is primarily composed of wax; internally, it is primarily composed of cutin; there is a continuous gradation of one into the other. At the cuticle–cell-wall interface, the cutin is in contact with the pectin and the pectin is in contact with the cellulose of the cell-wall.

There is a gradual transition in the polar nature of the cuticle–cell-wall complex from the surface wax to the cellulose. The cuticular wax is the most nonpolar (hydrophobic), followed by cutin, pectin, and cellulose, respectively. Cellulose is polar (hydrophilic). Therefore, polar compounds have

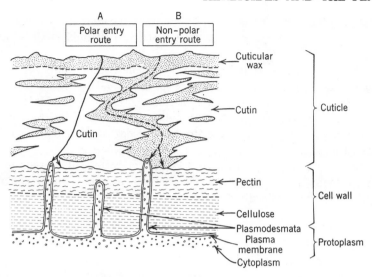

Figure 3-1. Hypothetical diagram representing the foliar absorption aspects of cuticle-cell wall-protoplasm structure. Hypothetical routes of entry of (*a*) polar and (*b*) nonpolar herbicides. [Adapted in part from Orgell (32).]

considerable difficulty entering the cuticular wax, but once they pass this barrier they enter each succeeding phase more readily. In contrast, nonpolar compounds readily enter the cuticular wax but have increasing difficulty passing into each succeeding phase. Thus the polar nature of the herbicide may have considerable influence on its rate of absorption.

Figure 3-1 shows the hypothetical routes of entry of both polar (*A*) and nonpolar (*B*) herbicides and their ultimate entry into the symplastic system via the plasmodesmata or the apoplastic system via the cell wall. The exact molecular requirements for entry into these two systems are unknown. However, atrazine, monuron, and chlorpropham primarily enter the apoplastic system whereas 2,4-D, amiben, and fenac primarily enter the symplastic system. Several herbicides, such as amitrole, dalapon, and picloram, readily enter both systems.

Any substance that will bring the polar herbicide into more intimate contact with the leaf surface should aid absorption. Wetting agents increase leaf absorption by (1) reducing interfacial tension to give better "wetting," and (2) modifying the wax- and oil-like substances of the cuticle, Therefore, wetting agents added to polar herbicides increase herbicide toxicity. Sodium and amine salts of 2,4-D were 5 times more toxic when a wetting agent was added to the spray (31). The amount of aminotriazole absorbed 24 hr after application was 13.1% without a surfactant (wetting agent) and 77.8% with a surfactant (15).

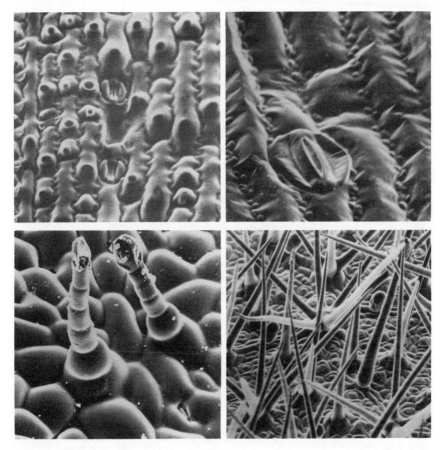

Figure 3-2. Leaf surfaces as they appear on a scanning electron microscope. Upper left: bermudagrass (450×). Upper right: nutsedge (1050×). Lower left: redroot pigweed (350×). Lower right: velvet leaf (170×). (D. E. Bayer and F. D. Hess, University of California, Davis.)

The addition of a wetting agent tends to equalize foliar herbicide absorption in all types of plants. Therefore, a wetting agent may reduce the selectivity of the herbicide if that selectivity is dependent on selective foliar absorption.

Increases in temperature may cause more rapid absorption. In many respects, the absorption process is a chemical process. Within biological limits, the *rate* of such chemical processes tends to double with each increase of 10°C or 17°F. Therefore, selectivity of postemergence herbicides dependent on selective foliar absorption may be reduced by excessive temperatures after treatment. This is true for postemergence treatments of dinoseb (21).

Root Absorption

Roots absorb many herbicides from the soil. Comparative studies have shown that roots absorb some herbicides (monuron, simazine) very rapidly and others (dalapon, amitrole) more slowly (9). Some herbicides are absorbed passively (monuron) while others (2,4-D) appear to be absorbed actively, requiring an expenditure of energy (12).

Herbicides appear to enter into roots by three routes: apoplast, symplast, and apoplast-symplast (see Figure 3-3). The apoplast route involves movement exclusively in cell walls to the xylem. This route appears to require that the herbicide pass through the casparian strip and enter the xylem. The casparian strip is a water-tight barrier in the cell walls of the endodermis separating the cortex and the stele.

The symplast route involves initial entry into cell walls and then into the protoplasm of the cells of the epidermis, cortex, or both. The herbicide remains in the protoplasm and sequentially passes into the endodermis, stele, and phloem by means of the interconnecting protoplasmic strands (plasmodesmata).

The apoplast-symplast route is identical to the symplast route but for the fact that the herbicide may reenter the cell walls after bypassing the casparian strip and then enter the xylem.

Although certain herbicides may be restricted to one route of entry, others

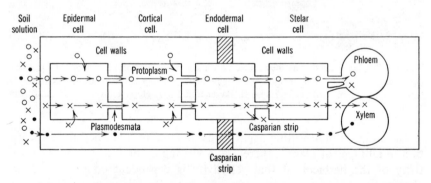

Figure 3-3. Hypothetical diagram representing herbicide absorption into roots. [Adapted from E. Epstein (13). Roots, SCIENTIFIC AMERICAN 228 (5): 48-58. 1973.

● Molecules able to enter cell walls (apoplast), diffuse through casparian strip, and enter xylem.

○ Molecules able to enter protoplasm (symplast), pass from cell to cell through the plasmodesmata, and enter the phloem.

× Molecules able to enter both cell walls (apoplast) and protoplasm (symplast), and enter both xylem and phloem.

may enter by more than one route. The chemical and physical properties of each herbicide primarily determine which route is followed.

Under most conditions, there is rapid translocation upward from roots in the xylem (transpiration stream), but only limited upward transport in the phloem. Therefore, in roots, herbicide entry into the xylem is more important than entry into the phloem. Roots, however, lack a cuticle, so that the root effectively absorbs polar compounds, but nonpolar compounds are absorbed with difficulty, if at all.

Shoot Absorption

Absorption by leaves and roots has been considered so far. In addition some compounds may be absorbed from soil by coleoptiles and young shoots as they develop and push upward through the soil following germination of seeds. Dawson (11), studying the response of barnyardgrass seedlings to soil-applied EPTC (S-ethyl dipropylthiocarbamate), found that exposure of primary roots of the seedlings of this weed to EPTC in soil gave little or no response, whereas exposure of the young shoot to concentrations above $\frac{1}{2}$ ppmw of EPTC (S-ethyl dipropylthiocarbamate) resulted in severe injury and, in most instances death. His 3-in. incorporation represents about $\frac{1}{2}$ lb/ acre dosage. Dawson concluded that the young shoot of barnyardgrass is the main site of EPTC uptake and also the prime site of injury.

Oliver et al. (23) continued studies with barley, wheat, sorghum, oats, and giant foxtail on the site of root uptake of and tolerance to EPTC. They were arranged so that (1) the upper shoot was exposed to treated soil; (2) the shoot and seed were exposed; (3) 2–4 mm of shoot and seed plus root were exposed; (4) the roots only were exposed, and (5) the entire shoot and root were exposed. Results of this test are shown in Figure 3-4. Barley and wheat seedlings were quite tolerant of shoot exposure to EPTC; sorghum, oat, and giant foxtail were susceptible. The roots were the major site of uptake by barley; in oats, root uptake gave more growth than any other treatment. These tests coupled with ^{14}C–EPTC studies indicate that differences in tolerance can be associated with the sites of uptake.

Stem Absorption

Direct applications of herbicides to stems of plants is not a common practice except for control of woody plants. However, the chemical may inadvertently come in contact with the stem in a foliar application or directed spray.

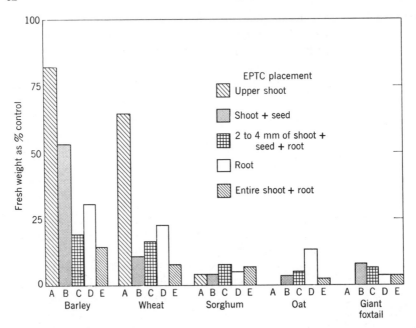

Figure 3-4. Effect of EPTC placement on fresh weight of five grass species after 14 days (23).

Generally, foliar absorption is much more important than stem absorption.

In stem treatment of woody plants the herbicide may be applied to (1) the bark, usually for small tree or brush control, (2) cuts through the bark (cut-surface method) for the control of large trees, and (3) stumps of cut trees or brush to prevent sprouting (see Chapter 27 for details of woody-plant control).

In bark treatments the herbicide must first penetrate the bark before it can be absorbed by the transport systems. The bark becomes thicker as the plant grows older and absorption decreases. The cut-surface method of treatment bypasses the bark barrier and places the herbicide in direct contact with the xylem and phloem.

TRANSLOCATION

Translocation of herbicides is of major importance in control of weeds. It is particularly important for those plants with below-ground reproductive organs. Herbicides are translocated within the plant through the symplastic system, the apoplastic system, and intercellular translocation. Some herbicides are primarily translocated in the symplastic system, others in the apoplastic system, and still others in both systems (see Table 3-1).

Herbicides move within the plant much the same as other solutes. Figure 3-5 shows the movement of herbicides through the symplastic and apoplastic systems.

Symplastic Translocation

When applied to leaves, symplastic mobile herbicides follow the same pathway as sugar formed there by photosynthesis. Such herbicides move from cell to cell in the leaf via the interconnecting protoplasmic strands (plasmodesmata) until they enter the phloem. Then they move out of the leaf and move both up and down the stem via the phloem, accumulating in those areas where sugar is being used for growth. Growth is most rapid in the apical growing point, expanding young leaves, rapidly elongating stems, developing fruits and seeds, and root tips.

Translocation through the phloem appears to involve a *mass flow* of solution. One explanation of this driving force is the difference in turgor pressure between the photosynthetic cells and the cells utilizing the products of photosynthesis, primarily sugar. Photosynthetic cells (high-pressure) are often called the *source* and using cells (low-pressure) the *sink*. The plasmodesmata and phloem are living; therefore, herbicides with great acute toxic properties kill these, stopping symplastic translocation.

2,4-D is rapidly absorbed through the leaves and may be quickly moved to other plant parts. Young plants move the materials faster than older plants (14). Calculations show that 2,4-D moves from the leaves to roots at rates up to 100 cm (40 in.)/hr. The 2,4-D moved as fast in poorly watered plants as in well-watered plants. Placement of the 2,4-D in a droplet over the midrib was more effective in moving the 2,4-D than placement near the edge of the leaf (7).

The translocation of herbicides downward through the phloem closely parallels the movement of food materials from the leaf to other plant parts (19). Thus, when a perennial plant with underground reproductive organs is being treated, translocation of the chemical to below-ground parts is most rapid and most effective when large amounts of food reserves are being moved toward the roots. This usually occurs after full leaf development. Little translocation takes place from the leaves if the readily available food supply has been diminished by continued darkness or by reduced light (25, 28, 29, 34). A high potassium and phosphorous content in the plant favors the translocation of 2,4-D from the leaves of tomato plants (27, 28).

Effective use of the phenoxy herbicides to kill the underground parts of perennial weeds largely depends on the maintenance of live phloem cells but

Table 3-1 Mobility of Herbicides in Plants. Mobility varies between compounds; it may also vary between plants and between various treatments (1)

Free Mobility			Limited Mobility			Little or No Mobility
In Apoplast	In Symplast	In Both	In Apoplast	In Symplast	In Both	
Atrazine	Amiben	Amitrole	Barban	2,4-D	Naptalam	DCPA
Bioxane	Chloramben	Dalapon	Dichlobenil	2,4-DP	Ammonium thiocyanate	DNBP
Bromacil	Fenac	Dicamba	Diquat	MCPA	$AsO_4^{\equiv}$	DNOC
Chlorpropham	Maleic hydrazide	Picloram	Fluorodiphen	2,4,5-T	Diallate	Endothall
Diuron		Pyriclor	Paraquat	TPA	EPTC	Nitrofen
Fluometuron		TBA			Ioxynil	PCP
Monuron					Propanil	Trifluralin
Propham						
Pyrazon						
Simazine						

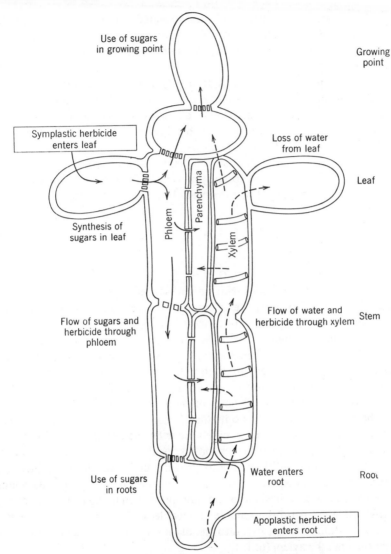

Figure 3-5. Diagram representing routes of translocation of herbicides in plants. [Adapted from PRINCIPLES OF PLANT PHYSIOLOGY by James Bonner and Arthur W. Galston. W. H. Freeman and Company. Copyright © 1952 (4).]

some farmers have figured—wrongly—that they would raise the dosage and get better results. This is not so. The excessive rate quickly immobilizes or kills the phloem cells, stopping translocation into the underground parts, and the tops quickly die. This method has much the same effect as cutting off the tops. The plant, however, may quickly resprout from below-ground buds.

Small dosages of the chemical, with repeated application, will usually bring about more desirable results. With such treatment, the plant is killed very slowly (20). The dosage-rate and timing of treatments need to be worked out for each plant species. With favorable growing conditions, most woody plants should show little or no yellowing 1 week after foliage treatment, some yellowing after 2 weeks, be completely yellowed after 3 weeks, and shed dead leaves after 4 weeks. This timing has worked especially well on such plants as honeysuckle, poison ivy, and blackberries, where 2,4-D is used.

The effective translocation of growth substances permits low-volume spraying. Spray volumes of 2–10 gallons of spray/acre are common with 2,4-D and related products. Uniform application is more important than the amount of water used in the spray. For more details, see several summaries (3, 7, 8, 18, 33, 35).

Apoplastic Translocation

Apoplasticly mobile herbicides which are absorbed by roots follow the same pathway as water. They enter the xylem and are swept upward with the transpiration stream of water and soil nutrients (see Figure 3-5). The driving force for this movement is the removal of water from the leaves by transpiration.

The xylem and cell walls are the principal components of the apoplastic system. They are considered nonliving. Therefore, all types of herbicides, including very toxic or poisonous chemicals, can be absorbed from the soil and quickly translocated to all parts of the plant. This movement does not injure the xylem because it is nonliving. Absorption and translocation may continue for a time even though acutely toxic herbicides have killed the root.

Actually, acutely toxic herbicides may move downward through the xylem tissue under special conditions. Plants growing under conditions of low soil moisture and high transpiration rates develop water deficits within the plant. When strongly acidic or basic solutions are sprayed on living leaves and stems, the cells of these structures are made permeable. If there is a moisture deficit, the toxic substances are drawn into the plant and may be drawn downward rapidly through the xylem (6, 17).

Interaction of Apoplast and Symplast Translocation

Translocation of most herbicides is not restricted to either the symplast or apoplast, but may involve both. However, many herbicides appear to be limited primarily to one or the other system. For example, chloramben is mainly translocated in the symplast, while monuron is mainly translocated in the apoplast. Amitrole is readily translocated in both systems and actually appears to circulate in the plant.

As the herbicide travels through long transport channels, xylem or phloem, some may move into adjacent cells. This may occur by simple diffusion or active uptake. From these adjacent cells, the herbicide may then move into the other long transport channel, phloem or xylem, and be translocated in it.

Intercellular Translocation

Nonpolar substances with low interfacial surface tensions may move through the plant's intercellular spaces. For example, oils may be absorbed by the plant through the cuticle, epidermis, bark, stomates, and even through injured roots. Oils move in any direction—up, down, radial, or tangential; the mechanism of movement is not well established. It is generally believed that the oil moves principally through the intercellular spaces (22, 30, 36, 37). Little or no movement of the oils takes place through the vascular system under normal conditions.

Movement of kerosene-like oils was studied in dandelions, carrots, and parsnips. Oil applied to *cut* roots moved up into the leaves, and oil applied to the leaves moved to the roots. Movement was confined to the intercellular spaces. In a large turgid dandelion root, the rate of diffusion was measured at 4–5 cm (2 in.)/hr (22). 2,4-D esters in kerosene can move in the intercellular spaces much the same as the kerosene (26).

MOLECULAR FATE

The *molecular fate* of a herbicide deals with the changes in the chemical structure of the herbicide molecule within the plant. Most of these changes reduce the phytotoxicity of the molecule (*inactivation*), that is, simazine to hydroxysimazine. However, with some changes the phytotoxicity is increased (*activation*), that is, 2,4-DB [4-(2,4-dichlorophenoxy) butyric acid] to 2,4-D. Since different species of plants may have varying degrees of ability to modify the chemical structure of a herbicide, this difference often determines their tolerance to a given herbicide. Some plants can inactivate a herbicide so rapidly that they are not injured while other plants are killed (e.g., atrazine in corn).

Molecular fate has also been referred to as degradation; however, many herbicides are not only degraded (from the Latin: *de*, down + *gradus*, a step) but may also form conjugates, a molecule containing the herbicide molecules combined with normal plant constituents (e.g., sugars, amino acids).

Higher plants have been shown to alter the molecular configuration of herbicides by a wide variety of chemical reactions. Most of these are probably catalyzed by specific enzymes; however some appear to be nonenzymatic. In most cases the specific enzyme or enzymes have not been isolated and characterized. The following reactions have been shown to be involved in herbicide degradation in higher plants: oxidation, decarboxylation, deamination, dehalogenation, dethioation, dealkylation, dealkyoxylation, dealkylthiolation, hydrolysis, hydroxylation, and conjugation. For more information on degradation, see references 1, 15, and 16.

PLANT METABOLISM

Plant metabolism includes the numerous biochemical reactions which occur in the protoplasm of living plant cells. Although most of these take place in all cells, for example, respiration, some occur only in specific cells, for example, photosynthesis in cells containing chlorophyll. A given herbicide may initially interfere with a single biochemical reaction (e.g., monuron rapidly inhibits the oxygen-evolving step of photosynthesis), or a herbicide may be relatively nonspecific and interfere with several reactions simultaneously (e.g., dalapon). Biochemical reactions are closely coupled and often when one reaction is altered by a herbicide, others are soon affected.

These biochemical reactions may be affected by herbicides: photosynthesis, respiration, carbohydrate metabolism, lipid metabolism, protein metabolism, and nucleic acid metabolism (1). By disrupting any of these biochemical reactions, herbicides may upset plant metabolism and thus injure or kill the plant.

GROWTH AND PLANT STRUCTURE

Growth and plant structure are the result of previous biochemical reactions. For example, seeds will not germinate and grow unless a source of energy is available; this energy comes from the metabolism of storage materials (carbohydrates, fats, or proteins) in the seeds produced by the parent plant. A plant containing chlorophyll and exposed to light produces its own energy by photosynthesis.

Herbicides may induce abnormal plant growth expressed in morphological, anatomical, and cytological effects. However, these effects are generally specific for a given herbicide on a particular plant species. Some of these abnormal responses are: (1) seed germination failure (Figure 3-6), (2) leaf

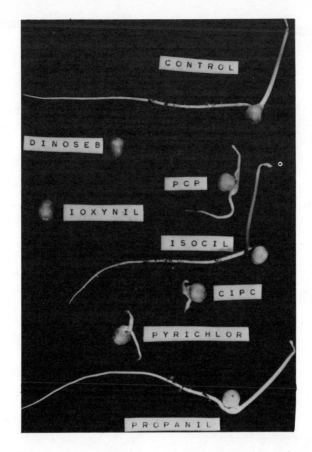

Figure 3-6. Effect of several herbicides on seed germination and root elongation in peas (32).

Figure 3 7.   The grape leaves on the right were exposed to 2,4-D; those on the left are normal.

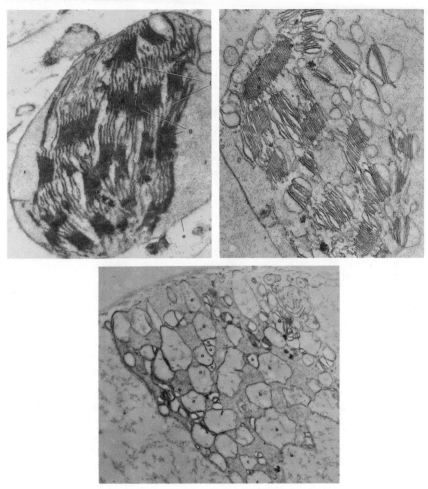

Figure 3-8. Effect of atrazine on the ultrastructure of bean chloroplasts. Upper left: control. Upper right: moderate injury. Lower: severe injury (1).

chlorosis, (3) leaf form (Figure 3-7), (4) stem swelling, (5) root elongation (Figure 3-6), (6) cell division, (7) and chloroplasts (Figure 3-8).

LITERATURE CITED

1. Ashton, F. M., and A. S. Crafts, *Mode of Action of Herbicides*, Wiley, New York, 1973.
2. Audus, L. J., *Physiology and Biochemistry of Herbicides*, Academic, New York, 1964.

3. Blackman, G. E., W. G. Templeman, and D. J. Halliday, *Ann. Rev. Plant Physiol.*, *2*, 199 (1951).

4. Bonner, J., and A. W. Galston, *Principles of Plant Physiology*, W. H. Freeman, San Francisco, 1952.

5. Casida, J. E., and L. Lykken, *Ann. Rev. Plant Physiol.*, *20*, 607 (1969).

6. Crafts, A. S., *Hilgardia*, *7*, 361 (1933).

7. Crafts, A. S., *Ann. Rev. Plant Physiol.*, *4*, 253 (1953).

8. Crafts, A. S., *Hilgardia*, *26*, 287 (1956).

9. Crafts, A. S., and S. Yamaguchi, *The Autoradiography of Plant Materials*, Division of Agricultural Science, Univ. of California, Berkeley (1964).

10. Currier, H. B., and C. D. Dybing, *Weeds*, *7*, 195 (1959).

11. Dawson, J. H., *Weeds*, *11*, 60 (1963).

12. Donaldson, T. W., *Absorption of the Herbicides 2,4-D and Monuron by Barley Roots*, Ph.D. Dissertation, Univ. of California, Davis (1967).

13. Epstein, E., *Sci. Amer.*, *228*, 48 (1973).

14. Fang, S. C., E. G. Jaworski, A. V. Logan, V. H. Freed, and J. S. Butts, *Arch. Biochem. Biophys.*, *32*, 249 (1951).

15. Freed, V. H., and M. Montgomery, *Weeds*, *6*, 386 (1958).

16. Kearney, P. C., and D. D. Kaufman, *Degradation of Herbicides*, Marcel Dekker, New York (1969).

17. Kennedy, P. B., and A. S. Crafts, *Plant Physiol.*, *2*, 503 (1927).

18. Leonard, O. A., and A. S. Crafts, *Hilgardia*, *26*, 366 (1956).

19. Linder, P. J., J. W. Brown, and J. W. Mitchel, *Botan. Gaz.*, *110*, 628 (1949).

20. Loomis, W. E., *Proceedings of North Central Weed Control Conference 6*, 101 (1949).

21. Meggitt, W. F., R. J. Aldrich, and W. C. Shaw, *Weeds*, *4*, 131 (1956).

22. Minshall, W. H., and V. A. Helson, *Proceedings of the Northeastern States Weed Control Conference*, pp. 8–12 (1949).

23. Oliver, L. R., G. N. Prendeville, and M. M. Schreiber, *Weed Sci.*, *16*, 534 (1968).

24. Orgell, W. H., Ph.D. Dissertation, Univ. of California, Davis (1954).

25. Rice, E. L., *Botan. Gaz.*, *109*, 201 (1948).

26. Rice, E. L., and L. M. Rohrbaugh, *Botan. Gaz.*, *11*, 76 (1953).

27. Rice, E. L., and L. M. Rohrbaugh, *Plant Physiol.*, *33*, 300 (1958).

28. Rohrbaugh, L. M., and E. L. Rice, *Plant Physiol.*, *31*, 196 (1956).

29. Rohrbaugh, L. M., and E. L. Rice, *Botan. Gaz.*, *11*, 85 (1949).

30. Rohrbaugh, P. W., *Plant. Physiol.*, *9*, 699 (1934).

31. Staniforth, D. W., and W. E. Loomis, *Science*, *109*, 628 (1949).

32. van Hoogstraten, S. D., Ph.D. Dissertation, Univ. of California, Davis (1972).

33. van Overbeek, J., *Ann Rev. Plant Physiol.*, *7*, 355 (1956).

34. Weaver, R. J., and H. R. DeRose, *Botan. Gaz.*, *107*, 509 (1946).

35. Woodford, E. K., K. Holly, and C. C. McCready, *Ann. Rev. Plant Physiol.*, *9*, 311 (1958).

36. Young, P. A., *J. Agr. Res.*, *49*, 559 (1934).

37. Young, P. A., *J. Agr. Res.*, *51*, 925 (1935).

4   Herbicides and the Soil

Many factors influence the effectiveness of a herbicide. Herbicides applied to the soil are directly affected by soil characteristics; however, herbicides applied to the foliage of plants are *not* directly affected by soil differences even though in foliar applications some of the herbicide may fall directly on the soil and usually minute amounts reach the soil later as the plants die and disintegrate.

The numerous soil factors, the many different kinds of herbicides, and the large number of plant species and climatic variations make the study of herbicides in soils very complex and diverse. There are at least 10 different soil variables of major importance, 125 different herbicides, and hundreds of different plant species involved. Thus, the complexity of herbicide–soil–weather–plant interactions is enormous.

We apply herbicides directly to the soil as (1) preplanting treatments, (2) preemergence treatments, or (3) postemergence treatments. The time of application may refer to the crop or to the weed.

Some preplanting treatments are mechanically mixed into the soil, while others are applied to the surface. When mechanically incorporated into the soil, the chemical is usually immediately effective on seeds germinating in the area—with no added moisture. When applied to the soil surface, most herbicides must be moved into the soil by rainfall or sprinkler irrigation to be effective.

The success of an incorporated preplanting treatment or a preemergence treatment depends largely on the presence of a high concentration of the herbicide in the upper 2 in. of soil. This is where most annual weed seeds germinate. Also, there must be a relatively low concentration of the herbicide in the zone where the crop seeds germinate, unless the crop seed is tolerant to the chemical (see Figure 4-1).

Soil surface

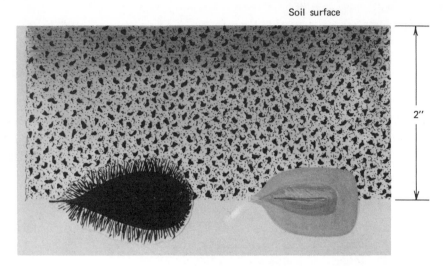

2"

Figure 4-1. Small-seeded weeds germinate principally in the upper 2 in. of soil. Most preemergence herbicides are applied to the surface and leached into the seed-germinating zone. There must be a low concentration of the herbicide where the crop seed germinates, or the crop seed must have true tolerance to the herbicide. (North Carolina State University.)

If the viable weed seeds in the surface soil are killed, the area may remain weed-free long after the chemical has disappeared. This is because most weed seeds will not germinate if buried deeply in the soil.

For effective soil sterilization, the chemical must remain active in the rooting zone to kill both germinating seeds and growing plants.

PERSISTENCE IN THE SOIL

The length of time that a herbicide remains active or persists in the soil is extremely important because it determines the length of time that weed control can be expected. Residual toxicity is also important because it relates to phytotoxic after-effects that may prove injurious to succeeding crops or plantings.

Herbicides may disappear faster with large amounts of water which provide leaching and with repeated cultivation or mixing of the soil. In some cases, fertilizers can be added to reduce the injurious after-effects. For example, nitrate nitrogen reduces the phytotoxic effects of sodium chlorate.

Seven factors affect the persistence of a herbicide in the soil: (1) microbial decomposition, (2) chemical decomposition, (3) adsorption on soil colloids,

(4) leaching, (5) volatility, (6) photo-decomposition, and (7) removal by higher plants when harvested. Each factor will be discussed later in this chapter.

Table 4-1 gives an approximate length of persistence for various herbicides under a given set of conditions. This list includes the most important herbicides, and the information is based on experimental work and general observations. In general, the conditions stated are favorable to rapid herbicide decomposition or a short period of herbicide toxicity. Those herbicides persisting 1 month or less may be used to control weeds present at the time of treatment. Those persisting 1–3 months will protect the crop only during a short period early in the growing season. Those providing 3 to 12 months control may provide protection to the crop for the entire growing season. Those providing more than 12-months control are used primarily for total vegetation control or where persistence is of little importance. If this group is used on crop land it is at low rates of application, with rotations of 2 or more years duration, and where crops in the rotation are known to be tolerant (see Figures 4-2 and 4-3).

A symposium (13) and book (6) are available which discuss the nature and fate of chemicals applied to soils, plants, and animals.

Microorganism Decomposition

The principal microorganisms in the soil are algae, fungi, actinomyces, and bacteria. They must have food for energy and growth. Organic compounds of the soil provide this food supply, the exception being a very small group of organisms that feeds on inorganic sources.

Microorganisms use all types of organic matter, including organic herbicides. Some chemicals are easily decomposed (easily utilized by the microorganisms) whereas others resist decomposition.

The microorganisms use carbonaceous organic matter in their bodies primarily through the process of aerobic respiration. In the process, oxygen is absorbed and CO_2 is released.

If an organic substance, that is, a herbicide, is applied to the soil, microorganisms immediately attack it. Those that can utilize the new food supply will likely flourish and increase in number. In effect, this hastens decomposition of that organic substance. When decomposed, the organisms decrease in number because their new food supply is gone.

Other factors beside food supply may quickly affect the growth and rate of multiplication of microorganisms. These factors are temperature, water, oxygen, and mineral nutrient supply. Most soil microorganisms are nearly dormant at 40°F, 75–90°F being the most favorable. Without water most

Table 4-1. Persistence of Biological Activity at the Usual Rate of Herbicide Application in Moist-Fertile Soils under Field Conditions and Summer Temperatures in a Temperate Climate[1]

1 Month or Less (Temporary Effects)	1–3 Months (Early Season Control)	3–12 Months[3] (Full Season Control)	Over 12 Months[4] (Total Vegetation Control)
Acrolein	Bentazon	Alachlor	Arsenic
Amitrole	Butachlor	Ametryn	Borate
AMS	Butylate	Atrazine	Bromacil
Barban	CDAA	Benefin	Chlorate
Cacodylic acid	CDEC	Bensulide	Fenac
Chloroxuron	Chloramben	Bromoxynil	Picloram
Dalapon	Chlorpropham	Chlorobromuron	Tebuthiuron
2,4-D	Cycloate	Cyprazine	Terbacil
2,4-DB	Diallate	DCPA	2,3,6-TBA
Dinoseb (DNBP)	2,4-DEP	Dicamba	
Diquat[2]	Diphenamid	Dichlobenil	
DSMA	EPTC	Dinitramine	
Endothall	Mecoprop	Diuron	
Fluorodifen	Naptalam	Fenuron	
Glyphosate	Pebulate	Fluometuron	
Metham	PCP	Isopropalin	
Methyl bromide	Propachlor	Linuron	
MCPA	Pyrazon	Metobromuron	
MH	Siduron	Metribuzin	
Molinate	Silvex	Monolinuron	
MSMA	TCA	Monuron	
Nitrofen	Triallate	Napropamide	
Paraquat[2]	2,4,5-T	Nitralin	
Phenmedipham	Vernolate	Norea	
Propanil		Oryzalin	
Propham		Prometryn	
		Pronamide	
		Propazine	
		Simazine	
		Terbutol	
		Terbutryn	
		Trifluralin	

[1] These are approximate values and will vary somewhat as the seven factors discussed in the text vary.

[2] Although diquat and paraquat molecules may remain unchanged in soils for long periods of time, they are adsorbed so tightly to many soils that they become biologically inactive.

[3] At higher rates of application, some of these chemicals may persist at biologically active levels for more than 12 months.

[4] At lower rates of application, some of these chemicals may persist at biologically active levels for less than 12 months.

Figure 4-2.   Monuron at rates of 4–20 lb/acre may provide annual-weed control up to 2 years. There is no mowing or fire problem around these posts and under the steel cable. (E. I. du Pont de Nemours and Company.)

organisms become dormant or die. Aerobic organisms are very sensitive to an adequate oxygen supply, and deficiency of nutrients, such as nitrogen, phosphorous, or potash, may reduce microorganism growth.

Thus, the herbicide may remain toxic in the soil for considerable time if the soil is cold, dry, poorly aerated, or if other conditions are unfavorable to the microorganisms. If the organisms are destroyed by soil sterilization (steam or chemical methods), decomposition of the herbicide may temporarily stop.

Soil pH also influences the microorganisms. In general, the bacteria and actinomyces are favored by soils having a medium to high pH, and their activity is seriously reduced below pH 5.5. Fungi tolerate all normal soil pH values. In normal soils, therefore, fungi predominate at pH 5.5 and below.

Figure 4-3. Simazine, at rates of 4–20 lb/acre, may provide annual-weed control up to 2 years. There is no mowing or fire problem in this gas-storage area. (Ciba-Geigy Chemical Corporation.)

Above pH 5.5, the fungi are reduced through competition with bacteria and actinomyces.

Thus a warm, moist, well-aerated, fertile soil with optimal pH is most favorable to microorganisms. Under these ideal conditions, the organisms can most quickly decompose organic herbicides. Microbial decomposition of many herbicides follows typical growth curves for bacterial populations (see Figure 4-4).

At the usual rate of herbicide application on farm lands, the total number of organisms is seldom changed to any great extent because the herbicide may benefit one group of organisms and injure another group. When the herbicides

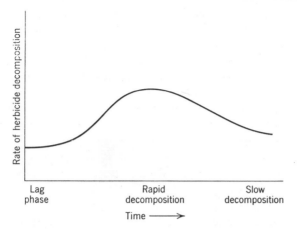

Figure 4-4. Rate of herbicide decomposition by microorganisms. (A. D. Worsham, North Carolina State University.)

are decomposed, the microorganism population returns to "normal." The biologic activity of most herbicides applied at rates recommended for cultivated crops disappears in less than 12 months (Table 4-1). Therefore, no long-time effect on the microorganism population of the soil is expected.

Chemical Decomposition

Chemical decomposition destroys some herbicides and activates others. Chemical decomposition may involve reactions such as oxidation, reduction, and hydrolysis. For example, potassium cyanate and dalapon will slowly hydrolyze in the presence of water, rendering both ineffective as herbicides.

In soil saturated with water, oxygen will likely become a limiting factor. Under such conditions, anaerobic degradation of organic compounds can be expected. It has not been established whether this is chemical or microbial, but it is probable that both are involved. Using trifluralin, and under standing water, degradation was complete in 7 days at 76°F in nonautoclaved soils, whereas only 20% had degraded at 38°F. For further details see page 193.

Adsorption on Soil Colloids

Colloids (from the Greek word meaning gluelike) have unusual adsorptive capacities. As a soils term, colloid refers to the microscopic (1 micron or less in diameter) inorganic and organic particles in the soil. These particles have an extremely large surface area in proportion to a given volume. It has been calculated that 1 cubic in. of colloidal clay may have 200–500 square ft of particle surface area.

Many clay particles resemble the negative radical of a weak acid, such as COO^- in acetic acid. Thus the negative clay particle attracts to its surface positive ions (cations) such as hydrogen, calcium, magnesium, sodium, and ammonium. These cations are rather easily displaced, or exchanged (from the clay particle) for other cations. They are known as *exchangeable ions.* This replacement is called *ionic exchange* or *base exchange.*

The base-exchange capacity of a soil is expressed as milliequivalents (meq) of hydrogen per 100 g of dry soil. A soil with a base-exchange capacity of 1 meq can absorb and hold 1 mg of hydrogen (or its equivalent) for every 100 g of soil. This is equivalent to 10 ppm or 0.001 % of hydrogen.

Adsorptive capacity, and thus exchange capacity, is closely associated with inorganic and organic colloids of the soil. Inorganic colloids are principally clay. There are two principal groups of clay: *kaolinite* and *montmorillonite*. Because of their chemical nature and lattice structure, the kaolinite colloids have a relatively low adsorptive or base-exchange capacity. Kaolinite clays have a tendency to predominate in areas of high rainfall and warm to hot temperatures. Thus clay soils of the tropics and the Southeastern United States are principally kaolinite. Aluminum and iron oxides are important constituents of kaolinic soils.

Montmorillonite clays have an adsorptive capacity of perhaps 3–7 times that of kaolin clays. The montmorillonite clays predominate in areas of moderate rainfall and moderate temperatures. They are typical of Corn Belt soils.

Organic colloids involve the humus of the soil. Organic colloids have a very high adsorptive capacity—about 4 times the base-exchange capacity of a montmorillonite clay, and perhaps 20 times that of kaolinite on a weight basis.

Scientists have studied the exchange of negatively charged particles (anions) much less than of cations. Anion exchange does occur. Anions such as OH^-, $PO_4^{\equiv}$, $HPO_4^=$, $H_2PO_4^-$, NO_3^-, HCO_3^-, and $CO_3^=$ may be involved. Anion exchange is probably more important in humus and in kaolinite clays than in montmorillonite clays. Because the active portion of many herbicide molecules is anionic, research designed to aid in the understanding of anionic adsorption should prove helpful.

Observations in research work as well as in the field have shown that:

1. Soils high in organic matter require relatively large amounts of most soil-applied herbicides for weed control.
2. Soils high in clay content require more soil-applied herbicide than sandy soils for weed control.

3. Soils high in organic matter and clay content have a tendency to hold the herbicides for a longer time than sand. The adsorbed herbicide may be released so slowly that the chemical is not effective as a herbicide.

The adsorption of monuron was studied in various soils; some of the results are shown in Table 4-2. These findings may help to clarify the principles of monuron adsorption (9, 10).

Table 4-2. Monuron Adsorption Correlated with Certain Soil Properties (9)

Soil Property (%)	Correlation Coefficient
Organic matter	0.991
Clay	0.209
Silt	0.358

The percent of organic matter had a high correlation with the amount of monuron adsorbed. The type of clay (see the previous discussion) may explain the relatively low adsorption of the clay in this study.

In another study (11), the toxicity of diuron to cotton and ryegrass in 12 different soils was measured. Organic matter content, cation exchange capacity, and total exchangeable bases gave a multiple regression coefficient (R) of 0.968 for cotton and 0.956 for Italian ryegrass.

Scientists have attempted to develop equations for predicting safe, effective rates of herbicides for various soil types (8, 12). Of the several chemical and physical properties of soils that were measured, organic matter gave the best prediction of performance. The use of other properties with organic matter in the equations did not greatly improve predictability (8). Such prediction equations are useful for many soils; however, the theoretical value of a given soil may be considerably different than that observed in practice. Therefore,

Table 4-3. Amount of Herbicide Required for 50% Grass Control in Soils Containing 1% and 40% Organic Matter; and Amount of Herbicide Required for Each 1% Increase in Organic Matter Content (12)

Herbicide	For 1% Organic Matter (lb/acre)	For 40% Organic Matter (lb/acre)	For 1% Increase in Organic Matter (lb/acre)
Simazine	0.13	18.12	0.46
Diuron	0.34	6.47	0.16
Chlorpropham	3.82	42.31	0.99

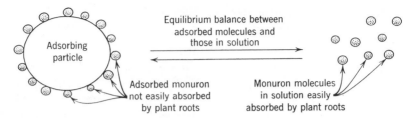

Figure 4-5. Plants absorb monuron in solution more easily than monuron which is adsorbed by the soil colloids.

such values should be used as guidelines but not as absolute recommendations.

"Activated" carbon is one of the most effective adsorptive materials known. It has been used to protect plants from herbicides. Roots of strawberry plants have been coated with "activated" carbon prior to setting the plants in herbicide-treated soil. Also, bands of "activated" carbon have been placed over previously seeded rows soon after planting and before a preemergence herbicide treatment. Well-decomposed organic matter could presumably have properties similar to "activated" carbon.

These studies indicate that a certain amount of herbicide is required to saturate the adsorptive capacity of a soil. Above this "threshold level" heavier rates will greatly increase the amount of herbicide in the soil solution, and thus increase the herbicide toxicity to plants.

Therefore, the nature and strength of the "adsorbed linkage" or "bonding" is of considerable importance for both cations and anions. Apparently the nature and characteristics of the colloidal organic matter, as well as the clay, may affect the tenacity of this bonding (see Figure 4-5).

In summary, the various soils show large differences in their adsorptive capacities. In practice, however, the *range* of herbicidal rates of application is much less than might be predicted from the very wide ranges in adsorptive capacity of the soils.

Leaching

Leaching is the downward movement of a substance by water through the soil. The movement of a herbicide by leaching may determine its effectiveness as a herbicide, may explain selectivity or lack of it, or may account for its removal from the soil. Preemergence herbicides are frequently applied to the soil surface. Rain leaches the chemical into the upper soil layers. Weed seeds germinating in the presence of the herbicide are killed. Large-seeded crops such as corn, cotton, and peanuts planted below the area of high herbicidal concentration may not be injured (see Figure 4-1). In addition to the protection offered by depth, crop tolerance to the herbicide is desirable.

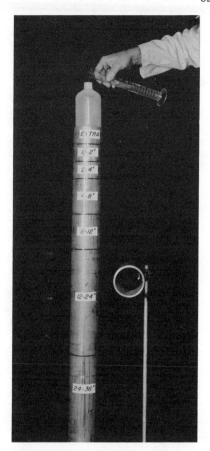

Figure 4-6. Laboratory study of herbicide leaching through a soil column. Simulated rainfall is being added at the top. The depth of leaching is determined by disassembling the steel columns and analyzing the soil enclosed. (R. P. Upchurch, formerly North Carolina State University.)

Some herbicides can be removed from the soil by leaching (see Figure 4-6). For example, sodium chlorate can be applied to shallow-rooted turf and the chemical leached to kill deep-rooted plants without injuring the turf.

The extent to which a herbicide is leached is determined principally by:

1. Adsorptive relationships between the herbicide and the soil.
2. Solubility of the herbicide in water.
3. Amount of water passing downward through a soil.

Solubility is sometimes offered as the principal factor affecting the leaching of a herbicide. Simple calculation of the amount of water in the usual rainfall disproves this assumption. A 4-in. rainfall weighs nearly 1,000,000 lb/acre. If you apply 1 lb of herbicide/acre, this equals 1 ppm of the herbicide in water; thus, if the herbicide is soluble to the extent of 1 ppm, you might expect a 4-in. rain to remove essentially all of the herbicide from the surface inches of soil.

1.5 lbs 2, 4–D/acre 6 lbs 2, 4–D/acre

Inches of water added

Figure 4-7.   Single-eye sugarcane seedpieces germinated in three soil types after receiving 2 levels of 2,4-D as a preemergence application and three quantities of water at a single irrigation. Photographed 7 weeks after planting (3).

84

An example is given to illustrate the point. Monuron, at 25°C is soluble in water to a concentration of 230 ppm. Monuron was applied at the rate of 2 lb/acre to a fine sandy soil, and 16 in. of water failed to remove the monuron from the surface inch of soil. This amount of water is capable of dissolving 400 times the amount of herbicide applied (7).

The interrelationship between binding of herbicides to the soil and water solubility can be demonstrated with 2,4-D. The salts of 2,4-D are water-soluble and readily leach through porous, sandy soils. Soils with high organic matter content adsorb the 2,4-D, reducing the tendency to leach. The ester formulations of 2,4-D have a low water solubility. Their tendency to leach is reduced by both low solubility and by adsorption by the soil.

Leaching of 2,4-D was tested in Hawaii with variables of three soil types and three amounts of water (3). Makiki, Kailua, and Monoa soils had relative adsorptions of 2, 8, and 75%, respectively, for 2,4-D in laboratory tests. One-eye seedpieces of sugarcane were planted 2 in. deep. Water was applied at 18, 10, and 2 in. after a preemergence application of 2,4-D.

Figure 4-7 shows that with 18 and 10 in. of water on the high adsorptive soil (Monoa), sugarcane was injured, but with 2 in. of water it was not injured. However, the relationship between water applied and injury was reversed with the low-adsorptive soils (Makiki and Kailua); plants were injured with the 2-in. application of water. Therefore, the 2,4-D injury suffered by the germinating sugarcane plant depended upon the adsorptiveness of the soil and the amount of water applied.

To restate the point, the strength of "adsorption bonds" is considered more important than water solubility in determining the leaching of herbicides.

Herbicides have been known to move upward in the soil. If water evaporates from the soil surface, water may move slowly upward and may carry with it soluble herbicides. As the water evaporates, the herbicide is deposited on the soil surface.

In arid areas where furrow irrigation is practiced, lateral movement of herbicides in soils also occurs.

Volatilization

All chemicals, both liquids and solids, have a vapor pressure. The evaporation of water is an example of a liquid and the vaporization of naphthalene (moth balls) is an example of a solid that will vaporize. At a given pressure, vaporization of both liquids and solids increases as the temperature rises (page 113).

Herbicides may evaporate and be lost to the atmosphere as gases. The gases may or may not be toxic to plants and the volatile gases may drift to susceptible plants. The *ester* forms of 2,4-D are volatile, and the vapors or fumes can cause injury to susceptible crops such as cotton or tomatoes.

The herbicide may move into a porous soil as a gas. EPTC is thought to move in this way. Adsorbed by the soil, the EPTC may effectively kill germinating seeds.

The importance of volatilization and the loss of a herbicide from the soil surface is often underestimated. In volatility studies, EPTC volatilized from a free-liquid surface at the rate of 57 μg per cm^2 (about 5 lb/acre) of surface area per hr at 86°F (1). This high rate of vaporization could easily explain the loss of the herbicide. EPTC, trifluralin, and other volatile soil-applied herbicides are usually mechanically mixed into the soil soon after application to reduce loss.

Codistillation with water evaporating from the soil surface (steam distillation) is another means by which a volatile herbicide may be lost. This process has been studied little but may be of considerable importance in view of the immense amount of water lost from the soil surface through evaporation.

Herbicides such as atrazine with very low vapor pressures, however, may be lost from a surface over an extended period of time, especially if exposed to high temperatures. Soil-surface temperatures have been measured as high as 180°F.

Rain or irrigation water applied to a dry or moderately dry soil will usually leach the herbicide into the soil, or aid in its adsorption by the soil. Once adsorbed by the soil, the loss by volatility is usually reduced.

Photodecomposition

Photodecomposition, or decomposition by light, has been reported for many herbicides (4). This process begins when the herbicide molecule absorbs light energy; this causes excitation of the electrons and may result in breakage, or formation of chemical bonds.

Most herbicides absorb radiation in the ultraviolet region (150 to 4000 mμ). However, solar energy below 295 mμ reaching the surface of the earth is considered to be negligible. The absorption maxima of a few are given in Table 4-4. The "sensitization" process may also be involved; light energy is absorbed by an intermediate molecule and transferred to the herbicide molecule by collision (4). Thus the effective wavelength of light could be outside the absorption spectra of the herbicide.

Some products of photodecomposition are similar to those produced by chemical or biological means.

Monuron solution containing 88.3 ppm in distilled water was sealed in quartz tubes and exposed to sunlight for 48 days. There was an 83% loss of the monuron as compared to the check (5).

Table 4-4. Ultraviolet Absorption Maxima of Selected Herbicides in Water (2)

Common Name	Chemical Name	Absorption Maximum, (mμ)
Simazine	2-chloro-4,6-bis(ethylamino)-s-triazine	220
2,4-D	(2,4-dichlorophenoxy)acetic acid	220, 230, 283
2,4,5-T	(2,4,5-trichlorophenoxy)acetic acid	220, 289
Propham	isopropyl carbanilate	234
Monuron	3-(p-chlorophenyl)-1,1-dimethylurea	244
Propanil	3',4'-dichloropropionanilide	248
Chloramben	3-amino-2,5-dichlorobenzoic acid	297
Dinoseb	2-sec-butyl-4,6-dinitrophenol	375
Trifluralin	α,α,α-trifluoro-2,6-dinitro-N,N-dipropyl-p-toluidine	376

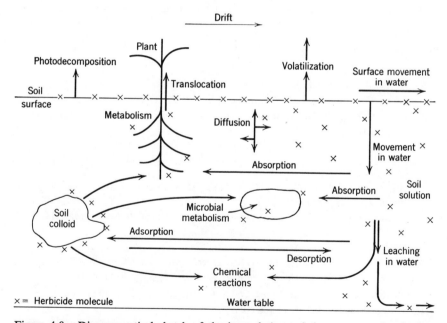

Figure 4-8. Diagrammatical sketch of the interrelations of the processes that lead to detoxication, degradation, and disappearance of herbicides. (From T. J. Sheets and D. D. Kaufman.)

Chemicals applied to soil surfaces are frequently lost, especially if they remain for an extended period without rain. It is entirely possible that photo-decomposition has taken place. However, do not overlook other factors that may account for the loss; volatilization accentuated by high soil-surface temperatures, biological and chemical deactivation, and adsorption are a few of the factors that should be considered.

Removal by Higher Plants

Higher plants may absorb herbicides from the soil in which they are growing. The absorbed herbicide may then be removed when the crop is harvested. This may not be a major factor in persistence of herbicides under most conditions; however, it has been used to help remove persistent herbicides from soils where they were applied as soil sterilants and the planting of ornamentals was desired (e.g., corn for simazine or atrazine removal).

Figure 4-8 shows the interrelations of the processes that lead to detoxication, degradation, and disappearance of herbicides in the environment.

LITERATURE CITED

1. Ashton, F. M., and T. J. Sheets, *Weeds, 7*, 88 (1959).
2. Bailey, G. W., and J. L. White, *Residue Rev., 10*, 97 (1965).
3. Burr, G. O., and F. M. Ashton, *Report Hawaiin Sugar Technology, Sixth Ann. Meet.*, 1948, p. 1.
4. Crosby, D. G., and M. Li, "Herbicide Photodecomposition", in *Degradation of Herbicides*, P. C. Kearney and D. D. Kaufman, Eds., Marcel Dekker, New York, 1969, p. 321.
5. Hill, G. D., J. W. McGahen, H. M. Baker, D. W. Finnerty, and C. W. Bingeman, *Agron. J., 47(2)*, 93 (1955).
6. Kearney, P. C., and D. D. Kaufman, Eds., *Degradation of Herbicides*, Marcel Dekker, New York, 1969.
7. Ogle, R. E., and G. F. Warren, *Weeds, 3*, 257 (1954).
8. Sheets, T. J., A. S. Crafts, and H. R. Drever, *Agric. Food Chem., 10*, 458 (1962).
9. Sherbourne, H. R., and V. H. Freed, *J. Agr. Food Chem., 2*, 937 (1954).
10. Sherbourne, H. R., V. H. Freed, and S. C. Fang, *Weeds, 3(1)*, 50 (1956).
11. Upchurch, R. P., *Weeds, 6*, 168 (1958).
12. Upchurch, R. P., F. L. Selman, D. D. Mason, and E. G. Kamprath, *Weeds, 14*, 42 (1966).
13. U.S.D.A., *The Nature and Fate of Chemicals Applied to Soils, Plants, and Animals— A Symposium*, United States Department of Agriculture, A.R.S.-20-9, Washington D.C., September 1960.

5   Selectivity of Herbicides

A selective herbicide kills or retards growth of one or more plants, whereas another plant is tolerant. Thus, a selective herbicide retards growth or kills one plant species (the weed), whereas another plant species (the crop) is tolerant to the same treatment. Ideally, the weed is killed, but sometimes it is necessary only to retard its growth long enough for the crop to become dominant. A herbicide is selective to a particular crop only within certain limits. These limits are determined by a complex interaction between *plants*, the *herbicide*, and the *environment*.

ROLE OF THE PLANT

Seven factors affect plant response (both weeds and crops) to a chemical: age, growth rate, morphology, physiology, biophysical processes, biochemical processes, and genetic inheritance.

Age

The younger the plant, the higher percentage of the plant that is meristematic (growing) tissue, thus creating higher total-biological activity of the plant. Thus, the age of a plant often determines its response to a particular herbicide; young plants are less tolerant than older ones. Preemergence treatments which kill germinating weed seeds or seedlings commonly have little or no effect on established weeds.

Growth Rate

Probably for this same reason, the growth rate of plants has a pronounced effect on their reaction to some herbicides. In general, fast-growing plants are more susceptible to treatment than are slow-growing plants.

Morphology

The morphology of a plant can determine whether or not it can be killed by a specific herbicide. Morphological differences are found in root systems, location of growing points, and leaf properties.

Root Systems

Annual weeds in a perennial crop can be controlled because most annual weeds have shallow root systems, whereas perennial crops (e.g., alfalfa) have deep, extensive root systems. Such deep root systems may escape injury through depth-protection, whereas the shallow roots may be killed. This way, it is possible to control many annual weeds among deep-rooted plants such as fruit trees and woody perennial ornamentals.

Location of Growth Points

The growing points of grasses are located at the base of the plant and below ground; thus, they are protected from overtop-contact herbicides (sprayed from above). Therefore, a contact spray may injure the leaves of cereals but not the growing points.

Most broadleaf plants have exposed growing points at the tips of the shoots and in leaf axils. These growing points are directly exposed to chemical sprays. If all the growing points are killed, the plant dies.

Leaf Properties

Certain leaf properties protect crops treated by selective herbicides. Liquid spray droplets can stick to only a small part of the surface of narrow, upright leaves (as in cereals and onions), or waxy leaf surfaces, or leaves that are corrugated or formed of small ridges. When sprays hit such leaves the droplets have a tendency to bounce off or wet the surfaces only in small spots, thus reducing the effect of the herbicide.

Broadleaf plants usually have wide, smooth leaf surfaces, extending horizontally from the plant stem. Such leaves intercept and hold more of the spray; droplets are less apt to bounce off. Therefore, when broadleaf weeds such as lambsquarters, wild radish, pigweed, or wild mustard are sprayed with contact herbicides, the spray solution tends to spread as a thin film, or the droplets spread so as to wet a large portion of the leaf, thus killing the weed. The same sprays on cereals or onions rebound and leave the plants uninjured.

Physiology

The physiology of the plant also determines the amount of herbicide taken up by the plant (absorption) and how it moves within the plant (translocation). Generally, plants which absorb and translocate the greatest amount of herbicide will be killed.

Absorption

Plants with penetrable cuticle or large stomata (minute pores on leaf surfaces) absorb more of the herbicide, increasing the toxicity of the herbicide. Suitable wetting agents increase absorption through both the stomata and cuticle; thus, a wetting agent may equalize absorption in different plants, losing a selectivity that previously existed (see Figure 5-1).

Translocation

Once a herbicide has entered a plant it may move from the point of absorption to other parts of the plant. Movement within the plant is called *translocation*. Translocation occurs both upward from the root to plant parts above the soil (primarily through the xylem), and downward from the leaves to the underground parts (primarily through the phloem). When movement occurs readily in both directions, the herbicide is said to be *completely systemic*. (see Figure 3-5 and Table 3-1).

The translocation rate and the amount of herbicide translocated vary both with different herbicides and plant species. The rate varies even within a given species under different environmental conditions. The rate and amount of translocation of 2,4-D, for example, are usually greater in susceptible species than in resistant species.

Biophysical Processes

Biophysical differences between plants, such as adsorption and membrane stability, may determine whether or not a plant is killed.

Adsorption

Adsorption of herbicides by plant-cell constituents inactivates herbicidal material, probably by a physical rather than biochemical process. Radioactive

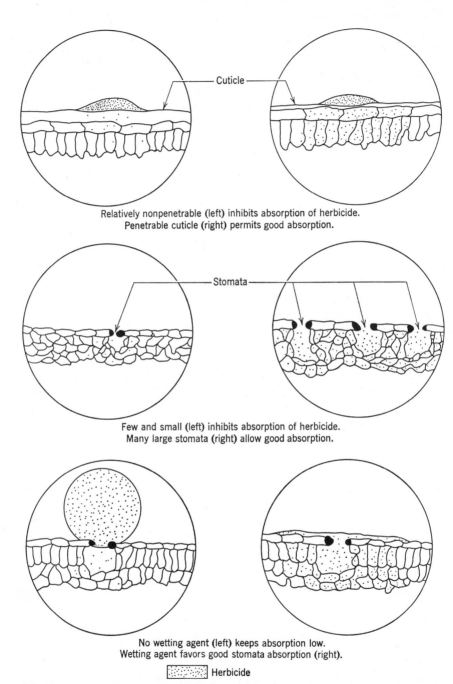

Relatively nonpenetrable (left) inhibits absorption of herbicide.
Penetrable cuticle (right) permits good absorption.

Few and small (left) inhibits absorption of herbicide.
Many large stomata (right) allow good absorption.

No wetting agent (left) keeps absorption low.
Wetting agent favors good stomata absorption (right).

Herbicide

Figure 5-1. Absorption. Cuticle penetrability, number and size of stomata, and wetting agents influence the foliar absorption of herbicides (1).

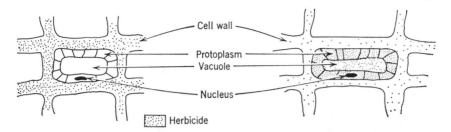

Figure 5-2. Biophysical. Left: The herbicide is absorbed by cell wall, and is prevented from reaching protoplasm. Right: The herbicide not absorbed by cell wall, and reaches protoplasm.

tracer studies have shown that the movements of herbicides are slowed down by the surrounding plant tissues. In extreme cases a herbicide may be bound so tightly to some plant constituent that it is not readily translocated from the point of application to the site of action; or it may even be held so tightly that it is unavailable for herbicidal action (see Figure 5-2).

Membrane Stability

Tolerance to oils, existing in carrots and other crops of the carrot family, is one of the oldest examples of biophysical selectivity. Selective oils used for weed control in carrots kill weeds by damaging the cellular membranes and allowing the cell sap to flow into the intercellular spaces (causing the leaves to appear water-soaked); this causes the cells to die and tissues to dry out later. Because their cellular membranes are resistant to this damage, the carrots are not killed (see Figure 5-3).

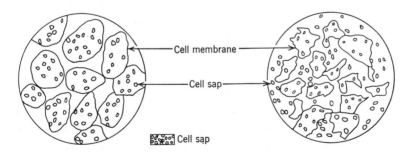

Figure 5-3. Biophysical. Carrot cell membranes (left) are resistant to selective oil. Membranes stay intact, keeping cell sap inside. Weed-cell membranes (right) are damaged by selective oil, allowing the cell sap to leak into intercellular spaces.

Biochemical Processes

Biochemical reactions in various plants may protect them from injury by certain herbicides. Their reactions include enzyme inactivation and herbicide activation or inactivation.

Enzyme Inactivation

Many herbicides reduce enzyme activity in one plant species but not in another, thereby selectively interfering with one or more of the plant's metabolic processes or perhaps with photosynthesis. This differential enzyme inactivation can kill certain plants and leave others unharmed (see Figure 5-4).

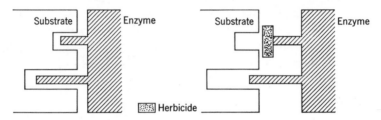

Figure 5-4.   Biochemical. Herbicide (left) does not interfere with enzyme reaction and metabolism. Herbicide (right) alters structure and attachment of enzyme and upsets metabolic processes.

Herbicide Activation

Activation of a harmless chemical into a herbicide sometimes can be used for selective weed control. For example, the relatively harmless compound 2,4-DB is changed in some sensitive plants into the weedkiller 2,4-D, in resistant plants (e.g., alfalfa) this reaction takes place very slowly. In the resistant alfalfa, lethal concentration of 2,4-D from the 2,4-DB treatment does not accumulate.

Genetic Inheritance

The genetic complement of a plant determines the extent to which it responds to its environment. Many of these responses are morphological, physiological, biophysical, or biochemical. These responses vary from genus to genus, but within a genus plant reactions to a given herbicide tend to be similar. However, there are exceptions—tolerance to a herbicide may vary considerably from species to species within a genus or even from variety to variety within one species. Thus, it is potentially possible to develop and select crop varieties that are tolerant to a specific herbicide.

ROLE OF THE HERBICIDE

Molecular Configuration

Variation in molecular configuration of a herbicide changes its properties, which in turn modifies its effects on plants. This is illustrated by Figure 5-5 which shows the herbicides trifluralin and benefin. The only difference is that a methyl ($-CH_2$) group is moved from one side of the molecule to the other. Trifluralin kills lettuce at rates required for weed control, but benefin at recommended rates will control weeds without harming lettuce.

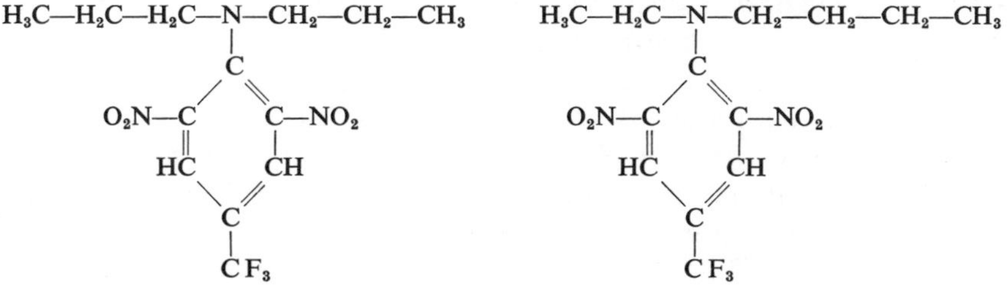

$$H_3C-H_2C-H_2C-N-CH_2-CH_2-CH_3$$

$$H_3C-H_2C-N-CH_2-CH_2-CH_2-CH_3$$

trifluralin   benefin

Figure 5-5.   Chemical structure of trifluralin and benefin are similar; however, there are differences in plant selectivity.

Types of Toxicity

Two types of toxicity to plant tissue, *acute* and *chronic*, were first noted during research with various petroleum oils (4). The word acute is used here to mean "intense" or "penetrating"; thus, acute herbicidal toxicity is an intense, rapid killing of the plant. However, the plant may survive if it is not immediately killed; it suffers only a temporary setback. Contact herbicides usually produce acute toxicity.

The word chronic as used here means "of long duration" or "continuing for a long time." Therefore, chronic herbicidal toxicity is slow acting. Under some conditions the plant may show little visible effect for a week or longer. Even then, it may die gradually 3–10 weeks after treatment.

Concentration of the Herbicide

Concentration may determine whether the herbicide inhibits or stimulates the plants. Strychnine is a very poisonous alkaloid, yet doctors may prescribe it in very minute quantities as a stimulant for the human nervous system. Some herbicides react on plants in a similar way. It has been clearly established that in low concentrations dinitrophenol may stimulate respiration, whereas stronger concentrations directly inhibit respiration (2, 3, 5). Under many conditions 2,4-D may speed up the rate of respiration and cell division, but in excessive quantities it may immediately slow down the rate.

The concentration of the herbicide *at a vital location* in the plant *at any one given time* may determine the herbicide effectiveness. The same amount of herbicide over an extended length of time may have little or no effect. If the structure of the herbicide must be altered within the plant to an active form, or if absorption or translocation is slow, the concentration of the chemical at any one moment may be less than the lethal dosage.

Crop yields often increase where herbicides have effectively controlled weeds. It is possible for some herbicides at very low rates to have a stimulating effect on growth of the plant. However, the lack of competition from weeds is probably far more important. Also, when some herbicides are decomposed they may leave needed essential mineral elements in the soil and thereby have a fertilizing effect. This phenomenon is most noticeable with herbicides that contain nitrogen and are applied at reasonably heavy rates. Calcium cyanamide and ammonium sulfamate often show a fertilizing effect after the chemical breaks down.

Formulation

The formulation of a herbicide is vital in determining whether it is selective or not with regard to a given species. Perhaps the most striking example is the solid, granular form which permits the herbicide to "bounce off" the crop and fall to the soil. Other substances known as adjuvants and surface-active agents (surfactants) are often added to improve the application properties of a liquid formulation; these additives may increase or decrease toxicity. The addition of nonphytotoxic oils or surfactants to atrazine or diuron induces foliar contact activity in these normally soil-active herbicides. These herbicides have little foliar activity in their usual form.

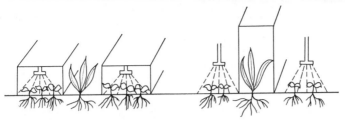

Figure 5-6. Shielded sprays. Shielded sprays protect crops from being sprayed with herbicide by having spray confined in shields (left) or crop covered by shields (right) (1).

How the Herbicide is Used

A herbicide can be applied so that most of it covers the weed but little of it contacts the crop. This can be done by using shielded or directed sprays.

Shielded Spray

Shields prevent the herbicidal spray from touching the crop while the weeds are covered by the spray. Simply place the spray nozzles under a hood or cover the crop with a shield (see Figure 5-6).

Directed Spray

This method is usually used where the crop is grown in rows and where the crop is higher than the weeds. Drop nozzles are used to spray weeds between the crop rows, and little herbicide contacts the crop.

Shielded sprays may not be needed if you choose nozzles that have little spray drift, if nozzle height and direction are carefully controlled, and if the crop is taller than the weeds (see Figures 1-7 and 5-7).

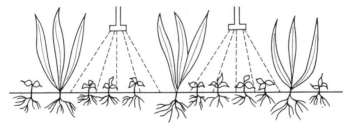

Figure 5-7. Directed sprays. Directed sprays are directed toward the base of the crop plant, favoring minimum coverage of crop and maximum coverage of weeds (1).

ROLE OF THE ENVIRONMENT

Dominant factors of environment which affect selectivity are *soil* texture (size of particles) *rain-fall or overhead irrigation,* and *temperature.*

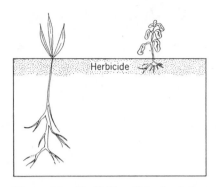 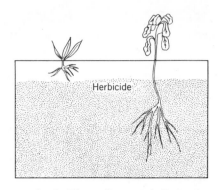

Figure 5-8. Herbicides differ in their tendency to leach. These diagrams indicate the differences that may result from differences in the location of the herbicide in the soil. Left: Deep-rooted crop (left) is not affected by herbicide which remains near soil surface. Shallow-rooted weed is killed by herbicide which stays near surface. Right: Shallow-rooted crop (left) remains alive if herbicide moves beyond its rooting zone. Deep-rooted weed is killed when the herbicide is leached into the deeper zones of the soil.

In general, soil type combined with the amount of rainfall determines the actual location of a specific herbicide in the soil. Among factors affecting movement of herbicides in soil are adsorption by soil particles, water solubility of the herbicide, amount of rainfall, and soil type. The intensity with which the chemical molecule is held by the soil particle will strongly affect movement or lack of movement in the soil. Also, high water solubility of the herbicide, high amounts of rainfall, and light soil types favor a deep penetration of the herbicide. Some herbicides are extremely resistant to leaching, whereas others readily move with the water. This movement is normally downward but the herbicide may move up as water evaporates from the surface.

Some herbicides not inherently selective may be made to function as such by their location in the soil. Such selectivity depends on different rooting habits of crop and weed. If you want to remove deep-rooted weeds while leaving shallow-rooted crops intact, you must use a herbicide that readily moves beyond the rooting zone of the crop into the rooting zone of the weed. But if you want to remove shallow-rooted weeds from a deep-rooted crop, you must choose a herbicide that remains near the soil surface (see Figure 5-8).

The temperature of the environment in which a plant is growing has considerable influence on the rates of its physiological and biochemical reactions. For example, the optimum temperature for germination of seeds of different species varies greatly; for example, spinach (41°F) and cantaloupe (77°F). The selectivity of various plants to herbicides also varies as the temperature differentially affects their processes.

The effect of temperature on these processes is often expressed as the temperature coefficient (Q_{10}). Within limits, the rate of a process increases, or decreases, with each 10°C (17°F) change in temperature. If the rate of a process is doubled, the Q_{10} is 2. Most chemical reactions have a Q_{10} between 2 and 3, while physical processes often have a Q_{10} only slightly greater than 1. A herbicide is a chemical and most of the reactions which the herbicide influences in the plant are chemical in nature. Thus, a change from 80 to 97°F may result in doubling or tripling the activity of the herbicide.

LITERATURE CITED

1. Ashton, F. M., and W. A. Harvey, *Circular 558, California Agricultural Experiment Station and Extension Service*, 1971.
2. Bonner, J., *Amer. J. Bot.*, *36*, 323 (1949).
3. Bonner, J., *Amer. J. Bot.*, *36*, 429 (1949).
4. de Ong, E. R., H. Knight, and J. C. Chamberlin, *Hilgardia*, *2*, 351 (1927).
5. Kelly, S., and G. S. Avery, Jr., *Amer. J. Bot.*, *36*, 421 (1949).

6 Formulations, Drift, and Calculations

FORMULATIONS

Herbicides are formulated to make their application easier and to improve their effectiveness under field conditions. The chemist can change the formulation of a chemical to affect its solubility, volatility, specific gravity, toxicity to plants, and numerous other characteristics. He can add other substances, such as surface-active agents, which can considerably alter the effectiveness of a herbicide. He may dissolve the active ingredient in a suitable solvent for a liquid formulation or mix it with an inert material such as clay for a wettable powder formulation, add it to a coarse particle such as vermiculite for a granular formulation, or combine the ingredients and, with pressure, produce a pellet.

For effective weed control, the quantity of chemical (active) may be as low as $\frac{1}{8}$ lb/acre (only 2 oz), while less active compounds may require several pounds per acre. The formulated herbicide is usually diluted with water for field applications and applied at 1–10 gal/acre by aircraft, or 5–20 gal/acre by ground sprayers. Uniform, even applications are required so that germinating seeds or tiny seedlings, perhaps only $\frac{1}{8}$ in. across, are killed. Hence, uniform application is extremely important.

Herbicides are formulated to be applied as (1) solutions of water or oil, (2) emulsions, (3) wettable powder suspensions, and (4) granules. Detailed information on pesticide formulations is presented in a recent book edited by van Valkenburg (15).

Solutions

A *solution is a physically homogeneous mixture of two or more substances.* The substances may be solids, liquids, or gases. However, when we think of a

100

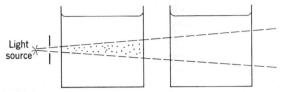

Figure 6-1.   Reflective properties of true solutions and colloidal suspensions, the Tyndall effect. Light is not reflected when it passes through a true solution (right). But light is reflected from colloidal suspensions (left) of either suspended liquid particles (emulsion) or suspended solid particles (wettable powder).

solution, we usually have in mind solids or liquids dissolved in a liquid. Every part of the solution is like every other part. The constituents cannot be seen separately, nor can they be separated by mechanical means. The solution is clear in appearance, even though it may have color. If darkly colored, the transparency of the solution may be lost (see Figure 6-1).

Solutions are formed when sugar or salt (solute) dissolves in water (solvent), when alcohol dissolves in water, or when kerosene dissolves in gasoline. The dissolved constituent is the *solute*, and the substance in which solution takes place is the *solvent*.

The salts of most herbicides are soluble in water. A few examples are: sodium and amine salts of 2,4-D, 2,4,5-T, MCPA, and silvex; amine salts of dinoseb; sodium salt of pentachlorophenol; sodium salt of TCA; and sodium salt of dalapon. These can be dissolved in convenient amounts of water and sprayed efficiently.

The "parent acid" formulations of some of these are oil-soluble. For example, DNBP and pentachlorophenol are soluble in oil. They are often used to increase the toxicity of oil sprays, or to *fortify* the oil. Ester formulations of 2,4-D and related products are also soluble in oil.

In a solution, a certain percentage of solute molecules are dissociated into ions. These ions are then free to combine with other ions of the solution. For example, "hard water" may have a high calcium or magnesium content. If a salt form of 2,4-D is added to hard water, it too becomes ionized. The 2,4-D ions are free to react to form calcium-2,4-D or magnesium-2,4-D salts. The calcium salt of 2,4-D is soluble in water to the extent of 2.5 g/liter and the magnesium salt to 17.4 g/liter at 68°F. Thus the calcium and magnesium 2,4-D molecules form precipitates if the quantity exceeds these amounts. These precipitates may clog filters and nozzles on the sprayer.

Emulsions

An emulsion is one liquid dispersed in another liquid, each maintaining its original identity. Without agitation the liquids may separate. Once mixed, some emulsions require very little agitation to prevent separation; others, however, require constant heavy agitation to prevent separation. Remixing following separation will usually reform the emulsion.

Oil-soluble herbicides are often formulated for mixture with water as an emulsion. For example, formulations of CIPC, CDEC, CDAA, trifluralin, and ester formulations of 2,4-D, 2,4,5-T, MCPA, and silvex are usually mixed with water to form emulsions, which appear milky. Emulsions of the *oil-in-water* (O/W) type are easily applied as sprays. The *water-in-oil* (W/O) type of emulsion varies in viscosity; some may be too viscous to be applied as a spray.

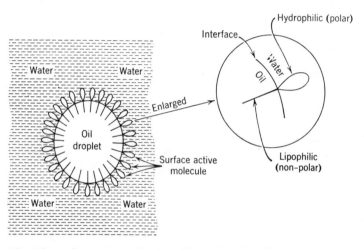

Figure 6-2. The surface-active molecule tends to bind the oil-water surfaces together, reducing interfacial tension.

To reiterate, an emulsion is a liquid dispersed in another liquid. The emulsifying agent binds and also has a tendency to "insulate" the surfaces of the two liquids. There is, however, no direct contact between the two liquids (see Figure 6-2). Therefore, little or no opportunity exists for chemical reaction between the two. The emulsion offers the opportunity, at least for a short period, of mixing reactive chemicals without the formation of pre-cipitates. The emulsifying agent must be stable in the presence of both. For example, ester formulations of 2,4-D are oil-like and form emulsions when mixed with water; even when mixed with "hard water," no calcium or magnesium precipitates are formed. The dispersed, oil-like droplets are suspended in the water but insulated by the emulsifying agent.

Emulsions are discussed more fully on page 109.

Wettable Powder Suspensions

A wettable powder suspension consists of finely divided, solid particles dispersed in a liquid. It usually consists of a mixture of the pesticide with clay, a dispersant, and a wetting agent. At the time of application it is suspended in water and sprayed.

Some herbicides are nearly insoluble in both water and oil-like substances. Therefore, concentrated solutions or emulsions cannot be prepared. In such cases, a wettable powder suspension may be the most practical formulation.

The suspension will usually give a "clouded" appearance to the liquid. It will also give reflection to a light cone (Tyndall effect) similar to an emulsion (see Figure 6-1). The addition of a surface active agent is needed to maintain the suspension in water.

Some gel-like substances are made by chemically or physically treating special clays. When water is added, a gel-like substance is formed, and various pesticides can be added to this gel-like substance. At the time of application the gel-like formulation is added to water and applied as a spray.

Most wettable powder suspensions require agitation to prevent settling of the solid particles. The smaller the solid particles, the slower the rate of separation. Similar density of the solid particle and the liquid will also slow the rate of separation.

Herbicides sold as wettable powders include simazine, atrazine, monuron, diuron, DCPA, nitralin, oryzalin, diphenamid, and tebuthiuron. They are marketed either as dry wettable powders or as "liquid suspensions."

Granules

Some chemicals are applied at rates high enough so that crystals of the chemical itself can be uniformly applied. This is true with many of the borate and sodium chlorate compounds. In other cases, the herbicide is applied by mixing with a "carrier" to provide sufficient bulk for even distribution. Pellets are granules that are extruded under pressure to yield a particular size and shape. Many carriers are used, including clay, sand, vermiculite, and finely ground plant parts (ground corn cobs, tobacco-leaf wastes, etc.). Granules are made in many different sizes. Granular materials can be spread by hand, or mechanical spreaders resembling seed or fertilizer spreaders are often used.

Granular materials have three advantages over sprays: (1) water is not needed to apply the material, (2) costlier spray equipment is not required, and (3) large granules have a tendency to fall off the leaves of valuable plants without injuring them.

However, granules also have disadvantages: (1) they are heavier, bulkier, and shipping charges may considerably increase their cost, (2) some granular materials are easily moved by wind or water, and (3) application is seldom as uniform as it is with sprays.

Dust

Dusts are a popular method of application for insecticides and fungicides. However, *herbicides* are not applied as dusts because of the drift hazard.

SURFACE ACTIVE AGENTS

Surface active agents (surfactants) include wetting agents, emulsifiers, detergents, spreaders, sticking agents, and dispersing agents. One or more of these types of surfactants are used in essentially all herbicide formulations. We are interested in the properties of the different surfactants because these properties may explain why some sprays adhere to plant surfaces and others bounce or run off. Properties also tell how a wetting agent may increase the toxicity of a herbicide in one case and decrease it in another.

Since certain surfactants may increase the phytotoxicity of specific herbicides and others decrease the phytotoxicity (10), the applicator should not add any surfactant to a commercial formulation unless recommended by the manufacturer, or unless independent research has shown that a specific surfactant herbicide combination increases herbicidal effectiveness.

Incidentally, the "mechanism of action" of surface-active agents used with herbicides is closely related to that of chemical detergents in the home laundry, textile dyeing and finishing, electrochemical plating, mining-ore flotation, paper toweling, and even fire fighting.

Basic Concepts

Water is not compatible with many chemicals used as herbicides, or with many plant surfaces (see Figure 6-3). For example, water does not mix with oil or oil-like chemicals; one usually repels the other. By adding a surfactant, in this case an emulsifying agent to the oil, the herbicide can be mixed with water to form an emulsion; as such, it can be easily sprayed through the sprayer.

Water is also repelled by the waxlike cuticle found on plant surfaces. By adding a surfactant, in the form of a wetting agent, the effectiveness of a

Figure 6-3. Water droplets on a waxed pane of glass. Left: Pure water with high surface tension. Right: Water droplets spread over the waxed surface as a thin layer when a wetting agent is added. (Atlas Powder Company, Wilmington, Delaware.)

herbicide may be completely changed, as shown in Figure 6-3. For example, an effective surfactant added to amitrole (in water spray) increased the amount of amitrole absorbed by bean leaves by 6 times over a 24-hr period (6).

Surface Relationships

Here are the four surface relationships:

    1. *Liquid to liquid*—oil dispersed in water by agitation to form an *emulsion*.
    2. *Solid to liquid*—clay suspended in water, or a wettable powder herbicide suspended in water.
    3. *Solid to air*—carbon in the air to form smoke, or pesticides applied as *dusts*. A spray droplet that dries in the air to leave a small solid particle suspended in the air is another example.
    4. *Liquid to air*—tiny water droplets suspended in the air to form fog, or spray droplets in the air.

In emulsions and suspensions, the water has a tendency to be repelled by the other liquid, or by the solid forming the suspension. This produces surface tensions or interfacial tensions. To "bind" the two surfaces, a substance is needed which has an affinity for both water and the other substance. Such a substance will have a molecule that orientes itself between the two surfaces so they are "bonded" in a more intimate contact. Thus, *a surface-active agent modifies the surface forces (interfaces) by orienting itself between the interfaces, providing a more intimate coupling*.

A molecule or ion will usually possess surface activity if it contains a strongly polar group that is attracted to water (hydrophilic), and a nonpolar group that is attracted to nonaqueous materials (lipophilic or hydrophobic). This nonpolar group is usually attracted to oils, fats, and waxes. Lipoid means fatlike; thus, the term oil-loving or lipophilic.

You might visualize surfactant molecules as tadpole shaped, with the head portion being hydrophilic, and the tail portion lipophilic. When added to a mixture of oil and water, the lipophilic portion orientes itself in the oil and the hydrophilic portion in the water (see page 102).

The distinct head–tail relationship is lacking in some of the newer synthetic detergents. However, their molecules do have hydrophilic and lipophilic regions that permit effective orientation between the two surfaces.

Surface Tension

The molecules of a liquid strongly attract other molecules that are close to it. Therefore, the surface molecules are attracted toward the center of a liquid body. Thus, a "free" water droplet appears as if held by a tense, elastic membrane. This can be illustrated by the use of a waxed or oiled surface (nonpolar) which has little attraction for the water molecules (polar). Because the water molecules are mutually attracted, causing surface tension, a nearly spherical drop is formed (see Figure 6-3). In summary, we can define *surface tension as the tendency of the surface molecules of a liquid to be attracted toward the center of the liquid body.*

A small steel sewing needle will float on the surface of pure water; the needle is not heavy enough to break the surface tension. But the needle will not float if a wetting agent is added. Surface tension is not strong enough to support the needle with the wetting agent added to the water.

Hydrophile–Lipophile Balance (HLB)

The HLB of a surfactant molecule or a mixture of two or more surfactant molecules is a quantitative value of their polarity. This concept, developed by Griffin (8), is particularly useful when surfactants are used as emulsifiers. It uses a scale of 0 to 20; the higher the value, the greater the hydrophilic properties. The HLB value is calculated from the nature of the molecule or molecules, or determined experimentally (1). A rough approximation of HLB can be obtained by observing the water solubility (see Table 6-1).

Although HLB values are valuable in the chemical formulation of herbicides, little information is available on the correlation of the HLB of a herbicide formulation with herbicidal effectiveness. However, in one study certain foliar-applied herbicides reached maximum phytotoxicity at an HLB value of about 14 (1,9).

Types of Surfactants

Surface active agents can be classed as *nonionic* and *ionic*, depending on their ionization or dissociation in water. Nonionic surface active agents have no

Table 6-1. Approximation of HLB by Water Solubility (1)

Behavior when Added to Water	HLB Range
No dispersibility in water	1–4
Poor dispersion	3–6
Milky dispersion after vigorous agitation	6–8
Stable milky dispersion (upper end almost translucent)	8–10
From translucent to clear	10–13
Clear solution	13+

particle charge, whereas ionic agents show either a positive or negative charge.

The *nonionic* surfactants ionize little or not at all in water. They are classed as nonelectrolytes and are usually chemically inactive in the presence of the usual salts. Thus, they can be mixed with most herbicides and still remain chemically inert. Many emulsifying agents are of this type, and they are usually liquids.

Ionic surfactants ionize when in an aqueous medium, some being anionic (−) and others cationic (+). *Anionic* agents are those in which the anion part of the molecule exerts the predominant influence. Wetting agents, detergents, and some emulsifiers fall in this group. *Cationic* agents (also known as reversed soaps) become ionized in water with the cation part of the molecule exerting the predominant influence. They have powerful bactericidal action but they are expensive and have been of only limited use in agriculture.

Most emulsifiers in commercial usage are blends of anionic and nonionic substances. Specific blends can give a specific result. In general, however, a high proportion of the anionic agents will improve performance in cold water, and work best in soft water. Usually a predominance of nonionic types will perform better than other types in warm water and in hard water. See Figure 6-4 for the chemical structure of various surfactants.

Surfactants Classed According to Use

Wetting Agents

A wetting agent increases a liquid's ability to moisten a solid. It lowers the interfacial tension, bringing the liquid into intimate contact with the solid (see Figure 6-5). The degree of effectiveness of a wetting agent is shown by the increase in spread of the liquid over a surface area. The spread over the surface area determines the "contact angle" of the liquid with the surface.

Nonionic

$$C_{12}H_{25} \vdots O(CH_2CH_2O)_{23}H$$

lauryl alcohol

$$C_{17}H_{35} \vdots COO—CH_2$$
$$| $$
$$CHOH$$
$$| $$
$$CH_2OH$$

glyceryl monostearate

Ionic

Anionic $C_{17}H_{35} \vdots COONa$ $C_{12}H_{25} \vdots OSO_3Na$

sodium stearate   sodium lauryl sulfate

Cationic $C_{16}H_{33} \vdots N(CH_3)_3Br$

cetyl trimethyl ammonium bromide   lauryl pyridinium bromide

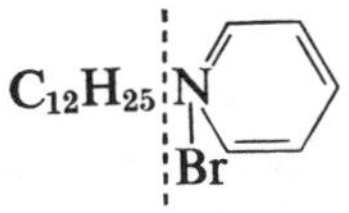

Figure 6-4. Structural formulas of the various types of surface-active agents. The portion of the molecule to the left of the dashed line is lipophilic; that to the right, the hydrophilic (1).

Wetting agents may increase or decrease the effectiveness of herbicide sprays. For example, the effectiveness of contact herbicides depends largely on both complete and uniform wetting of the plant. A contact chemical applied in a spray at the rate of 10 gal/acre without a wetting agent will not "wet" the vegetation. Lacking a wetting agent, the chemical remains as droplets, burning small "pinholes" in the leaf. The plant is not killed. The addition of a wetting agent causes the spray droplets to spread, and this action helps to cover the plant uniformly. At low gallonage, the wetting agent may be expected to increase the effectiveness of the spray.

Now, let us suppose that the same chemical is applied at the *same rate* per acre, but it is diluted in 200 gal of spray. Without a wetting agent this rate heavily "wets" the foliage, with no runoff. If a wetting agent is added, the spray forms a thin layer over the plant. The low surface tension permits approximately two-thirds of the spray to run off the plants to the ground. The chemical is lost from the surface of the plant and the herbicide loses

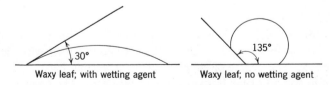

Waxy leaf; with wetting agent   Waxy leaf; no wetting agent

Figure 6-5. Water droplets containing a wetting agent spread in a thin layer over a waxed surface. Pure water will stand as a droplet, with a small area of contact with the waxed surface.

some of its effect. Thus, with high-gallonage sprays, a wetting agent may actually decrease the effectiveness of contact herbicides. This is seldom a problem, however, because excessive spray quantities are not often applied.

Herbicidal *selectivity* may be lost by the addition of a wetting agent. This is true if selectivity depends on selective wetting or selective absorption.

Emulsions (see discussion on page 102)

If oil is added to water and shaken vigorously, the oil is momentarily suspended as small droplets in the water, forming an emulsion. The water is a continuous body and is therefore referred to as the *continuous phase*; however, the oil is dispersed and is therefore referred to as the *discontinuous phase*.

If the mixture is allowed to stand, the oil and water will separate. A third material is added to decrease the tendency to separate or to increase the stability of the emulsion. This is called the *emulsifying agent* or *emulsifier*.

Emulsification is principally a "surface-action" phenomena. The emulsifier molecule orients itself between the oil–water interface, coupling the two (see Figure 6-2). This prevents the dispersed droplets from coalescing to form large droplets which quickly separate.

For example, milk is an emulsion of butterfat (dispersed) and water (continuous), with casein acting as the emulsifying agent. This type of emulsion is referred to as an *oil-in-water* (O/W) emulsion. Most herbicidal emulsions are of the O/W type. When mixed with water they have approximately the same viscosity as water, with a "milky appearance."

Butter and mayonnaise are emulsions with water in the dispersed phase and fat or oil in the continuous phase. This type of emulsion is called a *water-in-oil* (W/O) emulsion. Water-in-oil emulsions are usually viscous, too thick to pass through the usual spray equipment; they resemble whipped cream. The W/O emulsion may be referred to as an *invert*. Invert 2,4-D and 2,4,5-T emulsions have been prepared and have been used particularly for brush control.

Three factors largely determine the stability of an emulsion. These are: (1) size of dispersed particles, (2) relative density of the two liquids, and (3) viscosity of the emulsion. Size of particles is especially important. For example, in homogenized milk the fat globules have been broken into very tiny droplets. These separate much more slowly than large butterfat droplets in plain milk.

There are hundreds of detergents and emulsifying agents. McCutcheon (12) listed more than 700 detergents and gave the manufacturers, chemical formulas, main uses, concentrations, and types. Bennett (2) named about 600 emulsifying agents, giving the common name as well as the chemical name for many of the products. Cupples (4) prepared a list of wetting, dispersing, and emulsifying agents, and gave the properties and uses of more than 300 compounds. Gast and Early (7) rated the toxicity of 77 emulsifiers and 66 organic solvents when applied to beans, corn, cotton, cucumber, tobacco, and tomato plants. In general, the emulsifiers were more toxic than the organic solvents tested.

Detergents

Detergency indicates the cleaning power or the ability of a chemical to remove soil or grime. All detergents are surfactants. Many of the common detergent chemicals have been used with herbicides as wetting agents, spreaders, and emulsifiers; hence, the word detergent has appeared in herbicidal literature.

Spreaders

Spreaders and wetting agents, while sometimes considered independently, are closely related. When the wetting agent reduces surface tension, spreading naturally follows. Therefore, when used in connection with herbicides, spreaders and wetting agents should be considered together.

Adhesive or Sticking Agents

Adhesive or sticking agents, as the name implies, are substances that cause the herbicide to adhere to the sprayed surface. Many of the surfactants discussed above may also act as sticking agents.

Dispersing Agents

A dispersing agent is a substance that reduces the cohesion between like particles. It promotes separation of like particles and in some cases its reaction may be closely related to deflocculation. Some dispersing agents are good wetting agents, but others have little or no effect on the surface tension. Some wetting agents and dispersing agents are not compatible and have a tendency to interfere with each other if used together.

Effects On Plants

Surfactants generally intensify the action of a foliar applied herbicide. In some cases selectivity of the herbicide is lost. Surfactants probably have five important effects on the plant:

1. They favor uniform spreading of the spray, or uniform wetting of the plant.

Figure 6-6.   A wetting agent was added to an ammonium nitrate solution. The spray was directed to hit the Pennsylvania smartweed and corn stems without hitting the corn leaves. The wetting agent nearly doubled the burning and killing effect on the weeds. (North Carolina State University.)

2. Spray droplets usually stick to the plant—there is less bounce-off.

3. The chemical spray is brought into intimate contact with the plant surface. Droplets do not remain suspended on hairs, scales, or other projections.

4. The surface active agent may alter nonpolar plant substances. Substances such as the waxy cuticle or the lipoidal portion of the cell wall may be altered so the plant readily absorbs the chemical. Also, the cell membranes may be modified, thus permitting the cell sap to leak into the intercellular spaces. When this happens the surfactant may affect the plant as do oils (see Figure 6-6 and Chapter 5).

5. The detergents are believed to have many effects on the proteins. These may include, among others, protein precipitation, denaturization, and inactivation of enzymes, viruses, and toxins (14).

CHEMICAL DRIFT THROUGH THE AIR

Chemicals may drift through the air from the area of application and cause considerable injury if they contact susceptible plants. Movement through the air may result from *spray drift* or *volatility*. Yates and Akesson (15) have written a chapter on reducing pesticide drift.

Spray Drift

Spray drift is the movement of airborne spray particles from the target area. The amount of spray drift depends upon (1) size of the droplets, (2) amount of wind, and (3) height above the ground that the spray is released.

The size of the droplet will depend primarily on pressure, design of the nozzle, and surface tension of the spray solution. In general, low pressures have a tendency to produce large droplets and high pressures, small droplets. Small nozzles usually produce small droplets, and solutions possessing low-surface tension tend to produce small droplets.

Table 6-2 stresses the importance of large droplets in reducing spray drift. With a droplet size of $\frac{1}{50}$ in., or light rain, 1 gal of spray/acre would apply nine drops/in.^{2}. These would drift only 7 ft if released 10 ft above the ground in a 3-mph breeze.

Furthermore, if $\frac{1}{50}$-in. droplets (the size of light rain) are released in a 30-mph wind 2 ft above the surface, the spray will drift only 14 ft. The fog (5μ) size under similar conditions would drift over 3 miles.

As droplets fall, evaporation may occur, leaving an extremely small particle. Such particles may be nearly as small as smoke particles; the latter

Table 6-2. Spray-Droplet Size and Its Effect on Spray Drift (3)

Droplet Diameter (μ)	Type of Droplet	No. of Droplets/ In.^{2} from 1 Gal of Spray/Acre	Time Required to Fall 10 Ft in Still Air	Distance Droplet Will Travel in Falling 10 Ft with a 3-Mph Breeze
0.5	Brownian max	—	6,750 min	388 miles
5	Fog	9,000,000	66 min	3 miles
100	Mist	1,164	10 sec	409 ft
500 ($\frac{1}{50}$ in.)	Light rain	9	1.5 sec	7 ft
1000 ($\frac{1}{25}$ in.)	Moderate rain	—	1 sec	4.7 ft

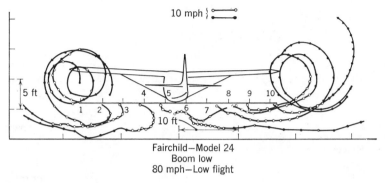

Figure 6-7. Circulation from a high-wing monoplane.

may be small enough to approximate the Brownian maximum. Such particles may drift as much as 388 miles under the conditions described in Table 6-2. If drift of the chemical is a hazard, the carrier liquid should have a low vapor pressure to slow the rate of evaporation, especially if applied by airplane.

Periods of strong wind velocities vary with locality. In general, winds are least turbulent just before sunrise, and another low period occurs just after sunset and throughout the night. As an average, winds are most gusty and turbulent between 2 and 4 p.m.

The height above ground that the spray is released is important for two reasons. First of all, the distance and thus the time required for the droplets to reach the ground is directly affected. Second, wind velocities are usually much lower close to the ground than at higher elevations.

Therefore, spray applications from aircraft present greater drift hazards than ground sprayers. Air currents produced by an aircraft have a major effect on the trajectory of fine particles released from it. Basically any aircraft, rotary (helicopter) or fixed wing, has updrafts produced by wing-tip vortices and downdrafts under the middle of the aircraft (see Figure 6-7).

Nozzles that produce uniform droplets between 500 μ and 1000 μ in diameter will deliver a spray with essentially no spray-drift hazard. Such sizes will permit adequate coverage for most herbicidal spray operations at relatively low gallonages per acre. Nozzle design is discussed further in Chapter 7.

Volatility

Volatility refers to the tendency of the chemical to vaporize or give off fumes. The amount of fumes or vapors emitted is related to the vapor pressure of the chemical (5,11,13).

Figure 6-8. These tomato plants were kept under a jar that was exposed to the vapor or fumes from different 2,4-D formulations. *1.* Sodium salt of 2,4-D caused slight to no injury; *2.* Diethanolamine salt of 2,4-D caused no injury; *3.* Triethanolamine salt of 2,4-D caused no injury; *4* and *5*, Butyl and ethyl esters of 2,4-D killed the plants (11).

Vapor drift is the movement of vapors or fumes. Vapor drift may damage susceptible crops, or may simply reduce, through loss, the effectiveness of the herbicide treatment.

The volatility of 2,4-D has perhaps received more attention than that of any other chemical. Figures 6-8 and 6-9 explain the differences found between the salts and esters. The amine and sodium salts of 2,4-D have little or no volatility hazard. The ester formulations vary from low to high volatility.

The length and structure of the alcohol portion of the 2,4-D ester molecule directly affect its volatility. In general the longer the carbon chain, in the part contributed by the alcohol, the lower the volatility. Those esters made from five carbon alcohols, or less, are usually considered volatile. Inclusion of an oxygen as an ether linkage in the alcohol portion of the molecule will also

$$\left(\text{H}-\text{O}-\underset{\underset{\text{H}}{|}}{\overset{\overset{\text{H}}{|}}{\text{C}}}-\underset{\underset{\text{H}}{|}}{\overset{\overset{\text{H}}{|}}{\text{C}}}-\underset{\underset{\text{H}}{|}}{\overset{\overset{\text{H}}{|}}{\text{C}}}-\text{O}-\underset{\underset{\text{H}}{|}}{\overset{\overset{\text{H}}{|}}{\text{C}}}-\underset{\underset{\text{H}}{|}}{\overset{\overset{\text{H}}{|}}{\text{C}}}-\underset{\underset{\text{H}}{|}}{\overset{\overset{\text{H}}{|}}{\text{C}}}-\text{H} \right)$$

reduce the volatility of an ester of 2,4-D.

Soil surfaces exposed to direct sunlight often reach 170°F. At this temperature some chemicals quickly volatilize and may be carried away in the wind, and therefore are hazards to susceptible plants. Also, the herbicidal effect of the treatment may be lost.

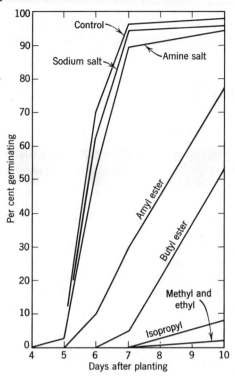

Figure 6-9.   Percent germination of pea seeds after being exposed to the vapors or fumes from different 2,4-D formulations (13).

CALCULATIONS

Active Ingredient

The active ingredient is that part of a chemical formulation which is directly responsible for the herbicidal effects. In some herbicides the entire molecule is considered to be the "active unit." Therefore, if the chemical was 99% pure, it would be considered 99% active ingredient.

In others, the herbicide activity is more accurately calculated on an acid-equivalent basis.

Acid Equivalent

The acid equivalent refers to that part of a formulation that theoretically can be converted to the acid. In this case, the acid equivalent is given as the active ingredient.

The percent active ingredient or acid equivalent is given on the label. To calculate the weight of the commercial product required, this formula may be convenient:

$$\frac{\text{Weight of the chemical to be applied (active ingredient)}}{\text{Percentage expressed as a decimal (active ingredient)}} = \text{Weight of commercial material required}$$

For example: You buy a herbicide with 80% (0.80) active ingredient as diuron. You want to apply 1.0 lb of diuron/acre. Therefore,

$$\frac{1.0}{0.80} = 1\tfrac{1}{4} \text{ lb of the commercial product needed to apply 1 lb of the active ingredient of diuron.}$$

In this case, $1\tfrac{1}{4}$ lb of the commercial product are added to the amount of water required to cover 1 acre.

Liquid formulations may show on their labels both the percent *active ingredient* or *acid equivalent*, and the *weight per gallon*. A label on a container of 2,4-D may read:

> 67% triethanolamine salt of 2,4-D by weight
> 40% 2,4-D acid equivalent by weight
> 4 lb of 2,4-D acid equivalent/gal

The above statements can be on the same label and still describe the contents accurately. The 40% and 4 lb represent the acid equivalent in this case; 1 gal of the herbicide would obviously weigh 10 lb. The 67% represents the total chemical formulation—27% triethanolamine and 40% 2,4-D acid equivalent. The other 33% may be additives such as wetting agents, but is principally inert ingredients.

Liquid formulations can be weighed in the same manner as dry forms, but it is usually easier to use liquid measure. For example, the above 2,4-D product contained 4 lb of 2,4-D acid equivalent/gal. Therefore, 1 qt will contain 1 lb of the 2,4-D acid equivalent, and if 1 lb of the 2,4-D is to be applied per acre, then 1 qt of the liquid should be included in the amount of spray required to cover 1 acre. Conversion factors that are useful in calculation are given in the appendix.

For applying to field conditions, the amount of the herbicide is usually given as pounds of active ingredient per given unit of volume. For example if a farm sprayer applies 10 gal of spray/acre, the farmer then adds the amount of herbicide per acre needed-to each 10 gal of spray. Or in making a mixture for basal spraying of brush he may add 12 lb of 2,4-D/100 gal of oil spray. If there are 4 lb of 2,4-D (active ingredient)/gal, this would mean adding 3 gal of the commercial 2,4-D to 97 gal of oil, to give a total of 100 gal of mixture. Each gallon of the mixture has 0.12 lb of 2,4-D (acid equivalent). In this case, 3.9 liquid oz of the 2,4-D concentrate would be added to enough oil to make 1 gal of the mixture.

Parts per Million

Parts per million (ppm) refers to the number of parts by weight or volume of a constituent in 1,000,000 parts of the final mixture, by weight or volume. The stated concentration should tell whether ppm is measured by weight or volume. For herbicidal purposes it usually refers to a given weight of a chemical in a given volume of the spray.

If we take the gram and the cubic milliliter (1 ml of water = 1 g) as the units, then 1000 ppm of 2,4-D would mean that 1000 g of 2,4-D are dissolved in sufficient water to make 1,000,000 ml of solution; or 1 g is dissolved in enough water to total 1000 ml of solution. If 1 qt of 2,4-D contains the equivalent of 1 lb (453.6 g) of 2,4-D acid, then to make up a spray of 1000 ppm the 1 qt is added to enough water to make up a spray weighing 453,600 g or nearly 120 gal.

Percent Concentration

Percent concentration is similar to ppm except that it is expressed as a percentage. For example, 1000 ppm is equivalent to 0.1%, and 5000 ppm to 0.5%. A 1.0% solution (10,000 ppm) contains 1 g of 2,4-D (acid equivalent) in 100 cc of spray. The conversion of percent to ppm, and vice versa, can be done according to the equation below.

$$\% = \frac{ppm}{10,000} \qquad ppm = \% \times 10,000$$

PRACTICE PROBLEMS. See page 414 for conversion factors.

1. You want to apply 1 lb of 2,4-D (acid equivalent)/acre. You buy a product that is 67% 2,4-D amine, 40% 2,4-D acid equivalent, and it has 4 lb 2,4-D acid equivalent/gal. How much will you apply per acre in terms of ———— pints; or ———— liquid oz; or ———— cc or ml?

2. You then see a product that has 3.34 lb/gal of 2,4-D acid equivalent. One pound of 2,4-D acid will be present in ———— pints; or ———— liquid oz; or ———— cc.

3. You see a granular material that is 10% active CDEC. You want to apply 8 lb/acre of active CDEC. How many pounds per acre will you need?

4. You want to apply $\frac{1}{2}$ lb/acre active ingredient of diuron to cotton. The label indicates that it is 80% active. You will need to apply ———— lbs; or ———— g of the commercial material/acre.

5. You want to apply $\frac{3}{4}$ lb/acre of the active ingredient trifluralin. The commercial product has 4 lb/gal of the active chemical. You will apply ———— pints; or ———— liquid oz; or ———— cc/acre of the commercial material.

6. You have a sprayer that applies 7 gal of spray/acre. Based on problem 5 above, you will add ———— pints; or ———— liquid oz; or ———— cc to each gallon of spray.

7. You have a sprayer that applies 25 gal of spray/acre. Using the rate given in problem 4 above, you will add ———— lb; or ———— g of commercial material to each gallon of spray.

8. From problem 6, your sprayer tank holds 55 gal. You need to refill the tank. You measure the amount of spray and find that 6 gal of spray remains. You will add ———— pints; or ———— liquid oz; or ———— cc of Treflan® to the tank prior to filling with water.

9. From problem 7, your sprayer tank holds 55 gal. You need to refill the sprayer, but 13 gal of spray remains. You will add ———— lb; or ———— g of the commercial diuron prior to refilling the tank.

10. Using the product described in problem 1, you will need ———— liquid oz; or ———— cc/gallon when mixed with water to give a concentration of 2000 ppm. This concentration is the same as ————%.

11. You have a nozzle that delivers approximately 3.8 gal/hr. The nozzle will treat a $3\frac{1}{2}$-ft middle in corn. Walking 3 mph, $1\frac{1}{4}$ acres are treated/hr; ———— gal of spray are applied/acre. If there are 4 lb of 2,4-D/gal, you will need ———— liquid oz of 2,4-D/gal of spray to apply $\frac{1}{2}$ lb of 2,4-D/acre.

12. You have a sprayer that sprays a swath $16\frac{1}{2}$ ft wide. Under regular operating conditions, it sprays 80 gal of water/hr. At a continuous 8 mph, ———— acres will be treated/hr; ———— gal of spray are applied/acre.

LITERATURE CITED

1. Becher, P., "The Emulsifier," in *Pesticide Formulations*, W. van Valkenburg, Ed. Marcel Dekker, New York, 1973, p. 65.

2. Bennett, H., *Practical Emulsions*, Chemical Publishing Co., New York, 1947.

3. Brooks, F. A., *Agr. Eng.*, *28*, 233 (1947).

4. Cupples, H. L., *U.S.D.A.*, *Bur. Entomol. Quarantine Publication E504*, 1940.

5. Day, B. E., E. Johnson, and J. L. Dewlen, *Hilgardia*, *28*, 255 (1959).

6. Freed, V. H., and M. Montgomery, *Weeds*, *6*, 386 (1958).

7. Gast, R., and J. Early, *Agr. Chem.*, *11*, 42 (1956).

8. Griffin, W. C., *J. Soc. Cosmetic Chemists*, *1*, 311 (1949).

9. Jansen, L. L., *J. Agr. Food Chem.*, *12*, 223 (1964).

10. Jansen, L. L., W. A. Genter, and W. C. Shaw, *Weeds*, *9*, 381 (1961).

11. Klingman, G. C., *North Carolina Research and Farming*, *4*, 3, (1947).

12. McCutcheon, J. W., *Soap Sanit. Chem.*, *25*:(8) 33, (9) 42, (10) 40 (1949).

13. Mullison, W. R., and R. W. Hummer, *Bot. Gaz.*, *111*, 77 (1949).

14. Putnam, F. W., in *Advances in Protein Chemistry*, Vol. 4, M. L. Anson and J. T. Edsall, Eds., Academic Press, New York, 1948, p. 79.

15. Yates, W. E., and N. B. Akesson, "Reducing Pesticide Chemical Drift," in *Pesticide Formulations*, W. van Valkenburg, Ed., Marcel Dekker, New York, 1973, p. 275.

7 Application Equipment

Herbicides are formulated as solutions, emulsions, wettable powders, and granular materials (see Chapter 6). The herbicide may be applied broadcast, in narrow bands, as individual spot treatments, or to a particular part of the plant. The quantity of the chemical may vary from a few ounces to several thousand pounds per acre. Obviously, to provide uniform coverage, special equipment is needed.

Solutions, emulsions, and wettable powders are usually applied as sprays. Water is the usual diluent or carrier but oils are frequently used, both as a herbicide and as a carrier. Granular materials are spread by special mechanical spreaders similar to those used for broadcasting seed or fertilizer.

In addition, herbicides are applied by (1) *mechanical incorporation* into the top 2 to 6 in. of soil, (2) *subsurface layering* (a horizontal layer a few inches below the soil surface), and (3) *injection* into (a) soil (usually methyl bromide, a phytotoxic gas), (b) irrigation water and drainage canals, and (c) lakes and reservoirs.

SPRAYING

Spraying is the most common method of applying herbicides. Sprays have many advantages over granular forms. For one thing, sprays can be applied more uniformly because extremely small quantities of a herbicide can be sufficiently diluted to permit even coverage. The amount of total spray can be varied from 1 gal to perhaps 500 gal/acre to suit the needs of the treatment. With properly designed and properly operated equipment, spray drift can be reduced to a minimum. Sprays can be accurately directed to a given area, either on the soil or on the plant.

Sprayers are classified as low-volume sprayers or as high-volume sprayers. Low-volume sprayers usually apply less than 30 gal of total spray/acre. As

120

shown in Chapter 6, it is theoretically possible to apply as little as 1 gal of spray/acre, with nine drops/in.². These drops are large enough so that there would be essentially no spray-drift hazard.

Systemic herbicides and *growth regulators* can be effectively applied in low-volume sprays. The entire plant need not be "wet", because the herbicide is translocated. However, with low-volume sprays, small droplets are usually more effective than large ones, probably a result of the spray sticking to the plant (less bounce-off) and better coverage, especially of small weeds. Reducing drop size in half increases the number of drops by eight-fold.

Contact herbicides usually require thorough "wetting" of the plant. The chemical is translocated little, if at all; therefore, complete coverage is important. With complete wetting of the plant, droplet size of the spray does not influence effectiveness of the herbicide.

Uniform application is extremely important with both low and high volumes.

Sprays can be applied from a sprinkler can, or a power or hand-pump-type sprayer. The power sprayer may be a small portable unit, a tractor-mounted sprayer, a jeep- or truck-mounted unit, or an airplane-type sprayer.

THE SPRAYER

Most spray units have some type of nozzle or nozzles, container (tank) to hold the spray, and pump to force the spray through the nozzles. You will also usually find one or more filters or strainers, pressure gauge, pressure regulator, shut-off valve, and connecting hoses (see Figure 7-1).

Nozzles

The nozzle is probably the most important part of the sprayer. Other parts exist only to help the nozzles operate properly. Uniformity of application, rate of application, and spray drift as influenced by droplet size are all largely determined by nozzle design and conditions of their operation.

The nozzle converts the spray liquid into spray droplets (see Figure 7-2). Nozzle design and the pressure of operation largely determine the size and uniformity in size of the droplets. The nozzle design and construction also determine the uniformity of the spray across the width of the spray pattern.

At low pressures, the liquid escapes from the nozzle tip as a liquid film. This film can be seen with droplets forming at the outer edge. As the film expands it forms ligaments, then droplets at the outer edge based on surface

Figure 7-1. A low-volume farm sprayer adapted to many different types of spray jobs. It can be removed from the tractor within a few minutes. (North Carolina State University.)

tension of the liquid. Liquids with low surface tension usually have small droplets.

As pressures increase, droplet formation occurs closer to the nozzle tip with the formation of small droplets. At high pressures, small droplets are formed directly from the nozzle tip as a result of escape from hydraulic force. With high pressures, droplets may be of fog and mist size, creating a drift hazard.

Most nozzles can be operated at a pressure which favors large-droplet formation. If nozzles are operated at this pressure, there will be a minimum amount of mist-size droplets to cause spray drift. At the selected pressure make sure that the nozzle is uniformly applying the herbicide. For nozzles, this pressure may range from as low as 5 psi to as high as 30 psi (pounds per square inch).

Nozzle tips are made of brass, aluminum, stainless steel, nylon, rubber, plastic, and ceramic-type material, and choice between these materials depends largely on corrosive and abrasive effects of chemicals and on cost. Some chemicals are chemically corrosive while others are abrasive; for example, TCA is chemically corrosive on brass, while most wettable powders are abrasive.

Nozzles that depend on finely milled edges for proper operation are easily damaged by either abrasion or corrosion. Damage to the sharp edges will ruin the spray pattern. Such damage may occur after a few hours' use with

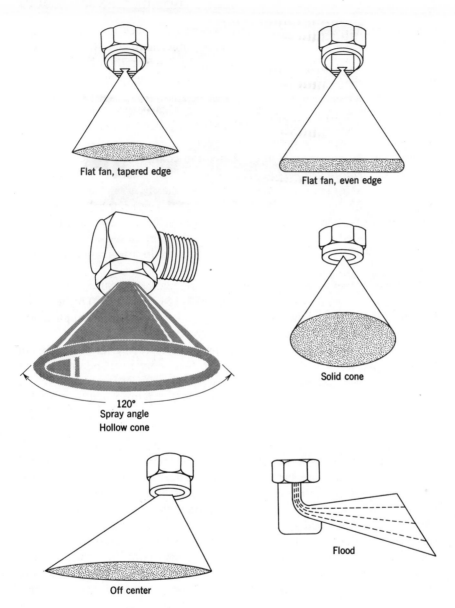

Flat fan, tapered edge

Flat fan, even edge

120°
Spray angle
Hollow cone

Solid cone

Off center

Flood

Figure 7-2. Nozzle designs used to apply herbicides.

sprays containing abrasive wettable powders. Also, such nozzles are easily damaged by cleaning with a wire, knife, or other hard object. Rinsing the clogged nozzle parts in water may loosen and remove the obstruction. A bristle brush, wooden peg, or wooden match will usually loosen lodged materials without damaging the nozzle.

Nozzles deliver many different spray patterns. No one pattern is best for all herbicide work. The nozzle must be chosen purely on the basis of satisfactory performance of a given job.

Spray patterns shown in Figure 7-2 are most commonly used for herbicide applications. *Flat-fan, tapered-edge* nozzles and the *wide-angle hollow-cone* nozzles are used for broadcast applications, and may be arranged in a series to give spray swaths up to 50 ft in width. The hollow-cone nozzle (120°), when tipped (30–45°) to spray to the rear (or forward), delivers a tapered-edge spray pattern as needed for uniform broadcast spray application. At 5–15 psi, a spray is delivered with large droplets and, as a result, it has minimum spray drift.

The flat-fan, even-edge, and the hollow-cone nozzles (70–80° spray angle—not shown) are used for band application of 8–14 in. in width. The herbicide is placed in a band over the crop row. Using this band-herbicide application, weeds between the rows are removed later by cultivation.

Flooding nozzles are widely used for broadcast spraying. Each nozzle can be set to give wide coverage—up to 10 ft. This nozzle can be operated at pressures of 10–15 psi with only moderate spray drift.

Pumps

Pumps are of two types—one delivers gas or air pressure and the other liquid pressure. Gas (or air) pressure is used mainly on research plot sprayers. If gas pressure is used, the entire system is under pressure; the spray tank is pressurized to force the liquid from the tank through the lines, boom, and nozzles. Compressed or liquified gases such as air, nitrogen, or carbon dioxide can be released through regulating valves to provide gas pressure for hand equipment. The spray tank must be strong enough to withstand the pressures. Because the tank is more expensive, the gas-pressure system often costs more than the liquid-pressure type and is usually limited to small hand-operated equipment.

In liquid pressure systems, pressure normally exists only from the pump through the lines and boom to the nozzles. The spray tank is not pressurized. Part of the liquid may be bypassed back to the tank for agitation, and for pressure control.

There are many designs of pumps for liquids, each with certain advantages and disadvantages. Because most herbicides can be applied at pressures of 5–40 psi, high-pressure pumps may be of no advantage. The pump must resist corrosion by the chemicals to be applied, provide relatively long, trouble-free service, and be priced so as to attract the user. Because pumps vary widely in design, cost alone is not necessarily a good measure for choosing a pump for a particular job.

Pumps having positive displacement require a pressure-relief system, if the sprayer has a cut-off valve. This will prevent broken hoses and possible damage to the pump. Pumps not having positive displacement may not need the pressure relief system.

Many ingenious methods can be used to develop liquid pressure. The more common types of pumps include rotary-impeller, centrifugal or turbine, gear, diaphragm, and piston pumps.

Rotary-Impeller Pumps

Probably more rotary-impeller pumps, especially the roller-type impeller, are used for agricultural spray application than any other type of pump. Rotary-impeller pumps consist of a rotor set to one side within the pump housing. The rotor maintains contact with the outer pump housing through flexible rubber vanes, or through rollers that adjust to the outer wall of the pump housing. The rollers are held to the outer wall by centrifugal force. Space between the rotor and the pump housing expands during half of each revolution, and contracts during the other half. The expanding phase creates vacuum, and the contracting phase creates pressure.

The principle of operation is the same for the flexible-rubber impeller and for the roller-type impeller. The flexible-rubber impeller may be built as shown in Figure 7-3, or the rubber impeller may consist of a vaned rubber disc which rotates at an angle to a flat surface (not pictured). In the roller-type impeller pump, the rollers are usually made of nylon or hard rubber.

With good care, especially when in storage, the roller-type impeller pump is moderately long lived. After use, rinse with clean water, and lubricate the inside of the pump with oil.

Centrifugal or Turbine Pumps

Centrifugal or turbine pumps develop pressure as a result of centrifugal force. Blades, fins, discs, or similar structures rotate rapidly and produce liquid velocity. The velocity of the liquid, combined with its weight, gives it pressure. The liquid is usually released at the outer edge of the pump housing. These pumps will usually handle coarse and abrasive materials; naturally, they are *not* self-priming.

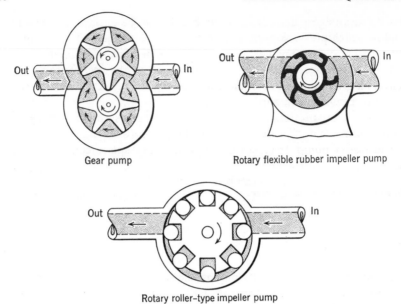

Figure 7-3.   Diagrams illustrating the method of developing pressure in the gear, rotary flexible rubber impeller, and rotary roller-type impeller pumps.

Gear Pumps

Gear pumps depend on the meshing of gear teeth to develop both vacuum and pressure. The principle of operation is clear in Figure 7-3. Also, one gear may be made to rotate internally to an external gear, with the same type of action. The gear pump is usually not long-lived, especially with abrasive materials, and as with most other pumps, the higher the pressure, the greater the wear. Some can provide up to 5000 psi; however, those used for herbicides usually operate in the range of 30–60 psi.

Diaphragm Pumps

Diaphragm pumps work on the same principle as the diaphragm fuel pump on an automobile. They can handle sprays that are both chemically and abrasively corrosive. The chemical touches only the valves, the diaphragm, and pump housing. Therefore, the pump can give long, trouble-free service with minimum upkeep.

Piston-Type Pumps

Piston-type pumps were used on nearly all sprayers before 1945 and they are still used where pressures over 100 psi are necessary. Piston-type pumps can produce very high pressures; the only limits are the structural strength of the pump and the power supply. They are usually reliable and long-life pumps, but usually expensive as well.

The piston pump is also used as a metering pump. Because the pump delivers a given amount of liquid with each stroke of the pump, the amount of spray per acre remains constant regardless of speed (within practical limits). The pump speed may be governed by the ground speed through special power take-off arrangements or by a wheel-driven mechanism.

The piston pump produces a significant pulsating effect, which can be partially corrected by the use of an air chamber.

Power Supply for the Pump

The most common sources of power to drive the pump include (1) the tractor power take-off; (2) gasoline engines or electric motors as direct drive, belt drive, or gear box drive; (3) a ground wheel traction drive; (4) on airplanes, a small propeller to drive the pump, and (5) water pressure from the water hydrant to provide power for small sprayers for home use.

Pump Capacity

Pump capacity needed is determined by total rate of discharge through the nozzles, plus the bypass needed for agitation. From 3 to 5 gal/min bypass is enough agitation for most farm sprayers holding up to 100 gal. The bypass liquid should be released at the bottom of the tank, so that the liquid force directly agitates the liquids at the tank bottom. To calculate the rate of discharge from the nozzles, use this equation:

$$\frac{\text{Pump capacity}}{\text{(gal/min)}} = \frac{\text{gal of spray/acre} \times \text{spray width of boom (ft)} \times \text{mph}}{495}$$

For example, if you want to apply 10 gal of spray/acre through a $16\frac{1}{2}$ ft boom, at 6 mph, the pump capacity will need to be a minimum of 2 gal/min. Additional capacity is needed for bypass agitation.

Occasionally, a farmer may want to increase the liquid volume discharged by his sprayer. For example, his sprayer may apply 5 gal/acre, but he wished to apply 30 gal/acre. He can accomplish this by using larger nozzles. If he does this the pump, lines, and boom capacity must be large enough to accommodate the increased volume. If the area is small, he may wish to accomplish the same purpose by reducing the speed to one-sixth his normal speed.

Filters or Strainers

Filters or strainers may be built as a part of the nozzle, installed as a line filter, or placed on the intake in the spray tank. A coarse strainer or funnel

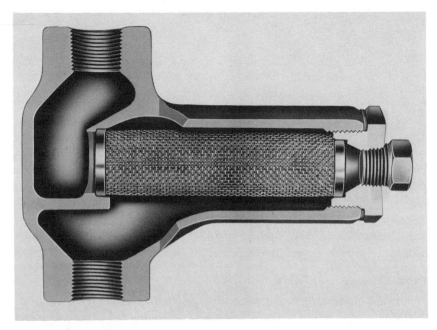

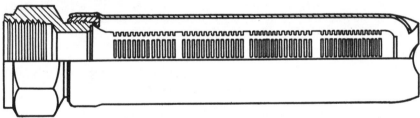

Figure 7-4.   Filters or strainers. Top: In the pressure line. Bottom: Suction strainer—
placed in the tank. (Spraying Systems Company, Bellwood, Ill.)

strainer is also desirable to filter the liquid as the spray tank is filled. This is
especially important if the water contains trash or dirt.

The size of the strainer openings is regulated by the size of the nozzle
orifices (openings). If the strainer openings are half as large as the nozzle
orifices, you will usually have little difficulty with clogged nozzles.

Wettable powders are seldom ground fine enough to pass through extremely fine filters. Thus, coarse filters as well as large nozzle openings are
usually needed for wettable powders. Most wettable powders will pass
through 50-mesh, or coarser, screens (see Figure 7-4).

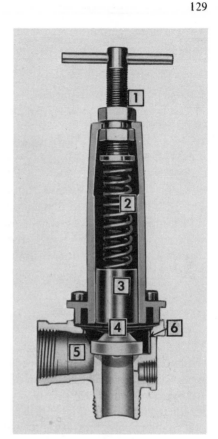

Figure 7-5. Pressure regulator with spring-regulated bypass. (Spraying Systems Company, Bellwood, Ill.)

Pressure Regulators

The pressure regulator reduces pressure to a relatively constant desired pressure.

The usual regulator on farm sprayers is a spring-loaded type that gives trouble-free service because of its simple construction. The regulator works only when you have enough liquid to require a bypass. The strength of the spring must be directly related to the chosen spray pressure. For example, with spray pressures of 15 psi, the regulator spring should be operative between the pressures of 10 and 20 psi.

Hoses

Choose hoses to fit the pressure or vacuum expected, and pick hose materials that will withstand the chemicals to be used.

Good-grade garden hose, rubber or plastic, can be used for pressures up to 65 psi. Heavy hoses are available for higher pressures. If the sprayer is built so that a vacuum may develop between the spray tank and pump, use a heavy walled hose or metal pipe in place of the hose. Also, hose is made with a metal interior to prevent collapse. A vacuum may develop from clogged suction strainers in the hose, the inlet opening held to the wall of the tank, or from collapsed or twisted hose. Insufficient liquid flow will damage some pumps and will reduce the efficiency of all pumps.

Oil-resistant hoses made of neoprene, plastic, or other oil-resistant materials may be most satisfactory with oils and oil-like herbicides. In addition to an increased life span, the oil-resistant hoses will resist absorption of the chemical making it less difficult to remove the herbicide.

Pressures in a hydraulic system are the same, regardless of hose size, minus any friction loss involved in movement. Therefore, hoses should be chosen that are large enough *not to restrict liquid flow* (see page 418).

Foaming

Herbicides often have surface-active agents as wetting agents, emulsifiers, and so on. Different surface-active agents have different foam-producing qualities. Agitation which tends to incorporate air may cause foaming, but returning the bypass solution *below* the surface of the liquid will reduce the amount of air incorporated and thus reduce the amount of foaming.

If foaming continues, it can usually be stopped by adding about 1 pint of kerosene or fuel oil/100 gal of spray. The excess surface-active agent is used in emulsifying the oil.

Calibrating the Sprayer

Several methods can be used to determine the number of gallons of spray applied per acre. Four methods are listed and briefly discussed.

Prepared Tables

Prepared tables which give nozzle spacing, pressures, speed, and various nozzle sizes giving various gallons of spray per acre are available from some nozzle manufacturers; the proper nozzle size can be selected from these tables. The disadvantage to this method is that there is no assurance that speeds and pressures used in spraying are correct.

Special Measuring Devices

Special measuring devices and prepared charts or graphs can be used. The spray is usually collected for a prescribed period of time or distance. The amount of spray collected is then converted, through tables or charts, into gallons per acre. Glass jars with the table printed directly on the jar are available to catch the spray. This has the same disadvantages as those given for prepared tables.

Measurement of Gallons of Spray Delivered

Measurement of gallons of spray delivered per hour and calculation of acres treated per hour at a given speed can be used to determine gallons per acre as follows:

$$\text{gal/acre} = \frac{\text{gallons applied/hr}}{\text{acres sprayed/hr}}$$

The rate of delivery of the spray can be determined by timing the rate of delivery of a known quantity of spray. From this timing, the gallons of spray per hour can be calculated.

The acres per hour can be calculated as follows:

$$\text{Acres/hr} = \frac{\text{mph} \times 5280 \text{ (ft/mile)} \times \text{spray width (ft)}}{43,560 \text{ (ft}^2\text{/acre)}}$$

For example: A sprayer traveling 6 mph, covering a swath $16\frac{1}{2}$ ft wide will spray 12 acres/hr of continuous operation. If the sprayer applied 60 gal of spray/hr, using the formula given first, the sprayer will apply 5 gal of spray/acre.

Spraying a Known Size Area and Measuring the Amount of Spray Applied

Perhaps the most accurate method is to actually spray an area of known size. By starting with a full tank and measuring the gallons required to refill the tank after spraying a specific area, the gallons per acre is quickly calculated. This method can be used by the aerial sprayer, the farm sprayer, or hand sprayer. The size of the area sprayed must be large enough to give an accurate measurement of the spray applied, and it is also imperative that all nozzles are the same size and that they operate properly. The airplane operator may need 5–10 acres, the tractor sprayer 1–2 acres, and the hand sprayer $\frac{1}{10}$–$\frac{1}{3}$ acre. (43,560 ft^2 = 1 acre).

Cleaning the Sprayer

Cleaning the sprayer before storage, even for short periods of time, will usually prove beneficial. Emptying the spray tank and rinsing it with water may be sufficient for short-time storage. Rinsing the pump and tank (both inside and out) with fuel oil will protect most metal parts from corrosion, but oil may injure parts made of natural rubber.

By cleaning the sprayer of leftover chemicals, you also reduce the possibility of injuring or killing sensitive plants in future sprayings; 2,4-D and related products have caused many problems through the neglect of this operation.

Farmer experience has shown that the barrel or tank is by far the most important source of contamination. It is especially difficult to clean barrels lacking a bottom drain. Therefore, if you expect contamination, change the barrel.

If the sprayer is to be cleaned, first rinse it with a material which acts as a solvent for the herbicide. Kerosene and fuel oils carry away herbicides known to be oil soluble (chemicals which form emulsions when mixed with water are oil soluble). Following the oil rinse, a rinse with a surfactant in water will help to remove the oil. The oil soluble herbicides, such as 2,4-D esters, are usually the most difficult herbicides to remove. 2,4-D salts are water-soluble and removed by thorough rinsing with water. Check the spray on susceptible plants, such as tomatoes, to make certain that the 2,4-D has been removed.

For wettable powder herbicides, examine the tank to see that *none of the wettable powder remains in the bottom*; otherwise, a thorough rinsing with water is usually sufficient.

GRANULAR MATERIALS

As shown in Figure 7-6, granular-application equipment should apply the granular material uniformly. Granular materials are spread in a manner similar to seed or fertilizer. (For further details see Chapter 6.)

SOIL INCORPORATION

Herbicides are incorporated (mixed) into the soil to (1) reduce volatility losses of relatively volatile herbicides, (2) place the herbicide close to the germinating weed seed, and (3) to improve dependability of weed control. The latter is particularly important in areas where rainfall is somewhat erratic,

Figure 7-6. Granular materials being applied in band over the row at planting time. (Gandy Company.)

and where there may not be sufficient water to leach surface-applied herbicides into the soil. Most preemergence herbicides must be in contact with the germinating weed seeds to be effective. Both liquid (spray) and granular applications may be incorporated into the soil (see Figure 7-7).

Virtually all types of cultivation equipment have been used to mix herbicides into the soil. However, all implements do not work equally well

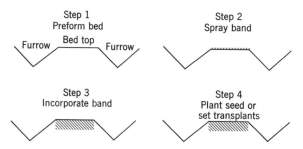

Figure 7-7. Steps in making a band application of a soil-incorporated herbicide to a preformed bed. Spray band must be wider than the width of the head of the incorporator. Selectivity may be influenced by the depth of seed placement within the treated soil. All of these steps can be performed in a single pass through the field by placing all the equipment required on one tractor.

for every situation or herbicide. For overall applications, a disc harrow or ground-driven rolling cultivator can be used. For band applications, power-driven rotary tillers are usually used. In addition to application rate, three major factors must be considered: depth of incorporation, soil conditions, and correct ground speed (4).

Depth of Incorporation

The depth of mixing is critical. If herbicides requiring shallow incorporation are placed too deep, they may lose some of their effect by dilution in a greater volume of soil (1). If volatile herbicides are mixed too shallow, some volatility loss may still occur. The more volatile herbicides (EPTC, pebulate, butylate, etc.) require deep mixing of 2–4 in. Less volatile herbicides (alachlor, trifluralin) may need to be mixed moderately deep, –2 inches. As a rule of thumb, the chemical is usually placed half as deep as the depth of discing.

The disc harrow is perhaps the most common incorporation tool. The second discing is commonly at right angles to the first. Speeds must be fast enough to effectively mix the soil. Also, a power-driven rotary tiller may be used for a band incorporation, with planting in the bands (4). Check the manufacturer's label for the depth of incorporation and equipment recommended for each herbicide.

Soil Conditions

Weed control may be erratic when many large clods are present. A fine seed bed with clods less than $\frac{1}{2}$ in. in diameter is recommended. The herbicide does not readily mix with excessively moist soil (4), which can also increase volatility losses by decreasing the absorption of herbicides on soil particles (2). The development of a plow sole, soil compaction, or loss of soil structure may occur by working soil with excessive moisture.

Ground Speed

Ground speeds of 4–6 mph are necessary to obtain adequate mixing of the herbicide into the soil with discs, harrows, and ground-driven rolling cultivators. Lower speeds (1–2 mph) can produce "streaking" and poor weed control. With power-driven rotary tillers, ground speeds of 1.5–3 mph are recommended. Good results have been obtained at a tiller speed of about 500 rev/min and 2-mph ground speed (4). The tillers must be close enough together to provide a "clean-sweep" of the soil. This may require L-shaped knives.

For an overall flat preplant soil-incorporation treatment followed by bedding (listing and bed forming), the lister shovels should always be set to run 1–2 in. shallower than the depth of incorporation. This setting prevents placing untreated soil on top of the bed. For a band application of a preplant soil-incorporation treatment to preformed beds, care must be taken not to remove treated soil or place untreated soil on top of the beds by subsequent planting or cultivating operations (see Figure 7-7).

SUBSURFACE LAYERING

The spraying of a horizontal layer of herbicide a few inches below the soil surface (subsurface layering) has produced effective control of several hard-to-control perennial weeds. One example is field bindweed control with trifluralin or dichlobenil. This method is also called the *spray-blade method* or simply *blading* or *layering*.

The spray-blade method consists of a blade with backward-facing nozzles attached under the leading edge. Figure 7-8 shows a straight blade that has been used in vineyards and orchards (3). Modified sweeps and V-shaped blades have also been used. The effectiveness of the control also can be seen in Figure 7-9. The concentrated layer of herbicide acts as a "protective wall", preventing the weed shoots from growing through it; thus, the deeper storage roots starve and the plant dies. Any disturbance of this layer by cultivation or natural cracking of the soil often permits the shoots to emerge, and control is therefore greatly reduced.

INJECTION

Herbicides are injected into soil or water to control terrestrial and aquatic weeds. The most common chemical injected into soil is methyl bromide. This herbicide is usually applied by a commercial applicator because of its toxicity and because the treated area must be covered with a gas-tight plastic tarpaulin for 24–48 hr. Methyl bromide is applied with chisel-type applicators which inject the chemical 6–8 in. below the soil surface. The chisel-injection units are spaced not more than 12 in. apart.

The injection of herbicides into water requires precise metering with adequate mixing. This is particularly important when applying herbicides through a sprinkle or furrow irrigation system. The precision of this rate of

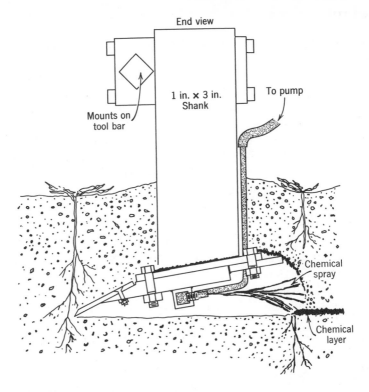

End view

1 in. × 3 in.
Shank

To pump

Mounts on
tool bar

Chemical
spray

Chemical
layer

Figure 7-8. Equipment used for placing a concentrated layer of herbicide below the soil surface.

Figure 7-9. Bindweed control is clearly evident in this research plot. Treflan-blade treated area is rectangle in foreground, with solid bindweed on all sides. (Elanco Products Company, a Division of Eli Lilly and Company.)

application is no better than the accuracy of the water distribution, and is usually less accurate than a spray application.

To control aquatic weeds, herbicides are injected into flowing waters, (irrigation and drainage canals) and injected from a boat into static waters (lakes and reservoirs). See Chapter 28 for details of aquatic weed control.

AIRCRAFT

Aircraft sprayers are less commonly used to apply herbicides than other pesticides because of the drift hazards (see Chapter 6). In addition, the precision of application is somewhat less than that of ground sprayers. Despite these limitations, aircraft are especially adapted for spraying or applying granular formulations to areas not readily accessible for ground sprayers, such as utility lines and fire breaks through remote woody areas, flooded rice fields, very large pasture or range areas and large cereal grainfields.

Both fixed and rotary (helicopter) wing aircraft are used. In general, the components of sprayers for aircraft are similar to those of ground sprayers. Included are a tank, agitators, pump, boom, valves, screens, nozzles, pressure regulator, and pressure gauge. Because the design of aircraft sprayers requires special engineering knowledge, the details of design are not discussed here.

CALCULATIONS

Calculations involving rate of application and various mixtures, in addition to practice problems, are presented in Chapter 6. Conversion factors useful in calculations are given in the appendix.

LITERATURE CITED

1. Ashton, F. M., and K. Dunster, *Weeds*, *9*, 312 (1961).
2. Ashton, F. M., and T. J. Sheets, *Weeds*, *7*, 88 (1959).
3. Lange, A. H., H. Agamalian, D. R. Donaldson, C. L. Elmore, W. D. Hamilton, O. A. Leonard, and H. M. Kempen, *Progress Report MA-41, Agricultural and Extension Services*, Univ. of California, 1972.
4. McHenry, W. B., and R. F. Norris, *Study Guide for Agricultural Pest Control Advisers on Weed Control*, Division of Agricultural Science, Univ. of California, 1972.

8 Aliphatics

The chemicals in this chapter are aliphatics (containing no rings). They are TCA, dalapon, acrolein, methyl bromide, glyphosate, and the organic arsenicals (cacodylic acid, MAA, MSMA, DSMA, MAMA). Certain amide and carbamate herbicides are also aliphatics, but their herbicidal properties place them more properly in Chapters 9 and 12, respectively. Except for the aliphatics, most organic herbicides contain one or more rings in their chemical structure.

TCA

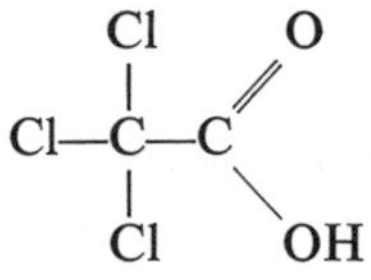

TCA, acid 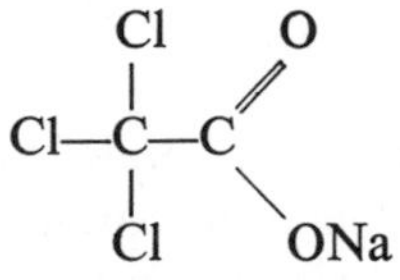   TCA, sodium salt

TCA is the common name for trichloroacetic acid. It is usually formulated as the sodium salt. Sodium TCA is a white solid, very soluble in water, 83.3 g/100 g of water. It is hygroscopic. Exposed to 90–95% relative humidity at 70°F, the chemical will absorb its equal weight in water in 8–10 days. Thus, the chemical must be stored in moisture-proof containers. The acute oral LD_{50} is about 5000 mg/kg.

Uses

Sodium TCA is primarily a grass killer. It has proven useful as a nonselective treatment on perennial weedy grasses such as johnsongrass, bermudagrass, and quackgrass at rates of 75–150 lb/acre (see Figure 8-1).

Figure 8-1. Sodium TCA was applied to the plot 4½ months before this picture was taken. (Texas Agricultural Experiment Station, College Station, Texas.)

As a selective, preemergence treatment in sugar beets to control annual grasses, TCA is applied at 8 lb/acre. In sugarcane, one preemergence or early postemergence treatment of 10–27 lb/acre controls seedling grasses, including johnsongrass seedlings.

Precautions

When handling and using sodium TCA, be especially careful to keep it away from your skin and eyes. Solutions of the sodium salt of TCA stronger than 10% may irritate and burn the skin and eyes unless rinsed off immediately, but solutions of 1% or less cause little or no irritation. However, if a dilute spray drift falls on your skin over a prolonged period, the water evaporates and the concentration of TCA increases to toxic levels.

A burning sensation will usually give sufficient warning to prevent more than peeling of the outer skin layer or the development of skin roughness. Ordinary glasses or sunglasses offer considerable eye protection from spray drift, and enclosed protective goggles provide more and better protection.

Mode of Action

When applied to foliage, TCA often causes rapid necrosis by contact action. It inhibits the growth of both shoots and roots, and causes leaf chlorosis and formative effects, especially to the shoot apex. At subherbicidal concentrations, however, growth may be stimulated (9, 10).

TCA is readily absorbed by leaves and roots (2). It is translocated throughout the plant from the roots, but only small amounts are translocated from leaves; therefore, it is primarily translocated via the apoplastic system. Perhaps its rapid contact action prevents symplastic movement. TCA is degraded slowly, if at all, by higher plants (2, 9).

Because TCA is used by the chemist to precipitate protein, its effectiveness as a herbicide has also been suggested. Foy (4) questions this, however, stating that "the acid or undissociated form is reasoned to be encountered rarely under physiological conditions." He further notes that halogenated acetates and propionates are theoretically able to alkylate the sulfhydryl or amino groups in enzymes. The suggestion has also been made that TCA induces conformational changes in enzymes (1).

At rates up to 100 lb/acre, TCA will usually disappear from a moist, loam soil with temperature of 70–100°F in 50–90 days. It persists somewhat longer in soils with a high organic-matter content and in dry or cold soils. It leaches readily in most soils.

DALAPON

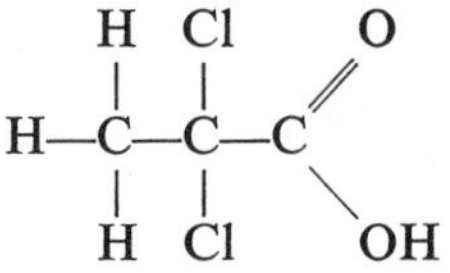

dalapon 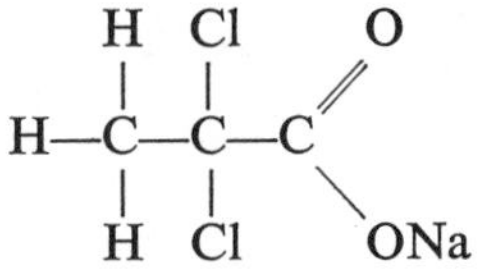   dalapon, sodium salt

Dalapon is the common name of 2,2-dichloropropionic acid. The trade names are Dowpon®, Radapon®, and Gramevin®. Its chemical structure is similar to TCA, except that one chlorine atom in TCA has been replaced by a methyl (—CH₃) group.

Dalapon is usually formulated as the sodium salt. This sodium salt is a white solid, quite soluble in water, 50.2 g/100 g of water at 25°C.

Dalapon is hygroscopic; in a 90% relative humidity, it will absorb enough water to dissolve itself. Exposed to 90% relative humidity at 70°F, the chemical will absorb its equal weight in water in 10–12 days. Therefore, storage requires moisture-proof containers.

When dalapon absorbs water, hydrolysis may occur and the chemical may lose its herbicidal effects. This equation shows the breakdown.

$$
\underset{\substack{|\quad| \\ H\quad Cl}}{\overset{\substack{H\quad Cl\quad O \\ |\quad|\quad /\!\!\!/}}{H-C-C-C}}\!\!\diagdown_{OH} + H_2O \xrightarrow{\text{heat}} \underset{\substack{| \\ H}}{\overset{\substack{H\quad O\quad O \\ |\quad\|\quad/\!\!\!/}}{H-C-C-C}}\!\!\diagdown_{OH} + NaCl + HCl
$$

dalapon + water $\xrightarrow{\text{heat}}$ pyruvic acid + sodium chloride + hydrochloric acid

At 25°C (77°F) this reaction is very slow, whereas at 50°C (122°F) conversion moves much faster (13). Therefore, prepare and apply spray solutions immediately, especially at high temperatures. Breakdown of dalapon is no problem if application is made within 24 hr after mixing with water.

The sodium salt of dalapon is relatively nonvolatile and nonflamable. It is less toxic to man and animals than is sodium TCA. The main precaution is to keep dalapon away from the skin and eyes. Glasses worn when spraying will aid in keeping the spray drift from the eyes. The acute oral toxicity, LD_{50}, is about 7000 to 9000 mg/kg.

Uses

Dalapon is an effective grass killer—considerably more effective than TCA in foliage application. It is more easily dissolved in water, somewhat less toxic to the skin and eyes, less corrosive to metals, and has a slightly shorter period of residual toxicity in the soil than TCA.

Dalapon controls perennial grasses such as bermudagrass, quackgrass, and johnsongrass. Two or three applications spaced 5–20 days apart at 5–10 lb/acre per application have usually given better control than one heavy application. Dalapon is also effective on cattails; one application of 5 lb/acre may be sufficient. An autumn application is most effective, about 3–4 weeks before leaves lose their green color (see Figures 8-2, 8-3, and 8-4).

Mode of Action

Dalapon inhibits the growth of both shoots and roots (7, 10). It also causes leaf chlorosis and formative effects, especially to the shoot apex, and interferes with cell division of root tips (12) and probably also shoot tips.

Dalapon may be absorbed through both foliage and roots. It enters the

Figure 8-2.   Spot treatment of johnsongrass clumps in a cotton field with dalapon spray.
Cotton seedling hit with the spray will likely be killed. (Dow Chemical Company, Midland
Mich.)

foliage through both cuticle and stomates. Surface-active agents aid movement into the stomates. The rate of dalapon absorption by plant leaves is
affected by the rate of treatment, type of plant surface, humidity, temperatures, and addition of surface-active agents. Absorption in the first 6 hr is by
far the most important, although absorption usually continues for at least
48 hr.

Translocation downward is primarily through the phloem and upward
movement is through the xylem (3). The amount of dalapon translocated
from the leaves is in direct proportion to the rate of application, provided
there is no burning or acute toxicity. With contact injury, translocation of
dalapon is greatly diminished or prevented (3). Translocation from leaves is
through living tissue.

Dalapon applied at high rates has acute plant toxicity. Quick killing of the
leaves, whether resulting from frost, heat, or a contact chemical, does not
favor translocation to other parts of the plant. Repeated application at low
to moderate concentrations gives the best control of deep-rooted perennial

Figure 8-3.   Taking a part of the hard work out of spot spraying. The cotton is the correct size for the first application of dalapon. (Dow Chemical Company, Midland, Mich.)

grasses. Applied in this way, the chemical shows chronic toxicity and kills very slowly. The plant may show little to no effect after the first treatment. But the chemical and method are highly effective after the second or third treatment (3).

Dalapon is readily absorbed by the roots and translocated to all parts of the plant. It was absorbed through the roots of cotton and sorghum and swept upward in the transpiration stream, reaching all parts of the plant within 1 hr (3). Under most circumstances, however, the chemical is applied more efficiently to the leaves for foliar absorption. At 40 lb/acre dalapon may remain toxic in moist, warm, loam soils for 20–60 days; cold or dry weather will lengthen the period of toxicity. Dalapon also leaches readily in most soil types.

Like TCA, dalapon probably acts by interfering with the activity of many enzymes by alkylating the sulfhydryl or amino groups (4), inducing conformational changes (1), or both. It has been shown that dalapon interferes with carbohydrate, lipid, and nitrogen metabolism (1).

Perhaps as a result of the enzyme disturbances, dalapon interferes with the plant's formation of pantothenic acid, one of the B vitamins essential to

Figure 8-4. Top: Dalapon applied at the rate of 6 lb/acre, 3 weeks prior to planting the corn, provided control of quackgrass. Bottom: Not treated with dalapon. Corn yields will be seriously reduced, perhaps resulting in a crop failure. (K. P. Buchholtz, University of Wisconsin.)

growth and development (6). Pantothenate artificially supplied to barley and oat plants partially overcame the toxic effects of dalapon (5).

ACROLEIN

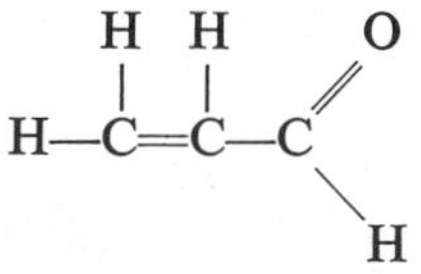

$$H-C=C-C\overset{\displaystyle O}{\underset{\displaystyle H}{<}}$$

Acrolein is the common name for acrylaldehyde or 2-propenal. The trade name is Aqualin®. It is a colorless liquid, moderately soluble in water, about 25 g/100 g of water, injected into water for the control of submersed and floating aquatic weeds. See Chapter 28 for details of aquatic-weed control.

Acrolein has an LD_{50} of 46 mg/kg. It is a flammable liquid subject to explosive reactions under certain conditions, and therefore may be applied only by applicators licensed by the manufacturer.

METHYL BROMIDE

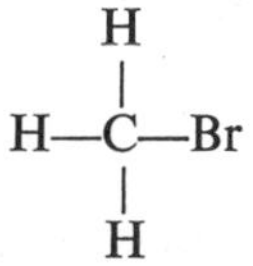

$$H-\overset{\displaystyle H}{\underset{\displaystyle H}{C}}-Br$$

Methyl bromide is a colorless, nearly odorless liquid or gas. At 1 atm of pressure and at 38°F (3.56°C), the liquid boils (turns to gas). The gas is 3.2 times heavier than air at 68°F. It is slightly soluble in water, very soluble in alcohol and ether, and is generally considered nonflammable and non-explosive. However, mixtures containing between 13.5% and 14.5% of the gas in air may be exploded by a spark.

Methyl bromide gas is poisonous to humans and animals, and the effects of exposures are cumulative. Because methyl bromide is nearly odorless (or has a very slight, sweetish odor), traces of other gases such as chloropicrin are sometimes added as warning agents. In prolonged storage with methyl bromide, the chloropicrin may decompose so that the warning odors are no longer present.

Methyl bromide is effective as a temporary soil sterilant. It controls all living plant, stem, or root tissue, most seeds, nearly all insects, and most disease organisms. As such, it makes an excellent soil treatment for seed beds of tobacco, flowers, vegetables, turf, and tree seedlings. It is also an effective

Figure 8-5. Methyl bromide being released under a sealed plastic cover. (Dow Chemical Company, Midland, Mich.)

treatment for propagating beds and for preplanting treatment of trees, shrubs, fruits, and gardens. Methyl bromide treatments, however, are expensive.

The chemical is applied under an airtight cover, usually plastic, at the rate of 1–2 lb/100 ft² of surface, or 100 ft³ of soil. It may be injected into the soil by power-driven equipment for large areas, or merely released under the plastic tarpaulin for small areas.

Methyl bromide may kill beneficial organisms in the soil as well as disease organisms, so that normal decomposition of organic matter into ammonia, then nitrite, and finally nitrate may be disturbed and toxic substances accumulate. This poisoning occurs if the organisms responsible for one of these decomposition changes are killed.

For example, in sterilized soils high in organic matter, ammonia or nitrite may build up to toxic levels. Such responses are seldom noted in sands or other low-organic-matter soils. Usually the beneficial microorganisms multiply sufficiently in 1–3 months in warm, moist soils so they restore a normal microorganism equilibrium and normal plant growth. That is why in preparing compost mixtures, it is usually desirable to sterilize the soil first and then add the peat, sawdust, or other organic matter.

GLYPHOSATE

$$\text{HO}-\overset{\overset{\displaystyle O}{\|}}{\text{C}}-\overset{\overset{\displaystyle H}{|}}{\underset{\underset{\displaystyle H}{|}}{\text{C}}}-\overset{\overset{\displaystyle H}{|}}{\text{N}}-\overset{\overset{\displaystyle H}{|}}{\underset{\underset{\displaystyle H}{|}}{\text{C}}}-\overset{\displaystyle}{\text{P}}\overset{\displaystyle O}{\underset{\displaystyle OH}{\diagup}}\text{OH}$$

Glyphosate is the common name for N-(phosphonomethyl) glycine. The trade name is Roundup®. It is a white solid, soluble in water, about 1 g/100 g water at 25°C, formulated as the isopropylamine salt for an aqueous spray. Glyphosate is relatively nontoxic to mammals, LD_{50} 4320 mg/kg.

Uses

Glyphosate is applied as a foliar spray. It gives outstanding control of perennial grasses such as johnsongrass, bermudagrass, and quackgrass, and appears to be superior to TCA or dalapon. It is also effective on certain perennial broadleaf weeds as well as annual grasses and broadleaf weeds. With such wide-ranging effects it is essentially nonselective, and therefore it is not used on crops except where it can be kept off the crop plant (selectivity by placement).

Mode of Action

Glyphosate, a very mobile compound, is translocated throughout the plant when applied to the leaves. It is rapidly degraded in the soil, but appears quite resistant to decomposition in higher plants. Toxicity symptoms develop slowly and may not be observed for 1–3 weeks. Jaworski (8) has suggested that glyphosate interferes with the biosynthesis of the essential amino acid phenylalanine, more specifically with the metabolism of chorismic acid in the aromatic amino acid biosynthetic pathway.

ORGANIC ARSENICALS

The organic arsenical herbicides include two similar compounds, cacodylic acid and MAA (methanearsonic acid), and their salts.

Cacodylic Acid

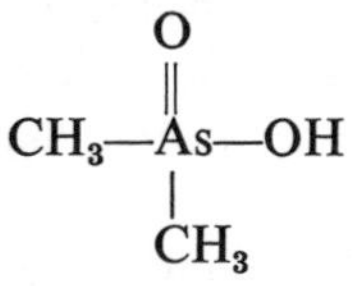

$$CH_3 - \overset{\overset{\displaystyle O}{\|}}{\underset{\underset{\displaystyle CH_3}{|}}{As}} - OH$$

The common name of hydroxydimethylarsine oxide is cacodylic acid. It was one of the first organic arsenicals to be introduced. It is a colorless, crystalline solid, soluble in water, 66.7 g/100 g of water. It is usually formulated as the sodium salt. Like other organic arsenical herbicides it has a much lower toxicity than elemental arsenic. Its acute oral LD_{50} is 830 mg/kg.

Cacodylic acid and its sodium salt are used as general contact sprays. They dessicate and defoliate a wide variety of plant species. They are used in noncrop areas (rights-of-ways, ditchbanks, fences, industrial sites), forests, lawn-renovation areas, and citrus orchards. Directed sprays are required in the citrus orchards.

Cacodylic acid is not thought to be translocated in plants. It is rapidly inactivated in soils, but continuous use over long periods of time may permit phytotoxic levels of elemental arsenic to accumulate.

MAA

$$
\begin{array}{cccc}
\overset{\displaystyle O}{\underset{\displaystyle OH}{CH_3-\overset{\|}{As}-OH}} & \overset{\displaystyle O}{\underset{\displaystyle ONa}{CH_3-\overset{\|}{As}-OH}} & \overset{\displaystyle O}{\underset{\displaystyle ONa}{CH_3-\overset{\|}{As}-ONa}} & \overset{\displaystyle O}{\underset{\displaystyle ONH_4}{CH_3-\overset{\|}{As}-OH}} \\
\text{MAA} & \text{MSMA} & \text{DSMA} & \text{MAMA}
\end{array}
$$

MAA is the common name for methanearsonic acid. It is usually formulated as monosodium methanearsonate (MSMA), disodium methanearsonate (DSMA), or monoammonium methanearsonate (MAMA). DSMA, the more widely used formulation, is a white, crystalline solid. It is soluble in water, 25.6 g/100 g of water. The acute oral LD_{50} is 1800 mg/kg.

The first major use of these compounds was for postemergence control of crabgrass in turf. They are now used to control certain other annual weeds in tolerant lawn grasses, cotton, citrus, and noncrop areas. They are also used to control perennials (johnsongrass and nutsedge) in noncrop areas, cotton, and citrus. Directed sprays are required for the crops.

In contrast to cacodylic acid, the methanearsonic acid salts appear to be translocated throughout the plant after foliar application. They are rapidly inactivated in soils, but elemental arsenic may accumulate with prolonged use.

Both cacodylic acid and the methanearsonic acid salts may act by interfering with phosphorus metabolism, complexing with sulfhydryl-containing enzymes, uncoupling oxidative phosphorylation, or both (1).

LITERATURE CITED

1. Ashton, F. M., and A. S. Crafts, *Mode of Action of Herbicides*, Wiley, New York, 1973.
2. Blanchard, F. A., *Weeds*, *3*, 274 (1954).
3. Crafts, A. S., and C. L. Foy, *Down to Earth*, *14*, 2 (1959).
4. Foy, C. L., "The Chlorinated Aliphatic Acids," in *Degradation of Herbicides*, P. C. Kearney and D. D. Kaufman, Eds., Marcel Dekker, New York, 1969, p. 207.
5. Hilton, J. L., J. S. Ard, L. L. Jensen, and W. A. Gentner, *Weeds*, *7*, 381 (1959).
6. Hilton, J. L., L. L. Jensen, and W. A. Gentner, *Plant Physiol.*, *33*, 43 (1958).
7. Ingle, M., and B. J. Rogers, *Weeds*, *9*, 264 (1961).
8. Jaworski, E. G., *Agric. and Food Chem.*, *20*, 1195 (1972).
9. Mayer, F., *Biochem. Z.*, *328*, 433 (1957).
10. Meyer, R. E., and K. P. Buchholtz, *Weeds*, *11*, 4 (1963).
11. Palmer, J. S. *A.R.S. Production Report No. 137*, United States Department of Agriculture, Washington D.C., 1972.
12. Prasad, R., and G. E. Blackman, *J. Exp. Bot.*, *15*, 48 (1964).
13. Report of Research Committee, *Southern Weed Conference Proceedings*, 7, 340 (1954).

FOR CHEMICAL USE, SEE THE MANUFACTURER'S LABEL AND FOLLOW THE DIRECTIONS. ALSO SEE THE PREFACE.

9　Amides

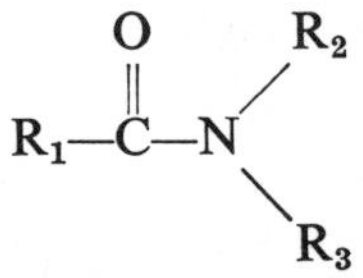

$$R_1-\overset{\overset{\displaystyle O}{\|}}{C}-N\overset{\displaystyle R_2}{\underset{\displaystyle R_3}{}}$$

The basic chemical structure of the amide-type herbicides is shown above. However, the substitutions on positions R_1, R_2, and R_3 vary greatly, making them a diverse group of chemicals. Likewise, the weeds that the different amide herbicides control, and the crops they are selective to, vary widely. Most amides are used as selective herbicides applied either preemergence or preplant. However, propanil is applied to the foliage of weeds to be controlled.

The amides include alachlor, CDAA, diphenamid, naptalam, pronamide, propachlor, napropamide, and propanil.

ALACHLOR

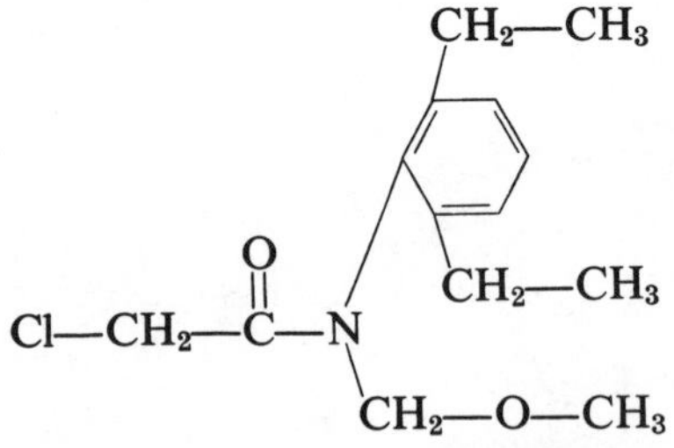

$$Cl-CH_2-\overset{\overset{\displaystyle O}{\|}}{C}-N$$

CH₂—CH₃

CH₂—CH₃

CH₂—O—CH₃

Alachlor is the common name for 2-chloro-2′,6′-diethyl-N-(methoxymethyl)-acetanilide, and the trade name is Lasso®. The chemical is formulated as an emulsifiable concentrate and granules. It is a cream-colored solid, relatively insoluble in water, 148 ppm at 25°C. The acute oral LD_{50} is 1800 mg/kg.

Figure 9-1. Barnyardgrass control in corn with alachlor. Upper left: untreated. (C. L. Elmore, University of California, Davis.)

Uses

Alachlor is applied principally as a preemergence treatment to control annual grasses and certain annual broadleaf weeds in corn, peanuts, and soybeans. It is also applied preplant and soil incorporated. It is sometimes combined with other herbicides to increase the spectrum of weeds controlled. See Figure 9-1.

Soil Influence

Because alachlor is adsorbed by soil colloidal particles, it is not subject to excessive leaching in most soils. In sandy soils low in organic matter, persistence is shorter than in heavy soils. In medium-textured soils, and with moderate moisture, herbicidal effectiveness usually lasts about 3 months.

Mode of Action

Alachlor inhibits the growth of shoots and roots, as well as inhibiting lateral root development (11).

CDAA

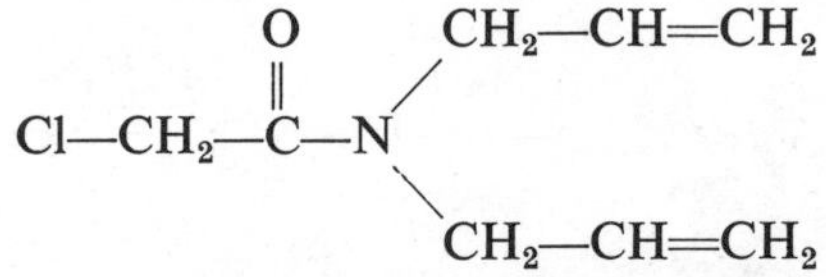

$$\text{Cl—CH}_2\text{—}\overset{\displaystyle\overset{O}{\|}}{\text{C}}\text{—N} \overset{\displaystyle CH_2\text{—CH}{=}CH_2}{\underset{\displaystyle CH_2\text{—CH}{=}CH_2}{}}$$

CDAA is the common name for *N,N*-diallyl-2-chloroacetamide. The trade name is Randox®. It is formulated as an emulsifiable concentrate and granules, and it is an oily, amber-colored liquid. The acute oral LD_{50} is 750 mg/kg. If spilled on the skin, it sensitizes the area to temperature changes; this may cause an aching or burning sensation for several hours. The historical development of CDAA, and related products, has been published (8).

Uses

CDAA is applied as a preemergence or preplant soil incorporation treatment for control of many grassy and broadleaf weeds in corn, sorghum, soybeans, and several vegetable crops. It is also formulated with trichlorobenzylchloride to increase its control of certain broadleaf weeds. Trade name of this formulation is Randox T®.

Soil Influence

Like most organic herbicides, the rate of leaching is correlated with the clay and organic content of the soil. The higher the content of these two factors, the less leaching. CDAA is especially effective on muck and clay loam soils. The persistence of CDAA in soil is relatively short, 4–6 weeks in moist soils, which is not long enough to give season-long weed control.

Mode of Action

CDAA inhibits cell division and root elongation (3). It is rapidly taken up by roots, translocated to the upper parts of the plant (9), and also readily absorbed by germinating seeds (9). CDAA is rapidly broken down in higher plants to glycolic acid and probably diallylamine; they are further metabolized to normal plant constituents (9, 10). CDAA has been shown to inhibit respiration, uncouple oxidative phosphorylation, and inhibit the development of certain hormone-induced hydrolytic enzymes. However, the biochemical basis for its action is still not fully understood (1).

DIPHENAMID

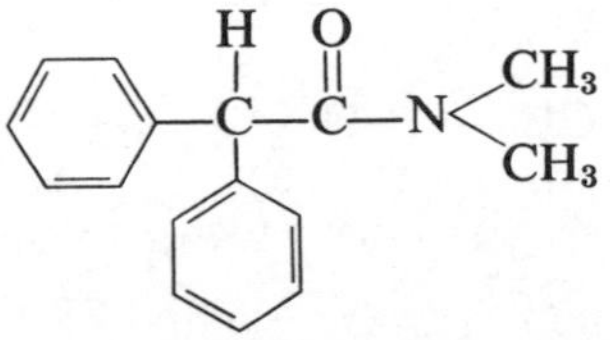

Diphenamid is the common name for *N,N*-dimethyl-2,2-diphenylacetamide. The trade names are Dymid® and Enide®. It is formulated as a wettable powder and granules. It is white to off-white, crystalline solid with a solubility in water of 261 ppm at 27°C. The acute oral LD_{50} is 686–776 mg/kg.

Uses

Diphenamid is primarily used as a selective herbicide to control annual grass and broadleaf weeds in cotton, peanuts, soybeans, tobacco, and several vegetable, horticultural, and orchard crops. These include apples, okra, peaches, peppers, potatoes, strawberries, sweet potatoes, and tomatoes. It may also be used on nonbearing blackberries, citrus, and raspberries. Dichondria and bermudagrass turfs, as well as many ornamentals, are tolerant to diphenamid. See Figure 9-2.

Diphenamid is applied as a preemergence or preplant soil-incorporation treatment. It is also formulated in combination with both trifluralin and dinoseb to increase the spectrum of weed species controlled. These products have the trade names of Trefmid® and Enide Dinitro®; but the combinations are not necessarily suitable for use on all the crops listed above.

Soil Influence

Diphenamid is leached fairly rapidly in sandy soils but more slowly in loam or clay soils. The soil colloids adsorb diphenamid, holding it from leaching. Under warm, moist conditions diphenamid normally persists from 1 to 3 months.

Mode of Action

Diphenamid appears to let seeds germinate, but kills the seedling plant before it emerges from the soil. At sublethal concentrations diphenamid inhibits root

Figure 9-2.   Right: tomatoes treated preemergence with diphenamid, followed by a lay-by treatment with trifluralin Left: not treated. (Elanco Products Company, a Division of Eli Lilly and Company.)

development in many species. In tolerant species such as tomato, marginal-leaf chlorosis may even occur after normal germination at relatively high application rates of diphenamid.

Diphenamid is readily absorbed by roots and rapidly translocated to the tops of plants with accumulation in the leaves (7, 13); this suggests apoplastic transport. Studies with tomatoes (13) and strawberries (7) show that diphenamid is metabolized by higher plants yielding N-methyl-2,2-diphenyla-cetamide as the major metabolite. Four additional minor metabolites were tentatively identified. Although diphenamid has been reported to inhibit RNA synthesis (2), this is not consistent with several studies showing that it does not inhibit protein synthesis (1). It appears that diphenamid inhibits the uptake of inorganic ions by roots and influences the distribution of calcium within the plant (15).

NAPTALAM

naptalam

Naptalam is the common name for *N*-1-naphthylphthalamic acid. The trade name is Alanap®. It is formulated as the sodium salt in water-soluble liquid and granule form. It is a purple, crystalline solid. The acid form has a water solubility of 200 ppm and the sodium salt 230,800 ppm. Increases in water solubility in herbicides of this magnitude are frequent when acid forms are converted to salts.

Uses

Naptalam is sold separately as a herbicide for use on crops such as cantaloupe, cucumbers, and watermelons. Formulated in combination with dinoseb, it is effectively applied preemergence to control annual broadleaf weeds and grasses in peanuts and soybeans. Naptalam also has growth-regulating properties and is used for blossom-thinning on peaches.

Soil Influence

Naptalam is subject to rapid leaching in porous soils. Thus, if heavy rains occur shortly after seeding and treatment, both crop injury and poor weed control may result. Weeds are usually controlled from 3 to 8 weeks after application and the herbicide presents no soil-residual problem.

Mode of Action

Naptalam has the unique property of acting as an antigeotropic agent; growing shoots and roots have a tendency to lose their ability to grow up or down, respectively. It has not been proved, however, that this is associated with its herbicidal action. Naptalam acts primarily as an inhibitor of seed germination, and also inhibits some growth responses induced by the normal plant hormones indole-3-acetic acid and gibberelic acid (1).

PRONAMIDE

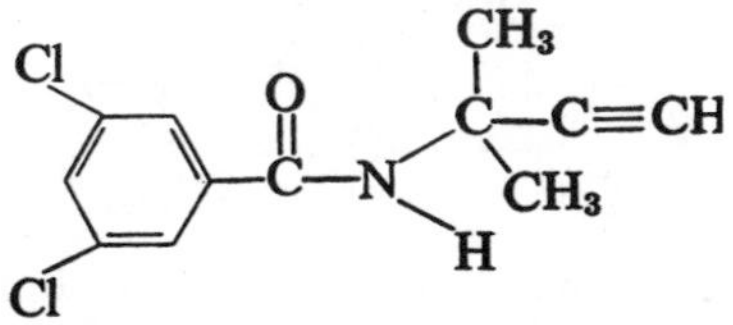

Pronamide is the common name for 3,5-dichloro-N-(1,1-dimethyl-2-propynyl)benzamide. The trade name is Kerb®. It is an off-white solid with a water solubility of 15 ppm, formulated as a wettable powder. The acute oral LD_{50} is 5620–8350 mg/kg.

Uses

Pronamide is a relatively new herbicide, and therefore its selectivities are not fully known. However it appears to be most useful for the control of annual grass and broadleaf weeds in small-seeded legumes, lettuce, certain orchard crops, turf, and some ornamentals. It may also control some perennial grasses.

Soil Influence

Pronamide is readily adsorbed on organic matter and other colloidal exchange sites and therefore leaches very little from most soils. It has intermediate persistence in soil, 3–8 months.

Mode of Action

Pronamide inhibits cell division and growth. It is readily absorbed by roots and translocated upward and distributed throughout the plant. Its translocation from leaves is not appreciable (4). This suggests translocation in the apoplastic system. Pronamide is slowly metabolized by higher plants by means of alterations of the aliphatic side chains (20).

PROPACHLOR

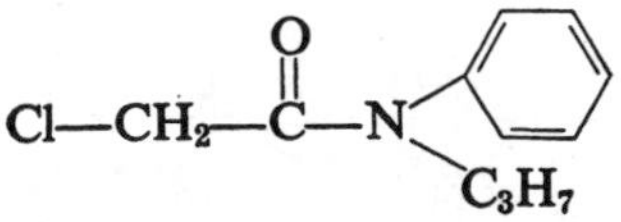

$$Cl-CH_2-\overset{\overset{\textstyle O}{\|}}{C}-N\diagdown^{C_6H_5}_{C_3H_7}$$

Propachlor is the common name for 2-chloro-*N*-isopropylacetanilide. The trade name is Ramrod®. It is a light-tan solid with a water solubility of 700 ppm at 20°C. It is formulated as a wettable powder and granules. The acute oral LD_{30} is 710 mg/kg.

Uses

Propachlor is used as a preemergence treatment to control many grass and broadleaf weeds in corn, cotton, peas, sorghum, and soybeans (seed crop only). It also has postemergence effects on small weeds and is applied in corn immediately after the crop emerges and before the weeds reach the two-leaf stage. It can be combined with atrazine as a preemergence treatment for annual-weed control in corn and sorghum.

Soil Influence

Propachlor is adsorbed by colloidal particles of the soil and is therefore not subject to excessive leaching. It primarily undergoes chemical breakdown in the soil, but the action of soil microorganisms also contributes to this breakdown. There are no soil persistence problems because the chemical is degraded in 4–6 weeks in most soils although it persists somewhat longer in organic soils.

Mode of Action

Propachlor inhibits root elongation. This is probably the result of inhibition of auxin-induced cell expansion (6). Although absorbed by roots (10), it is probably absorbed by the shoot as it emerges through the zone of treated soil (12, 16), and then translocated throughout the plant. Propachlor is very rapidly metabolized by corn and soybeans, and thus it has been suggested that the chemical forms a conjugate with some natural product (9). Propachlor inhibits protein synthesis, possibly by preventing the transfer of amino acids from amino-acyl-*t*-RNA to the polypeptide chain (5).

NAPROPAMIDE

Napropamide is the common name for 2-(α-naphthoxy)-N,N-diethylpropionamide. The trade name is Devrinol®. It is a white, crystalline solid with a water solubility of 73 ppm at 20°C. It is formulated as a wettable powder. The acute oral LD_{50} is greater than 5000 mg/kg.

Uses

Napropamide is a preemergence or preplant soil-incorporated herbicide used to control most annual grasses and many broadleaf weeds in several deciduous orchard crops including almonds, apricots, cherries, nectarines, peaches, plums, prunes, and oranges. It can be used on newly planted trees as well as established trees, and on grapes and tomatoes.

Soil Influence

Napropamide is quite resistant to leaching in most mineral soils. It is slowly decomposed by soil microorganisms. Its relatively long persistence, more than 9 months under some conditions, may cause injury to crops following use. When incorporated into moist, loamy sand and loam soils at 70–90°F, the half-life was 8–12 weeks.

Mode of Action

Limited basic research has been conducted on this herbicide. The following information was provided by the manufacturer. Napropamide inhibits the growth and development of the roots of grass weeds. Using radioactive napropamide, it was shown to be rapidly taken up by roots of tomatoes. Within 8 hr the radioactivity was translocated upward throughout the stems and leaves. Radioactive, ring-labelled napropamide is rapidly metabolized in tomato plants and fruit trees to water-soluble metabolites. The major metabolites appear to be hexose conjugates of 4-hydroxy napropamide.

PROPANIL

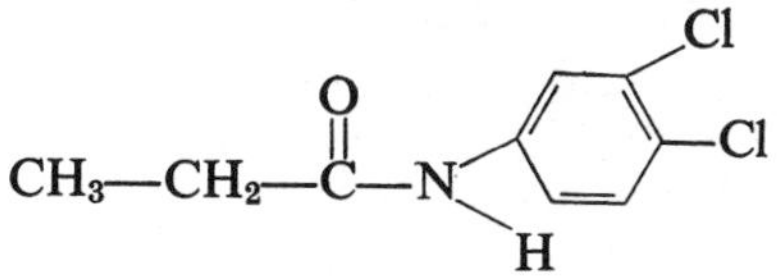

$$CH_3-CH_2-\overset{\overset{\textstyle O}{\|}}{C}-\underset{\underset{\textstyle H}{|}}{N}-\text{(3,4-dichlorophenyl)}$$

Propanil is the common name for 3',4'-dichloropropionanilide. The trade names are Stam F-34® and Rogue®. It is a light-brown to gray-black solid with a water solubility of about 500 ppm. It is formulated as an emulsifiable concentrate and as a liquid concentrate used without dilution. It has an acute oral LD_{50} of 1870–2270 mg/kg.

Uses

Propanil is applied postemergence in rice to control several annual grasses, especially barnyardgrass, and a limited number of annual broadleaf weeds and sedges. The time of application is critical, usually depending on the size of barnyardgrass. The ideal time is at the one- to three-leaf stage. It is used in both upland and flooded rice. Both aircraft sprayers and ground sprayers are used on upland rice, but only the former are used on flooded rice. Because propanil is essentially a contact spray, uniform coverage is essential and the drift of propanil onto other crops is a potential hazard.

Although rice is quite tolerant of propanil, the herbicide may severely injure this crop when certain insecticides (carbaryl or any organic phosphate) are applied within 14 days before or after propanil is used. These insecticides inhibit the action of the enzyme in rice which decomposes propanil and thus leads to its selectivity.

Because propanil is applied to the foliage, soil type has no effect on its action; it is rapidly broken down in the soil, therefore presenting no residual problem for later crops.

Mode of Action

Propanil causes chlorosis followed by necrosis to the leaves of susceptible species. When it remains as discreet, small droplets, a speckled pattern may be observed (1). Propanil is absorbed by leaves, but translocation within the leaf or from the treated leaves to the rest of the plant is very limited (18). The degradation of the chemical in higher plants has been studied rather extensively (1, 5, 14). Propanil is metabolized, through an intermediate, to 3,4-dichloroalanine and lactic acid, and the 3,4-dichloroalanine may then combine with glucose, lignin, or both (18, 19). Propanil inhibits a number of biochemical reactions, especially photosynthesis (1).

LITERATURE CITED

1. Ashton, F. M., and A. S. Crafts, *Mode of Action of Herbicides*, Wiley, New York, 1973.
2. Briquet, M. V., and A. L. Wiaux, *Meded. Rijksfac. Landbouwwetensch Gent.*, *32*, 1040 (1967).
3. Canvin, D. T., and G. Friesen, *Weeds*, *7*, 153 (1959).
4. Carlson, W. C., Ph.D. Dissertation, Univ. of Illinois, 1972.
5. Casida, J. I., and L. Lypken, *An. Rev. Plant Physiol.*, *20*, 607 (1969).
6. Duke, W. B., *Proceedings of the Twenty-Second Northeast Weed Control Conference*, 1968, p. 504.
7. Golab, T., R. J. Herberg, S. J. Parka, and T. B. Tepe, *J. Agr. Food Chem.*, *14*, 542 (1966).
8. Hamm, P. C. Weed Sci., *22*, 541 (1974).
9. Jaworski, E. G., *J. Agr. Food Chem.*, *12*, 33 (1964).
10. Jaworski, E. G., in *Degradation of Herbicides*, P. C. Kearney and D. D. Kaufman, Eds. Marcel Dekker, New York, 1969, p. 165.
11. Keeley, P. E., C. H. Carter, and J. H. Miller, *Weed Sci.*, *20*, 71 (1972).
12. Knake, E. L., and L. M. Wax, *Weed Sci.*, *16*, 393 (1968).
13. Lemin, A. J., *J. Agr. Food Chem.*, *14*, 109 (1966).
14. Matsunaka, S., *Residue Rev.*, *25*, 45 (1969).
15. Nashed, R. B., and R. D. Illnicki, *Proceedings of the Twenty-Second Northeast Weed Control Conference*, 1968, p. 500.
16. Nishimoto, R. K., A. P. Appleby, and W. R. Furtick, *Weed Society of America Abstr.*, 1967, p. 46.
17. Smith, G. R., C. A. Porter, and E. G. Jaworski, *Abstracts of the 152nd Meeting of the American Chemical Society*, 1966, p. A-42.
18. Yih, R. Y., D. M. McRae, and H. F. Wilson. *Plant Physiol.*, *43*, 1291 (1968a).
19. Yih, R. Y., D. M. McRae, and H. F. Wilson, *Science*, *116*, 37 (1968b).
20. Yih, R. Y., and C. Swithenbank, *J. Agr. Food Chem.*, *19*, 314 (1971).

FOR CHEMICAL USE, SEE THE MANUFACTURER'S LABEL AND FOLLOW THE DIRECTIONS. ALSO SEE THE PREFACE.

10 Benzoics

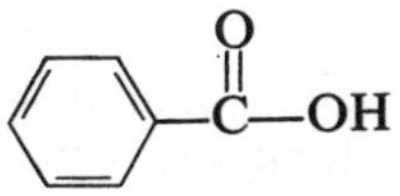

$$\text{(benzene ring)}-\overset{\overset{\displaystyle O}{\|}}{C}-OH$$

All benzoic herbicides are derivatives of benzoic acid, shown above. All contain at least two chlorine atoms, while one (dicamba) contains a methoxy ($-OCH_3$) group and another (chloramben) contains an amino ($-NH_2$) group. Zimmerman and Hitchcock (10) pointed out the growth-regulating properties of the substituted benzoic acids as early as 1942. In 1948 the Jealott's Hill Experiment Station in England evaluated the herbicidal properties of 2,3,6-TBA under field conditions, and similar investigations were simultaneously carried out in the United States. All the benzoic-acid herbicides are effective when applied to foliage or soil, except chloramben, which is used as a preemergence herbicide.

2,3,6-TBA

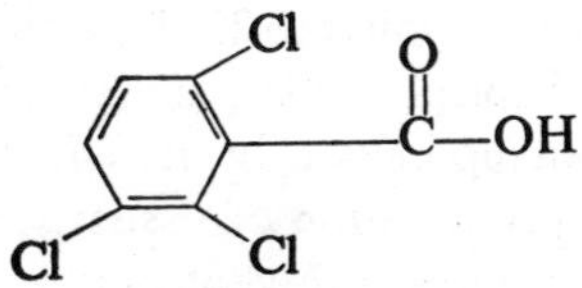

$$\text{(chlorinated benzene ring with Cl at 2,3,6 positions)}-\overset{\overset{\displaystyle O}{\|}}{C}-OH$$

2,3,6-TBA

2,3,6-TBA is the common name for 2,3,6-trichlorobenzoic acid. Trade names are Benzac®, Fen-All®, Tribac®, Trysben®, and Zobar®. Most commercial formulations contain a mixture of about 60% 2,3,6-TBA and 40% other chlorinated benzoic acids. Some of these latter derivatives are also phytotoxic, but the 2,3,6-chlorinated derivative is most phytotoxic. They are usually formulated as a liquid dimethylamine salt.

Pure 2,3,6-TBA is a white, crystalline solid, somewhat soluble in water, 0.84 g/100 ml. However, the dimethylamine salt is readily soluble in water. The acute oral LD_{50} of 2,3,6-TBA is about 750–1000 mg/kg and its dimethylamine salt approximately 1644 mg/kg.

Uses

2,3,6-TBA is considered a nonselective herbicide and is not used in crops, but it does control several broadleaf perennial weeds such as field bindweed, Canada thistle, and bur ragweed at 10–30 lb/acre, as well as certain woody brush species such as conifers, wild roses, and sassafras at 4–20 lb/acre. It is usually applied to the foliage, but is also taken up by the roots when leached into the rooting zone. Spray drift or possibly volatile fumes may injure susceptible crops such as beans, tomatoes, cotton, and various ornamentals, orchard, and vine crops.

Soil Influence

2,3,6-TBA is not markedly adsorbed by soil colloids and its high water solubility allows it to be readily leached into the rooting zone. It is not subject to rapid degradation in soil and often persists more than 1 year.

Mode of Action

2,3,6-TBA produces symptoms similar to 2,4-D, with epinasty of young shoots and inhibition of growth of apical meristems in dicotyledons. In monocots, it can result in increased lodging caused by its effect on nodal meristems. It promotes cell elongation, proliferation of tissues, and induction of adventitious roots (11). It also interferes with geotropic and phototropic responses (9).

This herbicide is readily absorbed by leaves and roots and translocated via both the symplastic and apoplastic systems with accumulation in areas of high metabolic activity, such as meristems. It has also been shown to readily "leak" or move out from roots into the surrounding medium when applied to the foliage. It appears to be very stable or resistant to degradation in plants (2, 5).

Figure 10-1. Modification of leaf structure of melons induced by dicamba. (A. H. Lange, University of California, Parlier.)

DICAMBA

dicamba

Dicamba is the common name for 3,6-dichloro-*o*-anisic acid. The trade name is Banvel®. The chemical structure of dicamba is similar to 2,3,6-TBA, except that the chlorine atom on the number 2 position has been replaced by a methoxy($-OCH_3$) group. It is formulated as a liquid in the dimethylamine salt form, and in granular form as the acid or amine salt.

Dicamba, as the acid, is a white, crystalline solid somewhat soluble in water, 0.45 g/100 ml. The dimethylamine salt is quite soluble in water. The acute oral LD_{50} is about 2900 mg/kg and the dimethylamine salt about 1028 mg/kg.

Uses

Dicamba is somewhat more selective than 2,3,6-TBA, but still applied only on grass crops including barley, corn, oats, sorghum, wheat, and pasture and range grasses. It is also used on noncropland and brush where it is often combined with 2,4,5-T, 2,4-D, or both, to broaden the spectrum of controlled weed species. It is usually applied to the foliage and stems of these plants, but also is active in the soil. Broadleaf perennial weeds and woody brush or trees are usually the target plant, but most annual broadleaf weeds are also controlled.

Soil Influence

It is relatively persistent in soils and phytotoxicity remains for several months.

Mode of Action

Dicamba affects plant growth in much the same way as 2,3,6-TBA, with epinasty of young shoots and proliferative growth. It is readily absorbed by leaves, stems, and roots and translocated throughout the plant, accumulating in areas of high metabolic activity (1). This indicates both apoplastic and symplastic transport. It also "leaks" from roots into the surrounding medium. Dicamba is relatively stable in higher plants, but it undergoes hydroxylation or demethylation with hydroxylation (1). These degradation products may then be conjugated with some endogenous metabolite. The rate of degradation varies greatly with species.

CHLORAMBEN

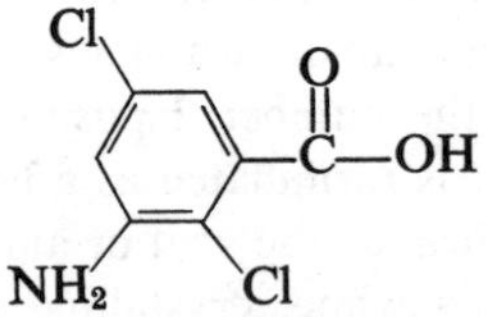

chloramben

Chloramben is the common name for 3-amino-2,5-dichlorobenzoic acid. The trade name is Amiben®. Chloramben is a white, amorphous, solid, somewhat soluble in water, about 700 ppm; usually formulated in liquid or granular form. As the ammonium salt it is quite soluble in water. A liquid formulation, Vegiben® 2E, contains the methyl ester derivative. The latter formulation has greater selectivity to certain vegetable crops. The acute oral LD_{50} of chloramben is about 3500 mg/kg.

Uses

Chloramben is much more selective than 2,3,6-TBA or dicamba. It is widely used as a preemergence herbicide to control many annual broadleaf weeds and grasses in soybeans. It also controls these same weeds in asparagus, beans, corn, peanuts, peppers, pumpkins, squash, sunflowers, sweet potatoes, and tomatoes. The herbicide is also formulated with linuron for annual-weed control in soybeans.

Soil Interaction

Chloramben is readily leached in sandy soils and subject to some leaching in other soils, especially after heavy rains. It lasts from 1 to 3 months in most soils, and so is much less persistent than 2,3,6-TBA or dicamba.

Mode of Action

Although chloramben is generally less phytotoxic than 2,3,6-TBA or dicamba, it still induces some auxin-like growth abnormalities. It is absorbed by seeds (4, 6, 7) and exhibits its phytotoxic properties on germinating seeds and young seedlings. Although it is also absorbed by leaves and roots, its translocation is quite limited. Apparently, interaction with some plant constituents (1) prevents its transport. The primary molecular fate of chloramben in higher plants apparently relates to the formation of a glucose conjugate (3,8).

LITERATURE CITED

1. Ashton, F. M., and A. S. Crafts, *Mode of Action of Herbicides*, Wiley, New York, 1973.
2. Balagannis, P. G., M. S. Smith, and R. L. Wain, *Ann. Appl. Biol.*, *55*, 149 (1965).
3. Colby, S. R., *Science*, *40*, 619 (1965).
4. Haskell, D. A., and B. J. Rogers, *Proceedings of the Seventeenth North Central Weed Control Conference*, p. 39, 1960.
5. Mason, G. W., Ph.D. Dissertation, Univ. of California, Davis, 1960.
6. Rieder, G., K. P. Buchholtz, and C. A. Kust, *Weed Sci.*, *18*, 101 (1970).
7. Swan, D. G., and F. W. Slife, *Weeds*, *13*, 133 (1965).
8. Swanson, C. R., R. E. Kadunce, R. A. Hodgson, and D. S. Frear, *Weeds*, *14*, 319 (1966).

9. Vander Beek, L. C., *Ann. N.Y. Acad. Sci.*, *144*, 374 (1967).

10. Zimmerman, M. H., and A. E. Hitchcock, *Contr. Boyce Thompson Inst.*, *12*, 321 (1942).

11. Zimmerman, M. H., and A. E. Hitchcock, *Contr. Boyce Thompson Inst.*, *16*, 209 (1951).

FOR CHEMICAL USE, SEE THE MANUFACTURER'S LABEL AND FOLLOW THE DIRECTIONS. ALSO SEE THE PREFACE.

11 Bipyridyliums

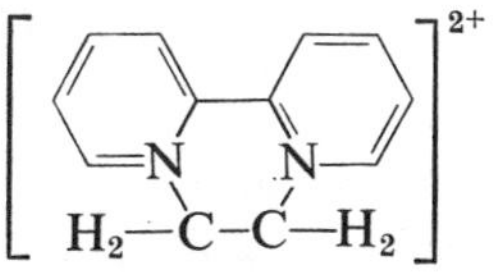

diquat ion

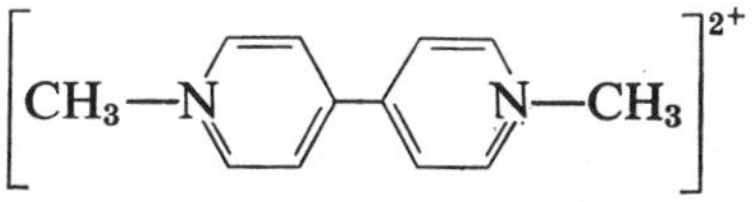

paraquat ion

The above heterocyclic (more than one type of atom in ring) organic compounds belong to the bipyridylium quaternary ammonium class. They were developed by Imperial Chemical Industries in England. Phytotoxic properties of diquat were discovered in 1955 (1) and paraquat a few years later. Among several reviews of these compounds, perhaps the most complete are by Calderbank (6) and Akhavein and Linscott (1).

DIQUAT

Diquat is the common name for 6,7-dihydrodipyrido[1,2-α:2′,1′-c]pyrazinediium ion. Trade names are Ortho® Diquat and Reglone®. The pure bromine salt is a yellow solid, but in aqueous solution it turns dark reddish-brown, soluble in water and is formulated as an aqueous solution. The acute oral LD_{50} is 400–440 mg/kg. Handle with care, and avoid breathing spray mist or getting the concentrate on skin. When diquat is used as an aquatic herbicide, there is a wide margin of safety between the recommended dosages and the rates necessary to cause toxic symptoms in fish.

PARAQUAT

Paraquat is the common name for 1,1′-dimethyl-4,4′-bipyridinium ion. Trade names are Ortho® Paraquat and Gramoxone®. The pure chloride salt is a white solid, but it forms a dark-red aqueous solution, soluble in water. The acute oral LD_{50} is about 150 mg/kg. Paraquat may also be absorbed through the skin; the acute dermal LD_{50} to rabbits is 240 mg/kg in a single dose. Avoid breathing the spray mist, and avoid prolonged skin or eye contact.

Oral ingestion of paraquat represents misuse, yet the serious damage from ingestion deserves special mention. Several poisonings and deaths have been reported as a result of accidentally or intentionally ingesting relatively small amounts of the liquid concentrate (13). Fatalities resulted mainly from progressive pulmonary fibrosis (of the lungs) with associated liver and kidney damage (11). Death may take 2-3 weeks following ingestion. No cure is known following ingestion of a lethal quantity.

Uses

Because many properties of diquat and paraquat are similar, they are discussed together. Both are considered general contact herbicides. Diquat is used as a preharvest desiccant for certain seed crops including alfalfa, clover, sorghum, and soybeans. It is also used for aquatic-weed control, weed control in sugarcane, and as a flower inhibitor in sugarcane.

Paraquat is used as a directed spray to control emerged weeds in many tree crops and as a preplant or preemergence (to the crop) treatment to emerged weeds in asparagus, corn, lettuce, melons, peppers, potatoes, sorghum, soybeans, sugar beets, and tomatoes. It is also used as a preharvest desiccant in cotton, potatoes, and soybeans.

Soil Influence

An important, unique property of diquat and paraquat is their speedy inactivation by soils (5). This results from reaction between the double positively charged cation of the herbicide and the negatively charged sites on clay minerals. In fact, the herbicide molecule becomes tightly held within the lattice structure of the soil itself, and is adsorbed by soil colloids (8). Therefore, these herbicides are essentially nonphytotoxic in most soils. Some phytotoxicity, however, has been demonstrated at high rates in very sandy soils. Although nonphytotoxic, the bound diquat or paraquat ion may persist in soils for long periods of time.

Mode of Action

Bipyridylium-type herbicides cause wilting and rapid desiccation of the foliage to which they are applied, often within a few hours. Mees (10) showed that light increased the rate of development of phytotoxic symptoms, but was not essential for herbicidal action. Best results in the field have often been obtained by a late-afternoon, rather than a morning or midday, application. This appears to allow some internal transport during the night, before development of acute phytotoxicity induced by light which could limit movement.

Translocation, following a foliar application, appears to be almost solely via the apoplastic system (2, 12, 14). However, after the loss of membrane integrity, induced by both herbicides, they do move into untreated leaves presumably along with the flow of other cellular contents. They are poorly translocated from roots (7) because they are tightly bound to cellular components.

These herbicides are not degraded in higher plants in the usual sense. However, they are reversibly converted from the ion form to the free radical form. This interconversion is cyclic and requires light, molecular oxygen, and the photosynthetic apparatus (3, 4, 9, 15). The reaction produces H_2O_2 or OH^- radicals which are proposed to be the actual toxicants (6, 9) (see Figure 11-1).

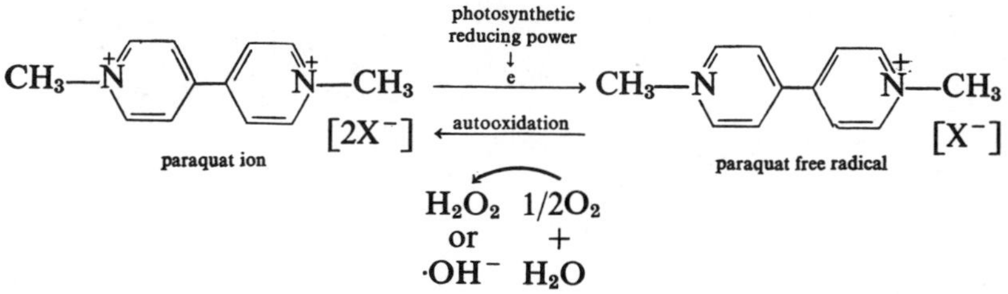

Figure 11-1. Free radical formation from paraquat ion and autooxidation of free radical yielding H_2O_2 or $\cdot OH^-$.

LITERATURE CITED

1. Akhavein, A. A., and D. L. Linscott, *Residue Rev.*, 23, 97 (1968).
2. Baldwin, B. C., *Nature*, 198, 872 (1963).
3. Black, C. C., *Science*, 149, 62 (1965).
4. Black, Jr., C. C., and L. Meyer, *Weeds*, 14, 331 (1966).
5. Brian, R. C., R. F. Homer, J. Stubbs, and R. L. Jones, *Nature*, 181, 446 (1958).
6. Calderbank, A., *Adv. Pest Control Res.*, 8, 129 (1968).
7. Damonakis, M., D. S. H. Drennan, J. D. Fryer, and K. Holly, *Weed Res.*, 10, 278 (1970).
8. Homer, R. F., G. C. Mees, and T. E. Tomlinson, *J. Sci. Food Agr.*, 11, 309 (1960).
9. Kok, B., R. J. Rurainski, and O. V. H. Owens, *Biochim. Biophys. Acta*, 109, 347 (1965).
10. Mees, G. C., *Ann. Appl. Biol.*, 48, 601 (1960).

11. Sinow, J., and E. Wei, *Bulletin of Environmental Contamination and Toxicology*, *9(3)*, 163 (1973).

12. Slade, P., and E. G. Bell, *Weed Res.*, *6*, 267 (1966).

13. Staiff, D. C., G. K. Irle, and W. C. Felsenstein, *Bulletin of Environmental Contamination and Toxicology*, *10(4)*, 193 (1973).

14. Wood, G. H., and J. M. Gosnell, *Proceedings of the South African Sugar Technological Association* 1965, p. 7.

15. Zweig, G., N. Shavit, and M. Avron, *J. Agr. Food Chem.*, *109*, 332 (1965).

FOR CHEMICAL USE, SEE THE MANUFACTURER'S LABEL AND FOLLOW THE DIRECTIONS. ALSO SEE THE PREFACE.

12   Carbamates

Carbamate herbicides derive their basic chemical structure from carbamic acid (NH_2COOH). Pure carbamic acid is not stable and quickly decomposes to NH_3 and CO_2. Beyond various substitutions for the hydrogen atoms of carbamic acid, the organic chemist also may replace the oxygen atom (or atoms) with one or two sulfur atoms. These are referred to as *thiocarbamates* and *dithiocarbamates*, respectively (*thio* = sulfur).

Friesen (11) first observed the effect of esters of carbamic acid on plants in 1929. Herbicidal properties of the carbamate herbicide, propham, were described by Templeman and Sexton (25) in 1945, following the reporting of 2,4-D. Carbamates thus followed closely the prime forerunner of modern chemical weed control. Among numerous carbamate compounds evaluated since 1945, at least 18 are now commercially available. These compounds will be discussed in the following sections.

CARBAMATES

As more carbamates were developed, the selectivity of specific compounds on particular crops has led to their increased use on a wide variety of crops. Selectivity also has given more effective weed control. Whereas the first carbamate herbicide, propham, is effective primarily as a preemergence herbicide, some of the newer compounds are applied almost exclusively to emerged weeds. The chemical structure of these compounds has become increasingly complicated (see Table 12-1).

Propham

Propham is the common name for isopropyl carbanilate. Although it has several trade names worldwide, the most common name in the United States is Chem-Hoe®. In pure form it is a white solid and has a water solubility of

171

Table 12-1. **Common Name, Chemical Name, and Chemical Structure of the Carbamate Herbicides**

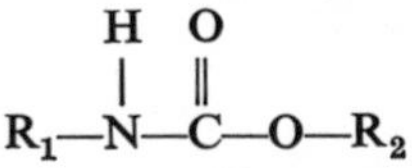

$$\begin{array}{c} H \quad O \\ | \quad \| \\ R_1-N-C-O-R_2 \end{array}$$

Common Name	Chemical Name	R_1	R_2
Barban	4-chloro-2-butynyl *m*-chlorocarbanilate	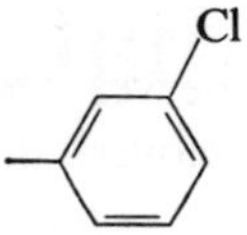 (m-chlorophenyl)	$-CH_2C{\equiv}CCH_2Cl$
Chlorpropham	isopropyl *m*-chlorocarbanilate	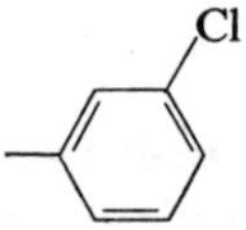 (m-chlorophenyl)	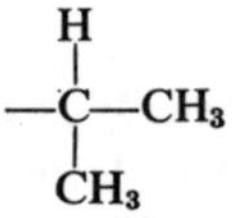 $-\overset{H}{\underset{CH_3}{C}}-CH_3$
Phenmedipham	methyl *m*-hydroxycarbanilate *m*-methylcarbanilate	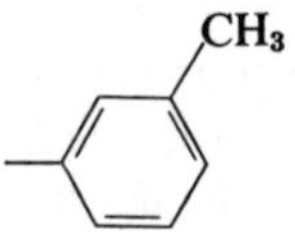 (m-methylphenyl)	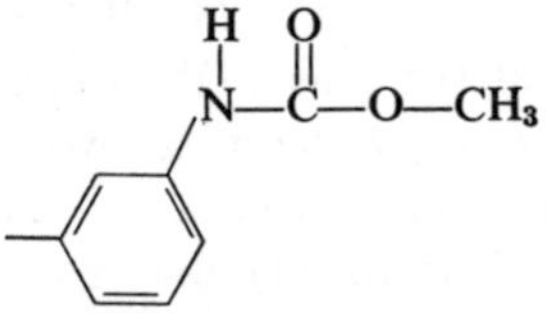 (m-substituted phenyl) $N-C-O-CH_3$ (H, O)
Propham	isopropyl carbanilate	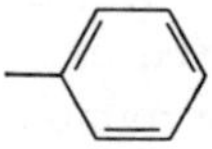 (phenyl)	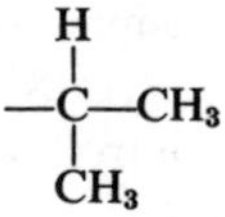 $-\overset{H}{\underset{CH_3}{C}}-CH_3$
Swep	methyl 3,4-dichlorocarbanilate	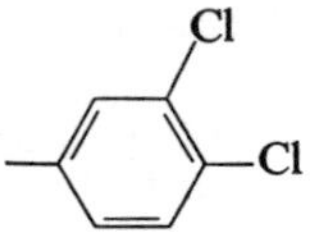 (3,4-dichlorophenyl)	$-CH_3$
Terbutol	2,6-di-*tert*-butyl-*p*-tolyl methylcarbamate	$-CH_3$	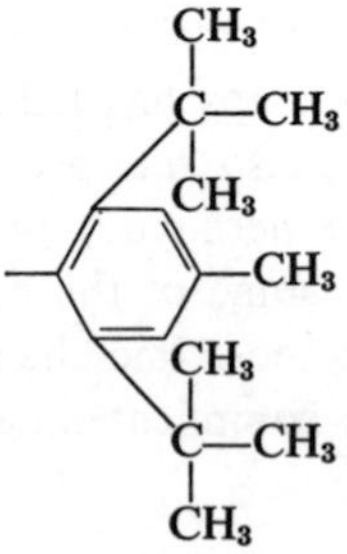 (2,6-di-tert-butyl-p-tolyl)

about 250 ppm; usually formulated as an emulsifiable concentrate, wettable powder, or granules. The acute oral LD_{50} is about 5000 mg/kg.

Uses

Propham is used primarily as a preemergence herbicide to control annual grasses including volunteer small grains. It also controls a few broadleaf weeds such as chickweed and burning nettle. Preplant and postemergence applications can be used in certain crops. However, when applied on emerged sensitive weeds, they should be very small, in the one- to two-leaf stage. Propham is used on alfalfa, clovers, flax, lentils, lettuce, peas, safflower, spinach, and sugar beets.

Soil Influence

Propham is moderately leached in most soils and moves slightly more than chlorpropham. When applied to moist soil followed by rainfall, it may penetrate to a depth of 5–8 in. It degrades rapidly in most soils, usually remaining phytotoxic for less than 1 month.

Mode of Action

When annual grasses are treated before emergence, they develop much thicker and shorter coleoptiles. The herbicide blocks cell division and induces polyploid nuclei (8). Propham is absorbed mostly by roots, but very little by leaves (4). It is translocated primarily in the apoplast (2) and appears to be rapidly degraded in higher plants.

Chlorpropham

Chlorpropham is the common name for isopropyl *m*-chlorocarbanilate. It was formerly called Chloro-IPC and now is trade-named Furloe® in the United States. Furloe®-124 contains an added "extender" (*p*-chlorophenyl *N*-methylcarbamate) which reduces the rate of microbial enzymatic degradation of chlorpropham in soil, thus extending the period of weed control. Pure chlorpropham is a white solid with a water solubility of 88 ppm; formulated as an emulsifiable concentrate and in granular form. The acute oral LD_{50} is from 5000 to 7500 mg/kg.

Chlorpropham was developed soon after propham was introduced. It has many of the same herbicidal characteristics, but it differs from propham in two ways: it persists longer in the soil, thus giving longer weed control. It is less selective than propham to certain crop species, such as lettuce.

Uses

Like propham, chlorpropham is applied primarily as a preemergence herbicide to control annual grasses and a few annual broadleaf weeds. It is also

Figure 12-1.   Shepherdspurse control in garlic with chlorpropham. Left: treated. Right: untreated.

used as a postemergence treatment because it has higher herbicidal efficacy than propham.

Chlorpropham is used in alfalfa, beans, blackberries, blueberries, carrots, clovers, cowpeas, cranberries, garlic, grass seed crops, lettuce, onion, peas, peppers, raspberries, rice, safflower, soybeans, spinach, sugar beets, and tomatoes. Chlorpropham also has growth-regulating properties and is used as a postharvest sprout inhibitor on potato tubers (see Figure 12-1).

Soil Influence

Chlorpropham is tightly bound to soil colloids and in general does not leach below the upper inch in nonsandy soil types. In part, this property is the basis of selectivity in certain large, deep-seeded crops such as beans. Chlorpropham persists in most soils at phytotoxic levels for 1–2 months. By adding the "extender" the period of persistence appears to be about doubled.

Mode of Action

Chlorpropham inhibits cell division, thus causing multinucleate root cells and inhibited root elongation (9). It is readily translocated throughout the plant when absorbed by roots (22) but not when applied to leaves, suggesting almost exclusive apoplastic transport. It also appears to be absorbed by emerging shoots as they pass through the treated soil (16). The vapors of chlorpropham have been shown to be absorbed by seeds (3) and emerged dodder plants (24). In the latter case, chlorpropham prevents the parasitic dodder from attaching to the host plant.

Chlorpropham is rapidly degraded in higher plants. It inhibits ATP (13), RNA (6, 20), and protein (19, 20) synthesis.

Barban

Barban is the common name for 4-chloro-2-butynyl *m*-chlorocarbanilate. The trade name is Carbyne®. It is a colorless solid with a water solubility of 11 ppm and formulated as an emulsifiable concentrate. The acute oral LD_{50} is 1350 mg/kg.

Uses

Barban is a postemergence herbicide used almost exclusively to control wild oats. Reed canarygrass is also highly sensitive. Time of application is critical to obtain control of wild oats. Barban must be applied at the two-leaf stage of wild oats—from the appearance of the second leaf until the appearance of the third leaf; earlier or later applications will give poor control. However, if wild oats are growing slowly because of cold weather, low soil moisture, or low fertility, and if it does not reach the two-leaf stage within 9 days after emergence, spray before the fourteenth day after emergence.

Barban is used in barley, flax, lentils, mustard, peas, safflower, soybeans, sugar beets, sunflower, and wheat.

Soil Influence

Since barban is applied to the foliage, its interaction with the soil is of limited significance. However, it is adsorbed by soil colloids and is rapidly degraded in less than 1 month.

Mode of Action

Substantial amounts of barban are absorbed by leaves within a few hours, but it continues to be absorbed for at least 1 week (10). Translocation is limited and appears to be mostly in the apoplast. It is degraded rapidly in most plant species (23). Barban may inhibit protein synthesis (18).

Phenmedipham

Phenmedipham, tradenamed Betanal®, is the common name for methyl *m*-hydroxycarbanilate *m*-methylcarbanilate. It has two carbamate groups in its chemical structure (see Table 12-1). It is a colorless solid with a water solubility of less than 10 ppm. Phenmedipham is formulated as an emulsifiable concentrate, and its acute oral LD_{50} is greater than 8000 mg/kg.

Uses

Phenmedipham is a postemergence herbicide which controls annual broadleaf weeds in sugar beets. Weeds should be small, with the rate of application depending on the size of the weeds: $1-1\frac{1}{4}$ lb/acre at the cotyledon to two-true-leaf stage and $1\frac{1}{4}-1\frac{1}{2}$ lb/acre at the two-to four-true-leaf stage; sugar beets should be past the two-leaf stage. Sensitivity of sugar beets to phenmedipham increases at temperatures over 85°F under moisture stress, or if preplant or preemergence herbicides are used before the phenmedipham application.

Soil Influence

Phenmedipham is not subject to excessive leaching. It has a half-life of about 25 days in most soils.

Mode of Action

Phenmedipham is readily absorbed by leaves and appears to be translocated via the apoplastic system. Studies conducted on sugar beets have shown that the chemical is degraded within several days after application and traces (0.1 ppm) of the breakdown product, 3-methylaniline, were detected at harvest upon hydrolysis of the tissue (14). Phenmedipham inhibits the Hill reaction of photosynthesis.

Swep

Swep is the common name for methyl 3,4-dichlorocarbanilate. The trade name is also Swep®. It is a white, crystalline solid, essentially insoluble in water, formulated as an emulsifiable concentrate. The acute oral LD_{50} is about; 522 mg/kg.

Swep is primarily used to control annual grasses in transplanted rice in the Far East. It is applied 10–15 days after transplanting. Broadleaf weeds are also controlled when MCPA is combined with Swep.

Terbutol

Terbutol is the common name for 2,6-di-*tert*-butyl-*p*-tolyl methylcarbamate. The trade name is Azak®. It is a white, crystalline solid with a water solubility of 6–7 ppm; formulated as a wettable powder. Terbutol has an acute oral LD_{50} of greater than 15,000 mg/kg.

Terbutol is used as a preemergence herbicide to control crabgrass in established turfs. Although it is not strongly absorbed by soils, its low water solubility limits its leaching. It persists in soils about 5 months in bluegrass turf in the spring and summer climates of the Northeastern and North Central states. Terbutol inhibits the growth of roots and rhizomes at the terminal meristems, and the seedling leaves of grasses are deformed. It is apparently absorbed through the roots.

THIOCARBAMATES

Thiocarbamate herbicides include butylate, cycloate, diallate, EPTC, molinate, pebulate, triallate, and vernolate. Their chemical structures are given in Table 12-2. As the name indicates, the carbamate molecule also includes *one* sulfur atom. All of the thiocarbamate herbicides except diallate and triallate, which are products of Monsanto, were developed by Stauffer Chemical Company.

Most thiocarbamate herbicides are relatively volatile. If not immediately mixed into the soil by tandem disc or power-driven cultivation equipment, much of the applied herbicide will be lost.

Butylate

Butylate is the common name for *S*-ethyl diisobutylthiocarbamate, with the trade name of Sutan®. This amber liquid has a water solubility of 45 ppm and is relatively volatile; it is formulated as an emulsifiable concentrate as well as in granular form. Formulations of butylate–atrazine combinations are also available. Atrazine improves the control of annual broadleaf weeds. The acute oral LD_{50} of butylate is from 3878 to 5431 mg/kg.

Uses

Butylate is used in a preplant, soil incorporation treatment to control annual weeds, especially grasses in corn (field, sweet, and silage). Purple and yellow nutsedge are also controlled.

Soil Influence

In sandy, dry soils, butylate is leached about 2.5 in. deep by 8 in. of water. Leaching is less in clay and organic soils. Butylate, however, is recommended for mineral soils only. It is not a persistent herbicide, lasting from 1 to 3 months under most field conditions.

Table 12-2. Common Name, Chemical Name, and Chemical Structure of the Thiocarbamate Herbicides

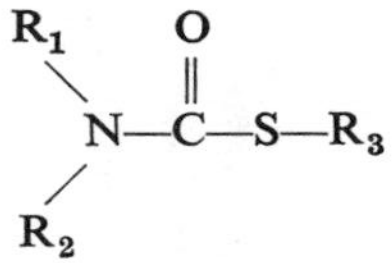

$$R_1\text{—}N(\text{—}R_2)\text{—}\underset{\underset{O}{\|}}{C}\text{—}S\text{—}R_3$$

Common Name	Chemical Name	R₁	R₂	R₃
Butylate	S-ethyl diisobutylthiocarbamate	$CH_3\text{—}CH(CH_3)\text{—}CH_2\text{—}$	$CH_3\text{—}CH(CH_3)\text{—}CH_2\text{—}$	$C_2H_5\text{—}$
Cycloate	S-ethyl N-ethylthiocyclohexanecarbamate	$C_2H_5\text{—}$	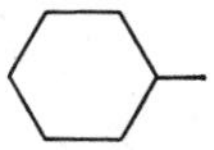 cyclohexyl	$C_2H_5\text{—}$
Diallate	S-(2,3-dichloroallyl)diisopropylthiocarbamate	$CH_3\text{—}CH(CH_3)\text{—}$	$CH_3\text{—}CH(CH_3)\text{—}$	$Cl\text{—}C(H)\text{=}C(Cl)\text{—}CH_2\text{—}$
EPTC	S-ethyl dipropylthiocarbamate	$C_3H_7\text{—}$	$CH_3\text{—}CH(C_3H_7)\text{—}$	$C_2H_5\text{—}$
Molinate*	S-ethyl hexahydro-1H-azepine-1-carbothioate	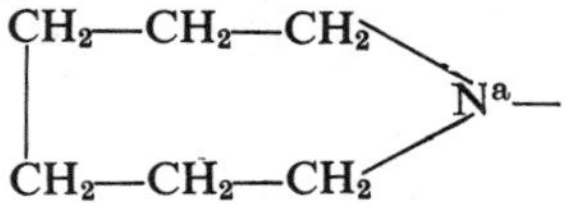 $CH_2\text{—}CH_2\text{—}CH_2\text{—}N^a\text{—}CH_2\text{—}CH_2\text{—}CH_2$ (ring)		$C_2H_5\text{—}$
Pebulate	S-propyl butylethylthiocarbamate	$C_2H_5\text{—}$	$C_4H_9\text{—}CH_3$	$C_3H_7\text{—}$
Triallate	S-(2,3,3-trichloroallyl)diisopropylthiocarbamate	$CH_3\text{—}CH(CH_3)\text{—}$	$CH_3\text{—}CH(CH_3)\text{—}$	$Cl\text{—}C(Cl)\text{=}C(Cl)\text{—}CH_2\text{—}$
Vernolate	S-propyl dipropylthiocarbamate	$C_3H_7\text{—}$	$C_3H_7\text{—}$	$C_3H_7\text{—}$

[a] The single nitrogen atom in the molinate ring structure is the nitrogen atom of the parent thiocarbamate acid molecule.

178

Mode of Action

Butylate inhibits growth in the meristematic region of the leaves of grasses. It is absorbed by both leaves and roots and is translocated upward throughout the plant following root uptake. Butylate is rapidly metabolized in corn to CO_2, diisobutylamine, fatty acids, conjugates of amines and fatty acids, and certain natural plant constituents (2).

Cycloate

Cycloate is the common name for S-ethyl N-ethylthiocyclohexanecarbamate. The trade name is Ro-Neet®. It is a colorless liquid with a water solubility of 85 ppm and relatively volatile. It is formulated as an emulsifiable concentrate and in granular form. The acute oral LD_{50} ranges from 2000 to 4100 mg/kg.

Uses

Cycloate controls most annual grasses including volunteer barley, several broadleaf weeds, and the nutsedges (yellow and purple) in sugar beets, table beets, and spinach. It is applied by preplant soil incorporation.

Soil Influence

Cycloate is resistant to leaching in heavy clay and high-organic soils. However, it is only used on mineral soils. It is less subject to leaching than EPTC or pebulate. In loamy sand, cycloate will leach 3–6 in. with 8 in. of water.

Mode of Action

Like other thiocarbamate herbicides, cycloate inhibits growth in the meristematic region of grass leaves. Absorbed by leaves and roots and readily translocated to leaves and stems of sugar beets following root uptake, cycloate is rapidly metabolized by sugar beets to ethylcyclohexylamine, CO_2, amino acids, sugars, and other natural plant constituents.

Diallate

Diallate is the common name for S-(2,3-dichloroallyl)diisopropylthiocarbamate. The trade name is Avadex®. It is an oily liquid with a water solubility of 14 ppm, and formulated as an emulsifiable concentrate as well as in granular form. The acute oral LD_{50} of diallate is 395 mg/kg.

Uses

Diallate is used primarily to control wild oats in sugar beets, lentils, peas, and flax, but is also registered for use in alfalfa, barley, clovers, corn, potatoes, safflower, and soybeans. This chemical can only be used in certain states on corn, potatoes, and soybeans, applied either as a preemergence or preplant soil-incorporation treatment.

Soil Influence

Diallate, as well as other thiocarbamate herbicides, competes with moisture for the adsorption sites on soil particles. It is adsorbed on clay and organic colloids, and therefore, less leaching occurs in clay and organic soils than in sandy soils. Under most conditions diallate persists 1–3 months in soil.

Mode of Action

Diallate appears to have more effect on cell division than on cell enlargement. At acute dosages lethal to wild oats, the first leaf does not emerge from the coleoptile, but at sublethal dosages, the first leaf emerges from the coleoptile and is somewhat distorted, dark green, and glossy in appearance (21) as well as brittle (5). In wild oats, absorption is primarily through the emerging coleoptile (1, 21).

EPTC

EPTC is the common name for S-ethyl dipropylthiocarbamate; its trade name is Eptam®. It is a light, yellow-colored liquid with a water solubility of 370 ppm; formulated both as an emulsifiable concentrate and in granular form. Formulations of EPTC containing the isooctyl ester of 2,4-D (Knotweed®) and with a plant-effect antidote (Eradicane®) are also available. The acute oral LD_{50} is 1652 mg/kg.

Uses

EPTC was the first thiocarbamate herbicide developed. Its volatile nature resulted in highly variable weed control when it was first used as a surface-applied preemergence herbicide. Soil incorporation corrected this deficiency and provided the first general use of that technique. From research done before 1960, it was well established that incorporation increased the efficacy of some herbicides. The power-driven rotary hoe operated to provide a "clean soil sweep" was the most efficacious. A second or repeated incorporation was suggested when the disc, sweep-type cultivator, or drag harrow was used (15). This method has since been used widely with many herbicides. Incorporation places the chemical in the soil in the seed-germination area. Thus the method is not dependent on rainfall or irrigation for leaching into the seed-germinating area of the weed.

EPTC, a preemergence herbicide, is commonly mixed mechanically to a depth of 2–3 in. In dry soil, EPTC may be incorporated by sprinkler irrigation immediately after application; it has also been applied through the sprinkler system and by metering into furrow-irrigation water for the treatment of crops such as deciduous or citrus orchards, alfalfa, mint, and clover. EPTC is used against a wide array of weeds including many grasses, and a variety of annual broadleaf weeds. It also controls yellow and purple nutsedge.

EPTC is used in alfalfa, almonds, certain beans, citrus (excluding lemons), clovers, cotton, flax, grapes, peas, potatoes, safflower, sugar beets, sunflower, sweet potatoes, and walnuts. Certain uses are restricted to specific geographical locations and special application techniques.

Two combination products involving EPTC may improve its results. Knotweed®, the EPTC formulation containing the isooctyl ester of 2,4-D, increases the control of certain broadleaf weeds in corn over EPTC alone. Eradicane® is an EPTC formulation containing an antidote or protectant which increases its selectivity or makes it less lethal to corn.

Soil Influence

EPTC is adsorbed by dry soil, but it is subject to some leaching. The amount of leaching decreases as the clay and organic content of the soil increase. It is used only on mineral soils. EPTC, like most other thiocarbamate herbicides, is readily volatilized from wet soil if not mixed in immediately after application. EPTC does not persist long in soils; its phytotoxic properties disappear in 1–3 months in most soils.

Mode of Action

EPTC inhibits growth of the meristematic region of grass leaves. The first leaf of grasses is often trapped in the coleoptile and emerges from the side of the coleoptile grossly distorted (7). Broadleaf weeds may develop cupped leaves with necrotic tissue around the edges of the leaf.

EPTC is absorbed by seeds, roots, and emerging shoots in contact with treated soil (12). The relative importance of these three sites of entry varies with the plant species. EPTC is rapidly degraded in higher plants (see Figure 12-2).

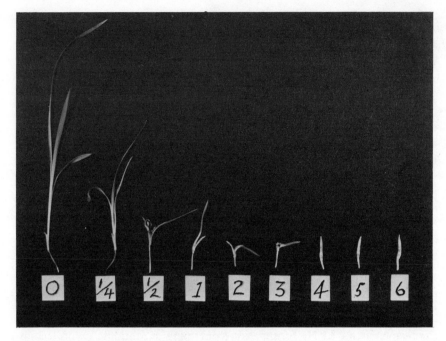

Figure 12-2. Barnyardgrass seedlings grown in sandy loam containing the indicated parts per million of EPTC. (J. H. Dawson ARS., U.S.D.A., Prosser, Washington).

Molinate

Molinate is the common name for S-ethylhexahydro-$1H$-azepine-1-carbothioate, with the trade name of Ordram®. It is a liquid with a water solubility of about 800 ppm; formulated as an emulsifiable concentrate and in granular form. The acute oral LD_{50} is 720 mg/kg.

Molinate is used as a preplant soil-incorporation treatment to control annual grasses, primarily barnyardgrass, in rice. It is adsorbed by dry soil, but subject to leaching. Its persistence in soils is relatively short, usually less than 1 month. Molinate is rapidly metabolized in higher plants.

Pebulate

Pebulate is the common name for S-propyl butylethylthiocarbamate, the trade name is Tillam®. It is a yellow liquid with a water solubility of 60 ppm, and formulated as an emulsifiable concentrate or in granular form. The acute oral LD_{50} is 921–1120 mg/kg.

Pebulate is used as a preplant, soil-incorporation treatment to control most annual grasses, yellow and purple nutsedge, and several broadleaf weeds. It is more effective on hairy nightshade than most other herbicides applied to tomatoes. Pebulate is used in sugar beets, tomatoes, and tobacco.

This herbicide is adsorbed on dry soil but is subject to leaching. Pebulate leaches less than EPTC but more than cycloate, both being thiocarbamates as well. Pebulate persists in most soils from 1 to 3 months. It is also readily metabolized by higher plants. Little is known about its mechanism of action, but it is probably similar to other thiocarbamate herbicides.

Triallate

Triallate is the common name for S-(2,3,3-trichloroallyl)diisopropylthio-carbamate. Avadex® BW and Far-Go® are its trade names. Triallate is an oily liquid with a water solubility of only 4 ppm; formulated as an emulsifiable concentrate. It has an acute oral LD_{50} of 1675–2165 mg/kg.

Triallate controls wild oats in barley, lentils, peas, and wheat. It is used as a preplant or postplant, soil-incorporation treatment. The postplant method is recommended for wheat, while either method can be used for barley or peas. In the postplant treatment, the seed is planted below the depth of incorporation.

The adsorption of triallate on soil colloids limits its leaching. It persists in soil up to 6 weeks under Northern Great Plains conditions. Triallate is mainly absorbed by emerging coleoptiles of grasses (21) and is most toxic during cell division. The chemical is metabolized by higher plants.

Vernolate

Vernolate is the common name for S-propyl dipropylthiocarbamate; the trade name is Vernam®. It is a liquid with a solubility in water of 90 ppm, and formulated as an emulsifiable concentrate and in a granular form. It has an oral LD_{50} of 1470–1780 mg/kg.

Vernolate is used to control annual grasses, many broadleaf weeds, as well as yellow and purple nutsedge in peanuts, potatoes, soybeans, sweet potatoes, and tobacco. It is applied as a soil incorporation treatment—preplant, preemergence, posttransplant, or postemergence (to the crop)—but does not control emerged weeds. A tank mix of vernolate and benefin, or vernolate and trifluralin, can be used on peanuts or soybeans, respectively, to control a wider variety of grasses and broadleaf weeds.

Vernolate is adsorbed by dry soil but it is leachable. It persists in soils from 1 to 3 months, inhibiting growth of the meristematic region of grass leaves. This herbicide is absorbed by roots of soybean and peanut plants and translocated throughout the leaves and stem.

It is rapidly metabolized by plants to CO_2 which in turn is incorporated into naturally occurring plant constituents.

DITHIOCARBAMATES

The dithiocarbamate herbicides were developed before the thiocarbamates. The two dithiocarbamate herbicides are CDEC and metham. (See Table 12-3 for their chemical structure.) As the name of this group indicates, the carbamate molecule has two sulfur atoms.

CDEC

CDEC is the common name for 2-chloroallyl diethyldithiocarbamate. As the trade name, Vegadex®, implies, it is primarily used in vegetable crops. CDEC is an amber, oily liquid with a water solubility of 92 ppm; formulated both as an emulsifiable concentrate and in granular form. Its acute oral LD_{50} is 850 mg/kg.

Uses

CDEC controls most annual grasses and several of the most common annual broadleaf weeds found in vegetable crops. These weeds include pigweed, henbit, chickweed, and purselane. It also controls dodder.

Table 12-3. Common Name, Chemical Name, and Chemical Structure of the Dithiocarbamate Herbicides

$$R_1 \diagdown \quad \overset{\displaystyle S}{\overset{\|}{}}$$

$$N{-}C{-}S{-}R_3$$

$$R_2 \diagup$$

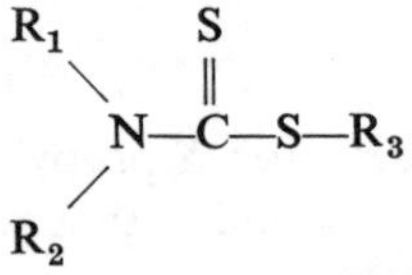

Common Name	Chemical Name	Chemical Structure		
		R_1	R_2	R_3
CDEC	2-chloroallyl diethyl-dithiocarbamate	C_2H_5-	C_2H_5-	$CH_2{=}\overset{\displaystyle Cl}{\overset{\|}{C}}{-}CH_2-$
Metham	sodium methyl-dithiocarbamate	CH_3-	$H-$	$Na-$

This herbicide can be used on 25 vegetables: beans (lima and snap), broccoli, brussel sprouts, cabbage, cantaloupe, cauliflower, celery, chicory, collards, corn, cucumbers, endive, escarole, hanover salad, kale, lettuce, mustard greens, okra, potatoes, soybeans, spinach, tomatoes, turnips, and watermelon. It is also used on some nursery stock, including hydrangea, euonymus, potentilla spirea, azaleas, junipers, yews, and privet. CDEC may also be combined with CDAA to broaden the spectrum of weed species controlled in celery and cabbage.

CDEC should be applied at seeding time or before crop and weed emergence. It does not control emerged annual weeds or perennial weeds. CDEC may also be applied to transplants or on certain of these crops following a clean cultivation. For best results, sprinkle-irrigate immediately after application. When CDEC is used with furrow irrigation, it must be incorporated into the soil.

Soil Influence

CDEC is most effective on "light" soils that are low in clay and organic content although with the use of sprinkle irrigation it can be used on muck soils. CDEC is reversibly adsorbed on soil colloids. Moderate rainfall or sprinkle irrigation, after application, enhances its performance, but heavy precipitation after application will cause excessive leaching in sandy soils and may produce substantial crop injury. CDEC is not persistent in soils and lasts from 4 to 6 weeks under most conditions.

Mode of Action

CDEC is readily absorbed by roots, but little, if at all, by foliage. Following root absorption it appears to be translocated throughout the plant in the apoplastic system. The chemical is rapidly metabolized by higher plants into CO_2 and other metabolites, including diethylamine as an intermediate.

Metham

Metham is the common name for sodium methyldithiocarbamate. The trade name is Vapam®. Formulated as a water-soluble solution, its acute oral LD_{50} is 820 mg/kg.

Metham is a temporary soil fumigant used to control nematodes, garden centipedes, soil-borne disease organisms, and most germinating weed seeds and seedlings. It may also be used to control certain shallow perennial weeds (e.g., nutsedge) limited to small patches.

This herbicide is used in the field as well as in potting soil. When used in the field, the soil should be cultivated before application to allow diffusion of the gaseous toxicant.

Metham may be applied in various ways depending on size of area to be treated and equipment available. For small areas, a sprinkling can or hose proportional diluter may be used. For large areas, soil injection, spray application with immediate rotary-tiller incorporation, or application through a sprinkler-irrigation system may be used. Metham is most effective when it is possible to confine the vapors with a plastic tarp; however, the water-seal method (saturating the top 2 in. of soil with water) may also be used. When using a tarp, keep the treated area covered for 48 hr or longer. Seven days after treatment, cultivate the area to a depth of 2 in. and wait at least 21 days after application before seeding the treated area.

Metham has also been used to kill roots in sewers.

LITERATURE CITED

1. Appleby, A. P., W. R. Furtick, and S. C. Fang, *Weed Res.*, *5*, 115 (1965).
2. Ashton, F. M., and A. S. Crafts, *Mode of Action of Herbicides*, Wiley, New York, 1973.
3. Ashton, F. M., and S. Helfgott, *Proceedings of the Eighteenth California Weed Conference*, 1966, p. 8.
4. Baldwin, R. E., V. H. Freed, and S. C. Fang, *J. Agr. Food Chem.*, *2*, 428 (1954).
5. Banting, J. D., *Weed Res.*, *7*, 302 (1967).
6. Briquet, M. V., and A. L. Wiaux, *Meded. Rÿksfac. Landb Wet. Gent.*, *32*, 1040 (1967).
7. Dawson, J. H., *Weeds*, *11*, 60 (1963).
8. Ennis, Jr., W. B., *Amer. J. Bot.*, *35*, 15 (1948).
9. Ennis, Jr., W. B., *Amer. J. Bot.*, *36*, 823 (1949).
10. Foy, C. L., *Research and Progress Report*, *Western Weed Control Conference*, 1961, p. 96.
11. Friesen, G., *Planta*, *8*, 666 (1929).
12. Gray, R. A., and A. J. Weierich, *Weed Sci.*, *17*, 223 (1969).
13. Gruenhagen, R. D., and D. E. Moreland, *Weed Sci.*, *19*, 319 (1971).
14. Kassenbeer, H., *Schering AG Conference, Berlin*, 1969, p. 5.
15. Klingman, Glenn C., *Southern Weed Conference Proceedings*, *14*, 63–68 (1961).
16. Knake, E. L., and L. M. Wax, *Weed Sci.*, *16*, 393 (1968).
17. Kossman, K., *Schering AG Conference, Berlin*, 1969, p. 16.
18. Mann, J. D., L. S. Jordan, and B. E. Day, *Plant Physiol.*, *40*, 840 (1965).
19. Mann, J. D., L. S. Jordan, and B. E. Day, *Weeds*, *13*, 63 (1965).
20. Moreland, D. E., S. S. Malhotra, R. D. Gruenhagen, and E. H. Schoraii, *Weed Sci.* *17*, 556 (1969).
21. Parker, C., *Weed Res.*, *3*, 259 (1963).

22. Prendeville, G. N., Y. Eshel, C. S. James, G. F. Warren, and M. M. Schreiber, *Weed Sci.*, *16*, 432 (1968).

23. Riden, J. R., and T. R. Hopkins, *J. Agr. Feed Chem.*, *10*, 455 (1962).

24. Slater, C. H., J. H. Dawson, W. R. Furtick, and A. P. Appleby, *Weed Sci.*, *17*, 238 (1969).

25. Templeman, W. G., and W. A. Sexton, *Nature*, *156*, 630 (1945).

FOR CHEMICAL USE, SEE THE MANUFACTURER'S LABEL AND FOLLOW THE DIRECTIONS. ALSO SEE THE PREFACE.

13 Dinitroanilines

The herbicidal properties of 2,6-dinitroanilines were first reported by Eli Lilly and Company scientists in 1960 (1), and other companies have also developed herbicides from this class of compounds. Most pure dinitroanilines are yellow-orange crystalline solids with a low water solubility. They are generally volatile and more or less susceptible to photodegradation. Most are selective herbicides used as a preplant soil-incorporation treatment.

The chemical structures shown below are those basic to anilines and 2,6-dinitroanilines. Table 13-1 also gives specific structures of ten dinitroaniline herbicides.

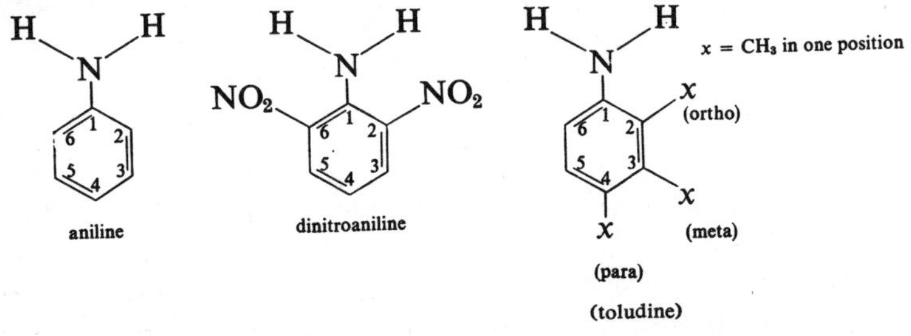

aniline

dinitroaniline

x = CH₃ in one position

(ortho)

(meta)

(para)

(toluidine)

TRIFLURALIN

Trifluralin is the common name for α,α,α-trifluoro-2,6-dinitro-N,N-dipropyl-p-toluidine. Trade names include Treflan®, Treficon®, Trefanocide®, Trim®, and Elancolan®. Trifluralin is a yellow-orange, crystalline solid with a water solubility of less than 1 ppm. The compound is formulated as an emulsifiable concentrate and as granules. The acute oral LD_{50} in mice is about 10,000 mg/kg.

188

Uses

Trifluralin, the first 2,6-dinitroaniline herbicide developed and marketed, is the most widely used herbicide of this class and one of the most important herbicides used selectively in crops. Although the major use has been in cotton and soybeans, trifluralin is used on more than 40 other crops including established alfalfa, several types of beans, barley, cole crops, carrots, celery, collards, cucurbits, grapes, guar, hops, kale, mint, mustard greens, okra, peanuts, peas, peppers, potatoes, safflower, southern peas, sugar beets, sugar-cane, sunflower, tomatoes, turnip greens, wheat, and many fruit and nut trees.

In most of these crops, trifluralin is applied as a preplant or preemergence, soil-incorporation treatment. In some crops, such as tomatoes, potatoes, sugar beets, cantaloupe, cucumbers and watermelon, the herbicide is too phytotoxic to be applied preplant on seeded crops, but can be used on transplants or on established plants.

Trifluralin controls most weed seeds as they germinate. This includes nearly all grass weed seeds and also many broadleaf weeds including pigweeds, purslane, lambsquarters, kochia, Russian thistle, henbit, knotweed, and chickweed to mention a few.

Trifluralin also controls certain perennial weeds (e.g, field bindweed and johnsongrass from rhizomes) when used in accordance with special rate recommendations and application techniques.

Behavior in Soil

Four factors are responsible for the disappearance or degradation of tri-fluralin in soil. These include volatilization, photodecomposition, microbial decomposition, and chemical decomposition. The relative importance of each factor is influenced by soil type, moisture content, temperature, microflora, rate of application, and method of incorporation. In warm moist soils, recommended rates of application are decomposed in less than 12 months (see Figure 13-1).

Studies on loss of trifluralin from soil surfaces have shown that volatilization and photodecomposition are important sources of herbicide loss. The rate of loss was highest in the few hours immediately after application. Wet soil surfaces and high soil temperature increased the rate of loss. The more adsorptive soils showed the longest retention time for the herbicide. Losses through volatilization and photodecomposition are minimized when tri-fluralin is mixed into the soil after application.

Table 13-1. Common Name, Chemical Name, and Chemical Structure of the Dinitroaniline Herbicides

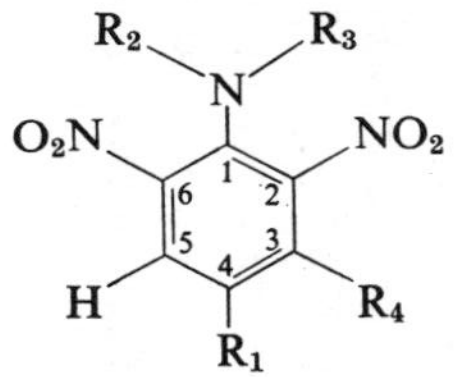

Common Name	Chemical Name	R_1	R_2	R_3	R_4
Benefin	N-butyl-N-ethyl-α,α,α-trifluoro-2,6-dinitro-p-toluidine	$-CF_3$	$-C_2H_5$	$-C_4H_9$	$-H$
Butralin	4-(1,1-dimethylethyl)-N-(1-methylpropyl)-2,6-dinitrobenzenamine	CH_3 $-\overset{\displaystyle CH_3}{\underset{\displaystyle CH_3}{C}}-CH_3$	$-H$	$-C_4H_9$	$-H$

Name	IUPAC name				
Dinitramine	N^4,N^4-diethyl-α,α,α-trifluoro-3,5-dinitro-toluene-2,4-diamine	$-CF_3$		$-C_2H_5$	$-NH_2$
Ethalfluralin	N-ethyl-N-(2-methyl-2-propenyl)-2-6-dinitro-4-(trifluoromethyl)benzenamine	$-CF_3$	$-C_4H_7$	$-C_2H_5$	$-H$
Fluchloralin	N-(2-chloroethyl)-2,6-dinitro-N-propyl-4-trifluromethyl)aniline	$-CF_3$	$-C_3H_7$	$-C_2H_4Cl$	$-H$
Isopropalin	2,6-dinitro-N,N-dipropylcumidine	$-CH(CH_3)_2$	$-C_3H_7$	$-C_3H_7$	$-H$
Nitralin	4-(methylsulfonyl)-2,6-dinitro-N,N-dipropylaniline	$-SO_2-CH_3$	$-C_3H_7$	$-C_3H_7$	$-H$
Oryzalin	3,5-dinitro-N^4,N^4-dipropylsulfanilamide	$-SO_2-NH_2$	$-C_3H_7$	$-C_3H_7$	$-H$
Penoxalin	N-(1-ethylpropyl)-2,6-dinitro-3,4-xylidine	$-CH_3$	$-H$	$-C_5H_{11}$	$-CH_3$
Profluralin	N-(cyclopropylmethyl)-α,α,α-trifluoro-2,6-dinitro-N-propyl-p-toluidine	$-CF_3$	$-C_3H_7$	$-CH_2$-cyclopropyl	$-H$
Trifluralin	α,α,α-trifluoro-2,6-dinitro-N,N-dipropyl-p-toluidine	$-CF_3$	$-C_3H_7$	$-C_3H_7$	$-H$

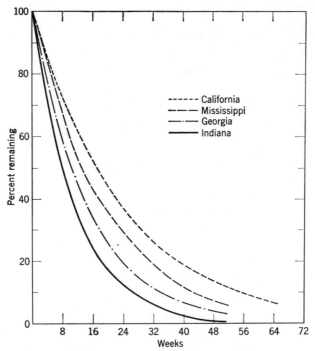

Figure 13-1.   Field persistence of trifluralin under different climatic conditions. Note that after 48 weeks, the trifluralin was nearly gone, with the longest persistence in dryland conditions in California. (Lilly Research Laboratories.)

In greenhouse studies, the rate of trifluralin degradation in soil was compared under anaerobic and aerobic conditions. After 40 days, with moisture at 200% of the field's water-holding capacity (anaerobic), 98% of the trifluralin had degraded. With aerobic moisture conditions (0.0%, 50%, and 100% of field capacity) less than 25% of the trifluralin had decomposed in the same time period.

In soil saturated with water, oxygen will most likely become a limiting factor. Under such conditions, anaerobic degradation of organic compounds can be expected. It has not been established whether this is chemical or microbial, but both are probably involved. In nonautoclaved, underwater soil, trifluralin degradation was complete in 7 days at 76°F, whereas only 10% had degraded at 38°F (11) (see Figure 13-2). Under warm anaerobic conditions, the trifluralin was rapidly degraded, whereas under aerobic conditions degradation was much slower (11). Under field conditions, trifluralin degradation was nearly complete after 48 weeks under four different climatic conditions (see Figure 13-1). More detailed discussions on degradation of trifluralin in soil have been published (6, 11, 12).

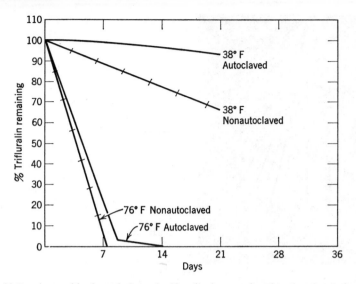

Figure 13-2.   Anaerobic degradation of trifluralin in autoclaved and nonautoclaved soil as a function of temperature in soil saturated with water (200% field capacity) (11).

Trifluralin is strongly adsorbed onto the soil and is extremely resistant to movement by water. By mixing the herbicide with soil, an effective concentration of herbicide is formed in the zone of weed-seed germination. Despite heavy rainfall, trifluralin is not leached from this weed-seed germination zone. With furrow irrigation, there is little, if any, lateral movement.

The availability of trifluralin to germinating weed seeds is influenced by its adsorption onto soil. The content of sand, silt, and clay influences the amount required for weed control, with the most trifluralin required on the clay (heavy) soils. Adsorption on organic matter appears to strongly restrict availability of trifluralin, thus limiting its efficacy as a herbicide. Trifluralin is not usually recommended for use on peat and muck soils of greater than 10% organic matter.

Mode of Action

Most research on mode of action of dinitroaniline herbicides has been conducted with trifluralin. Although these herbicides are basically similar in activity, varying degrees of tolerance by several crop species can be demonstrated (3).

Many studies have shown that trifluralin inhibits growth of roots. Characteristically, the root increases in diameter or swells in the active meristematic region near the root tip. Lateral or secondary root development is also

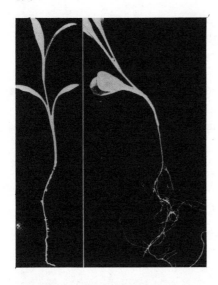

Figure 13-3. Inhibition of lateral root formation is typical of the dinitroaniline herbicides. Left: treated. Right: untreated. (D. E. Bayer, University of California, Davis.)

inhibited (2). Trifluralin disrupts cell division causing multinucleate cells or polyploidy (4), and probably by preventing microtubule development (microtubules are functional parts of the spindle) (7). (see Figure 13-3).

Trifluralin appears to be absorbed primarily by emerging grass shoots as they pass through the treated soil; however, some root absorption may also occur (8, 9, 10, 14). There is no significant translocation of trifluralin into the stem or foliage of higher plants, nor is it found in the harvested seed crop (12).

More detailed information regarding the metabolism of trifluralin in plants, animals, and soil has been published (6, 11, 12).

BENEFIN

Benefin is the common name for N-butyl-N-ethyl-α,α,α-trifluoro-2,6-dinitro-p-toluidine. Trade names include Balan®, Bonalan®, Benalan®, Carpidor®, and Quilan®. Pure benefin is a yellow-orange, crystalline solid, and its solubility in water is less than 1 ppm at 25°C. Benefin is susceptible to decomposition by ultraviolet light. In the adult rat, the LD_{50} was greater than 10,000 mg/kg. The herbicide is formulated as a $1\frac{1}{2}$ lb/gal concentrate and as granules.

Benefin controls most annual grasses and many broadleaf weeds including carpetweed, chickweed, knotweed, lambsquarters, pigweeds, purslane, redmaids, and Florida pusley. Benefin is recommended for alfalfa, birdsfoot trefoil, clovers, lettuce, peanuts, and tobacco. It is applied as a preplant soil-incorporated application in all of these crops except tobacco, in which it is soil-incorporated prior to transplanting. Granular benefin is also surface-applied to turf grasses for annual-grass weed control.

Behavior in Soil

Degradation of benefin in soil generally follows a pattern similar to trifluralin. Usually, benefin persists in soils for 4–5 months at recommended application rates.

BUTRALIN

Butralin is the common name for 4-(1,1-dimethylethyl)-N-(1-methylpropyl)-2,6-dinitrobenzenamine. The trade name is Amex® 820. It is a yellow-orange solid, with a water solubility of about 1 ppm. It is formulated as an emulsifiable concentrate. It has an acute oral LD_{50} of 2500 mg/kg.

Butralin controls most annual grasses and several broadleaf weeds including lambsquarters, pigweed, and purslane. It is applied in cotton and soybeans as a preplant, soil-incorporation treatment. As with other dinitroanilines, the rate used depends on the soil texture—low rates are applied on sand and sandy loams, and higher rates on silt-clay loams and clay loams.

DINITRAMINE

Dinitramine is the common name for N^4,N^4-diethyl-α,α,α-trifluoro-3,5-dinitrotoluene-2,4-diamine. The trade name is Cobex®. It is a yellow, crystalline solid with a water solubility of 1 ppm. It is formulated as an emulsifiable concentrate, and the acute oral LD_{50} is 3000 mg/kg.

In soybeans and cotton, dinitramine controls most annual grasses and several broadleaf weeds including lambsquarters, pigweed, and purslane, when applied as a preplant soil-incorporation treatment. The herbicide is strongly adsorbed by soils, and is therefore only slightly leached.

FLUCHLORALIN

Fluchloralin is the common name for N-(2-chloroethyl)-2,6-dinitro-N-propyl-4-(trifluoromethyl)aniline. The trade name is Basalin®. It is a yellow-orange, crystalline solid. Water solubility is probably less than 1 ppm. The compound is formulated as an emulsifiable concentrate and as granules. The acute oral LD_{50} is about 1550 mg/kg.

Fluchloralin controls annual grasses sprouting from seed as well as from several broadleaf weeds. At the same time, it has shown tolerance for many crops. Ultimate uses have not been determined, although the herbicide has been tested for use on cotton and soybeans.

ISOPROPALIN

Isopropalin is the common name for 2,6-dinitro-N,N-dipropylcumidine. The trade name is Paarlan®. This red-orange oillike chemical has a water solubility of less than 1 ppm. Isopropalin is formulated as an emulsifiable concentrate. The acute oral LD_{50} is about 5000 mg/kg.

Isopropalin controls most annual grasses and several annual broadleaf weeds starting from seed. These include pigweed, lambsquarters, purslane, carpetweed, Florida purslane, and poorjoe. It is used on transplant tobacco, tomatoes, and peppers, as well as direct-seeded peppers. It is applied as a preplant soil-incorporation treatment. Isopropalin is strongly adsorbed by soil and shows negligible leaching.

NITRALIN

Nitralin is the common name for 4-(methylsulfonyl)-2,6-dinitro-N,N-dipropylaniline. The trade name is Planavin®. Nitralin is a light yellow to orange, crystalline solid with a water solubility of 0.6 ppm. The compound is formulated as a water dispersible liquid and as a wettable powder. The acute oral LD_{50} is greater than 2000 mg/kg.

Nitralin can be used on alfalfa, certain beans, several cole crops, cotton, certain cucurbits, peanuts, peas, southern peas, transplant peppers, safflower, soybeans, and transplant tomatoes. On most of these crops it is applied as a preplant soil-incorporation treatment. In tomatoes, peppers, and cole crops nitralin may be applied before or after transplanting. It is applied to established alfalfa. Nitralin controls most annual grasses and several broadleaf weeds.

Nitralin is relatively immobile in soil and leaches very slowly. Nitralin like other dinitroanilines inhibits cell division and causes swelling of cells in the meristematic region of the root and inhibits root growth (13). It either prevents or limits spindle formation in dividing cells, resulting in a polyploid condition (5). Nitralin is absorbed during seed germination and by roots but only slightly, if any, by leaves.

ORYZALIN

Oryzalin is the common name for 3,5-dinitro-N^4,N^4-dipropylsulfanilamide. The trade name is Surflan®. Oryzalin is a yellow-orange, crystalline solid

Figure 13-4. Oryzalin was applied to this row of grapes for 3 years, consecutively. Shallow cultivation or the use of a contact herbicide further increases the effectiveness of the treatment. (Elanco Products Company., a Division of Eli Lilly and Co.)

with a water solubility of 2.5 ppm. The compound has a low vapor pressure and is relatively stable in sunlight; thus, oryzalin can be applied to the soil surface for subsequent movement into the upper soil zone by rain. It is formulated as a wettable powder. The acute oral LD_{50} is greater than 10,000 mg/kg.

Oryzalin alone, and in combination with other herbicides, is cleared for use on soybeans. In addition, the herbicide is used on tobacco, potatoes, fruit and nut trees, grapes, and woody ornamentals (see Figure 13-4).

Behavior in Soils

Laboratory studies indicate that oryzalin is biodegradable in soil. In laboratory leaching studies, oryzalin applied to a soil surface remained primarily in the top 2 in. when 6 in. of water was passed through an 8-in. column.

In field studies, at least 0.5 in. of rainfall or overhead irrigation water was necessary to position oryzalin in the weed-seed germinating layer of the soil. Excessive rainfall or irrigation did not leach oryzalin out of the weed-seed germination zone.

In areas with more than 25 in. of rainfall, oryzalin applied at the recommended rate has not been persistent from one year to the next. Cultivation does not reduce, and in fact may increase, the herbicidal efficacy of the compound.

PENOXALIN

Penoxalin is the common name for N-(1-ethylpropyl)-2,6-dinitro-3,4-xylidine. The tradename is Prowl®. It is an orange-yellow crystal, with solubility in water of less than 0.5 ppm. The compound is formulated as an emulsifiable concentrate, and its acute oral LD_{50} for the male rat is 1250 mg/kg. It is being tested for use as a herbicide for soybeans, cotton, peanuts, and in several combinations as a herbicide for corn.

PROFLURALIN

Profluralin is the common name for N-(cyclopropylmethyl)-α,α,α-trifluoro-2,6-dinitro-N-propyl-p-toluidine. The trade name is Tolban®. It is a yellow-orange, crystalline solid, or deep-orange liquid with a water solubility of 0.1 ppm. The compound is formulated as an emulsifiable concentrate, but granular formulations are also available. The acute oral LD_{50} is 2200 mg/kg.

Profluralin controls annual grasses from seed and several broadleaf weeds and has shown tolerance for many crops. Ultimate uses have not been determined.

LITERATURE CITED

1. Alder, E. F., W. L. Wright, and Q. F. Soper, *Proceedings of Seventeenth North Central Weed Control Conference*, 1960, p. 24.
2. Ashton, F. M., and A. S. Crafts, *Mode of Action of Herbicides*, Wiley, New York, 1973.
3. Ashton, F. M., and R. D. Kukas, *Abstracts of the Weed Science Society of America*, No. 36, 1974.
4. Bayer, D. E., C. L. Foy, T. E. Mallory, and E. G. Cutter, *Amer. J. Bot.*, *54*, 945 (1967).

5. Gentner, W. H., and L. G. Burk, *Weed Sci.*, *16*, 259 (1968).

6. Golab, T., R. J. Herberg, S. J. Parka, and J. B. Tepe, *Agr. Food Chem.*, *15(4)*, 638 (1967).

7. Hess, D., M.S., thesis. Univ. of California, Davis, 1973.

8. Knake, E. L., A. P. Appleby, and W. R. Furtick, *Weeds*, *15*, 228 (1967).

9. Negi, N. S., and H. H. Funderburk, Jr., *Abstracts of the Weed Science Society of America*, 1968, p. 37.

10. Parker, C., *Weeds*, *14*, 117 (1966).

11. Probst, G. W., T. Golab, R. J. Herberg, F. J. Holzer, S. J. Parka, C. Van der Schans, and J. B. Tepe, *Agr. Food Chem.*, *15(4)*, 592 (1967).

12. Probst, G. W., and J. B. Tepe, "Trifluralin and Related Compounds," in *Degradation of Herbicides*, P. C. Kearney and D. D. Kaufman, Eds., Marcel Dekker, New York, 1969, pp. 255–282.

13. Schieferstein, R. H., and W. J. Hughes, *Proceedings of the Eighth British Weed Control Conference*, 1966, p. 377.

14. Swann, C. W., and R. Behrens, *Abstracts of the Weed Science Society of America*, *No. 222*, 1969.

FOR CHEMICAL USE SEE THE MANUFACTURER'S LABEL AND FOLLOW THE DIRECTIONS. ALSO SEE PREFACE.

14 Nitriles

Nitriles are organic compounds containing a $-C\equiv N$ group. Those used as herbicides are benzonitriles with OH^-, Cl^-, and/or Br^- substitutions on the benzene ring. Herbicidal properties of these compounds were discovered in the late 1950s and early 1960s. Dichlobenil is applied to the soil, while bromoxynil and ioxynil are applied to the foliage of weeds to be controlled.

DICHLOBENIL

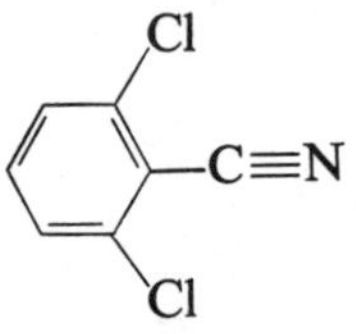

dichlobenil

Dichlobenil is the common name for 2,6-dichlorobenzonitrile. The trade name is Casoron®. It is a white, crystalline solid with a water solubility of 18 ppm. It is formulated as a wettable powder and in granular form. Its acute oral LD_{50} is greater than 3160 mg/kg.

Uses

Dichlobenil inhibits germination of seeds of both grass and broadleaf plants but it does not control emerged weeds. It controls annual weeds in 13 tree fruit and nut crops including almonds, apples, avocados, cherries, citrus, figs, filberts, mangoes, nectarines, peaches, pears, plums, and English walnuts. It is also used on five small fruits (blackberries, blueberries, cranberries, grapes, and raspberries) as well as many woody ornamentals. And it can be used to control aquatic weeds in ponds, reservoirs, and lakes.

In none of these crops can dichlobenil be applied immediately after transplanting. The waiting period after transplanting ranges from 4 weeks for most species to 6 months for pecans and 1 year for citrus. In certain crops there is a minimum number of days time restriction before harvest that dichlobenil must not be applied. Do not graze treated orchards.

Dichlobenil is relatively volatile and therefore needs to be leached into the soil by rainfall or sprinkler irrigation soon after application to prevent volatility losses. It can be mixed in by shallow incorporation.

Recent research with dichlobenil has shown promise for the control of certain perennial weeds (e.g., field bindweed) in orchards and vineyards when applied to the soil as a subsurface layer. The emerging weed shoots are not able to penetrate the concentrated layer of dichlobenil.

Soil Influence

Dichlobenil appears to be adsorbed tightly by soil colloids. This, together with its relatively low water solubility greatly restricts its leaching. Dichlobenil is persistent in soils and phytotoxicity often can be detected for as long as 8 months after application; as a concentrated subsurface layer it probably persists even longer.

Mode of Action

Dichlobenil acts primarily on apical growing points and root tips. This inhibition of growth is followed by a gross disruption of tissues mostly in the meristems and phloem. This may cause swelling or collapse of stem, root, and leaf petioles. The apical meristems and leaves may show dark discoloration (3).

Because dichlobenil is applied to soil, its rapid absorption by roots and seeds is of particular significance. However, it is also absorbed by leaves in vapor form (2). Translocation is limited and appears to occur primarily in the apoplast. Dichlobenil has been reported to greatly restrict the translocation of photosynthate from leaves to other plant parts (1), and is metabolized by higher plants (4). Some of this chemical, however, is also lost as dichlobenil in vapor form (2).

BROMOXYNIL

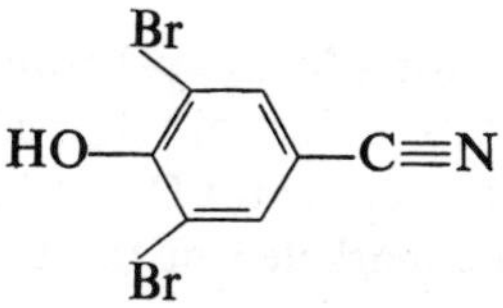

$$HO-\underset{\underset{Br}{|}}{\overset{\overset{Br}{|}}{\bigcirc}}-C\equiv N$$

bromoxynil

Bromoxynil is the common name for 3,5-dibromo-4-hydroxybenzonitrile. The trade names are Brominal® and Buctril®. It is a light buff solid with a water solubility of less than 0.02%; formulated in liquid form as the octanoic acid ester. The acute oral LD_{50} is 250 mg/kg.

Uses

Bromoxynil is used on barley and wheat as a postemergence spray to control annual broadleaf weeds not readily controlled by 2,4-D or MCPA (e.g., fiddleneck, resistant mustards, and wild buckwheat). It is applied when the grain is in the two-leaf to boot stage. It is more effective on small weeds than large weeds.

Bromoxynil is often mixed with MCPA to control a greater variety of broadleaf weeds, but this combination must be applied at the two- to four-leaf stage of the grain.

Mode of Action

The contact action of bromoxynil appears as blistered or necrotic spots on leaves within 24 hr. Later, extensive leaf tissue is destroyed and the plant dies. Bromoxynil is both a photosynthetic and respiratory inhibitor, readily absorbed by leaves, but with limited translocation. Therefore, complete coverage of the foliage is essential to good weed control.

IOXYNIL

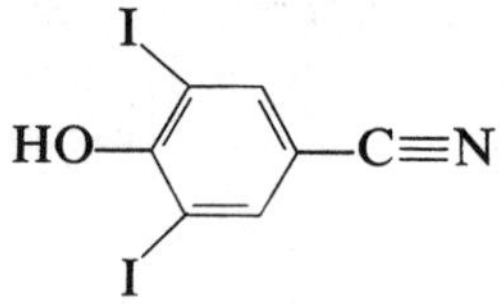

ioxynil

Ioxynil is the common name for 4-hydroxy-3,5-diiodobenzonitrile. The trade names are Actril®, Bantrol®, and Certrol®. It is a light buff solid with a water solubility of 50 ppm, and an acute oral LD_{50} of 110 mg/kg. Ioxynil is similar to bromoxynil and it is not marketed in the United States, and therefore additional details are not given.

LITERATURE CITED

1. Glenn., R. K., O. A. Leonard, and D. E. Bayer, *Research and Progress Reports, Western Society of Weed Science*, 1971, p. 150.
2. Massini, P., *Weed Res., 1*, 142 (1961).
3. Milborrow, B. V., *J. Exp. Bot., 15*, 515 (1964).
4. Verloop, A., and W. B. Nimmo, *Weed Res., 10*, 65 (1970).

FOR CHEMICAL USE, SEE THE MANUFACTURER'S LABEL AND FOLLOW THE DIRECTIONS. ALSO, SEE THE PREFACE.

15 Phenols

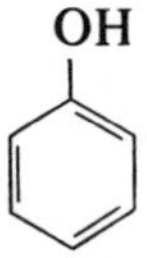

$$OH$$

Phenol is the monohydroxy (—OH) derivative of benzene. Herbicides of this class now registered for crop use in the United States have two nitro (—NO_2) groups on the ring in the number 4 and number 6 positions. These are dinitrophenols or simply dinitros.

Historically, PCP (pentachlorophenol) was used as a herbicide in certain crops, but it is not registered now for this use in the United States. While it can be used as a preharvest desiccant in small-seeded legume crops grown for seed, it is largely used as a wood preservative.

Substituted phenolic herbicides are contact herbicides. Some have been used for more than 40 years. The sodium salt of DNOC (4,6-dinitro-*o*-cresol) was first used to remove broadleaf weeds from small grain in France about 1933. Soon after, it was introduced into the United States and promptly became the major material for annual broadleaf weed control in cereal crops, flax, and peas. However, it is not now registered for these uses in the United States. It is presently used as a plant growth regulator for blossom thinning in apples.

DINOSEB

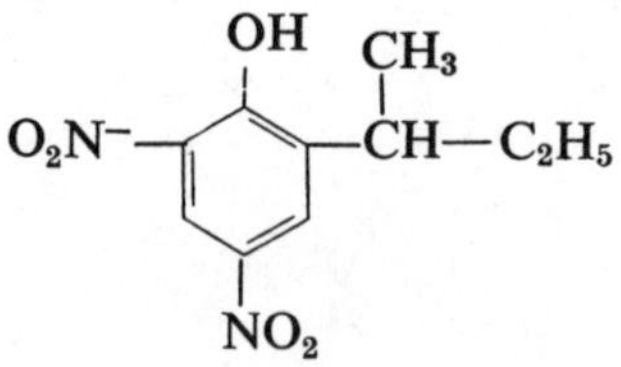

$$O_2N-\underset{NO_2}{\underset{|}{\bigcirc}}-\underset{|}{\overset{CH_3}{\underset{}{\overset{|}{CH}}}}-C_2H_5$$

dinoseb

Dinoseb is the common name for 2-*sec*-butyl-4,6-dinitrophenol. It is also often called DNBP. It has several trade names. The phenol form is usually formulated as an emulsifiable concentrate; it is also soluble in oil. It may also be formulated as water-soluble salts; the most common are the ammonium, triethanolamine, and a mixture of ethanol and isopropanolamine salts.

Dinoseb is a dark-brown solid or a dark-orange liquid, depending on temperature. As the phenol, it has a water solubility of 52 ppm, but the salts are quite soluble in water.

Toxicity to Humans and Animals

Dinoseb and its salts are dangerous poisons if taken internally, if inhaled as dusts, or if considerable quantities are absorbed through the skin. Therefore, avoid prolonged breathing of the spray drift or dusts, and avoid wearing contaminated clothing or shoes. If your skin is contaminated, wash it immediately with soap and water. Symptoms of poisoning are excessive fatigue, sweating, thirst, and fever. If these develop, send for a physician.

With normal precautions, the chemical can be applied routinely with little or no hazard to the applicators. Daily bathing and change of clothing is recommended whether the applicator thinks he is contaminated or not.

Residues on foliage normally constitute little or no hazard to livestock. In an unpublished study, a milk cow was given 1.7 g of chemical/kg body weight/day for 3 days with no ill effects. Dinoseb did not appear in the milk. If there is question, keep livestock away from sprayed foliage until a rain has removed much of the herbicide.

The acute oral LD_{50} for rats ranges from 5 to 60 mg/kg. The maximum amount tolerated in the diet for a 6-month period was 100 ppm. It is considered to be quite toxic.

Fish are sensitive to dinoseb; 1.0 ppm killed trout and sea lamprey in 14 hr, and bluegills in 5 hr (1).

Uses

Dinoseb is very toxic to growing plants, so it is used as a general contact herbicide. It is so toxic to all leaves that it lacks the selectivity of its salt derivatives. Dinoseb is valuable where mowing is impractical; for example, along fencerows, ditchbanks, and roadsides. It kills most annual weeds and removes the tops from perennial weeds. Underground parts of perennial plants are not killed except by repeated treatments. Thus, dinoseb can be used in dormant alfalfa to kill annual weeds.

It is also used for preemergence soil treatments in beans, corn, cucumbers, peanuts, potatoes, and soybeans. In most small-seeded legumes it can be used immediately after cutting and before new growth appears; however, do not graze treated areas or feed treated forage to livestock.

Dinoseb is also used as a directed spray to weeds in the fall after harvest or in early spring before bloom in currants, gooseberries, grapes, and raspberries. Directed sprays are also used in many fruit and nut crops. It is also used as a preharvest desiccant in small-seeded legumes, peas, and soybean seed crops, as well as for preharvest vine killing in potatoes.

Dinoseb is not soluble in water, but is soluble in oil. Dinoseb-oil mixtures are very potent contact herbicides.

Salts of dinoseb are soluble in water and insoluble in oil—just the opposite of dinoseb itself. They are more selective in toxicity to plants than dinoseb itself.

Figure 15-1. Soybeans treated with dinoseb (amine salt) just after the soybeans emerged (crook stage). This treatment controlled weeds for 1 month without harm to the soybeans. The nearly weed-free plot in the background was treated similarly—but 9 days later. Treatment at this stage was much less effective. (Kentucky Agricultural Experiment Station.)

For simplicity, the uses of all three common salt formulations (ammonium, triethanolamine, and the mixture of ethanol and isopropyl amine salts) are considered together. However, these salts may not be registered for use on all crops mentioned, and it is advisable to check the manufacturer's label, Federal registration, or both, before using.

Salts of dinoseb are used for selective postemergence weed control in alfalfa, beans, small grain, small-seeded legumes, corn, garlic, onions, peas, and peanuts (see Figure 15-1). They are also applied as a directed spray to emerged weed ground cover in certain tree and nut crops while the crop is dormant and before bloom; 1–3 lb of the active ingredient/acre are usually applied as a spray. The most favorable conditions for application are dry, sunny weather, temperatures of 70 to 85°F, and no rain for about 12 hr after treatment. At strong concentrations the salts resemble dinoseb in their toxic contact effects on plants.

Salts are also used for preemergence weed control; effects usually last for 3–5 weeks. Salts are applied preemergence to beans, corn, cucumbers, mint, peanuts, peas, potatoes, pumpkins, soybeans, and squash. From $4\frac{1}{2}$ to 12 lb/acre of the active ingredient are applied in enough water to give uniform soil coverage.

Light rain after treatment is beneficial to carry the chemical into the soil and to reduce volatility losses. Heavy rains may cause leaching which may injure the crop or prevent effective weed control by leaching the chemical below the weed seeds in the surface soil. No rain soon after treatment will most likely cause poor preemergence weed control, especially in hot weather. This is a result of volatilization (loss of the chemical as a vapor) under several days' exposure to high soil temperatures.

Soil Influence

Dinoseb leaches readily, but there is some evidence of partial adsorption in certain organic and clay soils. Persistence of dinitros in soil depends on many of the same factors discussed in Chapter 4. At 6–9 lbs of active ingredient/acre, and under warm, moist conditions, you can expect dinitros to last only for 3–5 weeks. No residual carry-over is expected from one season to the next.

Mode of Action

Dinoseb is translocated little, if at all, within the plant. It acts almost completely as a contact herbicide; therefore, uniform coverage of all foliage is important.

Dinoseb stimulates respiration in low concentration and powerfully inhibits respiration at high concentrations (2,4,5), and it also inhibits the coupling during phosphorylation and oxidation of pyruvate. Experimental evidence indicates that dinoseb acts on a basic mechanism in the cell by which the phosphate bond is coupled to oxidative reaction (5). Dinitrophenols act as protein coagulants (3), as well as causing the respiratory response.

To reiterate—dinoseb and its salts apparently affect respiratory responses and protoplasmic toxicity. Perhaps both are associated with protein coagulation of enzymes and other protoplasmic constituents.

LITERATURE CITED

1. Applegate, V. C., J. H. Howell, and A. E. Hall, Jr., *U.S.D.I.*, *Special Scientific Report—Fisheries No. 207*, Washington D.C., 1957.

2. Bonner, J., *Amer. J. Botany*, *36*, 429 (1949).

3. Crafts, A. S., and W. A. Harvey, *Agr. Chem. 5(3)*, 28–41, 89, 91 (1950).

4. Kelly, Sally, and G. S. Avery, Jr., *Amer. J. Botany*, *36*, 421 (1949).

5. Loomis, W. E., and F. Lyman, *J. Biol. Chem.*, *173*, 807 (1948).

FOR CHEMICAL USE, SEE THE MANUFACTURER'S LABEL AND FOLLOW THE DIRECTIONS. ALSO SEE THE PREFACE.

16 Phenoxys

Phenoxy compounds have the phenyl ring attached to an oxygen, which in turn is attached to a carboxylic group. Using the phenyl ring, oxygen, and butyric acid, this assumes the structural formula of:

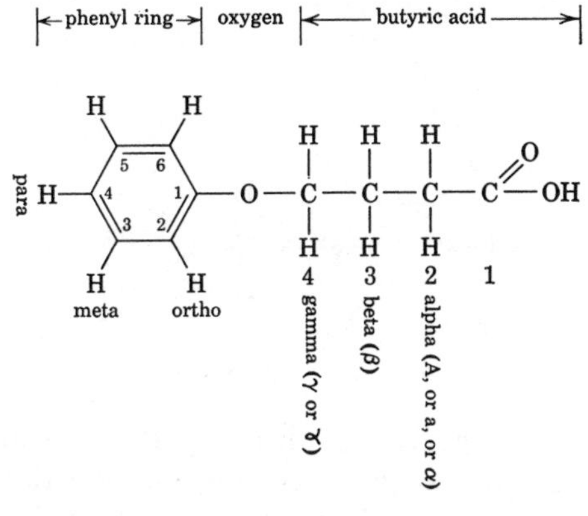

The carbon atoms in the benzene ring are numbered, starting at a given point (usually a point of attachment) as shown above. The terms ortho, meta, and para may also be used. The ortho position is equivalent to positions 2 and 6, the meta position is equivalent to positions 3 and 5, the para position to position 4.

The butyric acid, in this case, represents the carboxylic acid portion of the molecule. A carboxylic acid with one carbon atom would be named formic or methanoic acid; two carbons, acetic or ethanoic acid; three carbons, propionic acid; four carbons, butyric acid; and so on. The longest continuous chain of carbon atoms is usually selected to determine the name.

Besides choosing the appropriate name of the carboxylic acid, each carbon atom can be numbered or named. The "Geneva system" starts numbering from the end, which allows the smallest possible numbers to be used. See the formula above, starting with number 1 on the carboxylic carbon.

Greek letter names are also used. The *first carbon next to the carboxyl group* is named alpha (α), the second carbon beta (β), the third carbon gamma (γ), and so on.

These are only the preliminary rules used in naming chemical compounds. Chemistry textbooks and handbooks give further rules.

Some examples of phenoxy compounds are 2,4-D, MCPA, 2,4-DB, dichlorprop, 2,4,5-T, and silvex. We will discuss each, with the major emphasis on 2,4-D. Scientists believe that most other phenoxy materials have effects similar to 2,4-D.

Because of World War II security regulations, research on selective herbicides was not reported as it progressed. Technical papers were often published several years after the work was actually done.

Slade, Templeman, and Sexton (20) of Imperial Chemical Industries, Ltd., reported in 1945 that they used α-naphthaleneacetic acid to control yellow charlock (wild mustard) with slight injury to oats. They later used MCPA without injury.

2,4-D

2,4-D or (2,4-dichlorophenoxy)acetic acid

2,4-D is the common name for (2,4-dichlorophenoxy)acetic acid. There are numerous trade names. It is a white, crystalline solid with a water solubility of about 600 ppm. Its salts (sodium, lithium, amine), however, are quite soluble in water, and the acute oral LD_{50} of its various formulations range from 300 to 1000 mg/kg.

The first reference to 2,4-D in the literature is an article by Pokorny in 1941 (18). He outlined a very simple method of preparing 2,4-D but made no reference to its use. In 1942 Zimmerman and Hitchcock (27), of Boyce Thompson Institute, first described the use of 2,4-D as a plant-growth regulator. In 1944 Marth and Mitchell (13), U.S. Department of Agriculture, reported that 2,4-D killed dandelion, plantain, and other weeds from a bluegrass lawn, and Hamner and Tukey (8) described successful field trials using 2,4-D as a herbicide.

An excellent paper on the discovery and development of 2,4-D has been written by Peterson (17).

Figure 16-1.   Scientists studying the selective killing of giant ragweed in corn in Henderson County, Kentucky, 1947. This was one of the first large-scale tests made with 2,4-D in the United States. (Sherwin-Williams Company.)

Common Forms

2,4-D is formulated as an emulsifiable acid, amine salts, mineral salts, and esters. The esters are oil-soluble and emulsifiable with water.

2,4-D Acid

Because the acid form of 2,4-D is only slightly soluble in water, it is not commonly used in commercial formulations. However, at least one emulsifiable concentrate formulation of 2,4-D acid is available. It is used mostly to control hard-to-kill perennial broadleaved weeds such as field bindweed, Canada thistle and Russian knapweed.

The concentration of essentially all phenoxy herbicide formulations is expressed as *acid equivalents* in pounds per gallon. Recommendations are also made on this basis. *Acid equivalent* refers to that part of a formulation that theoretically can be converted to the acid.

2,4-D Amines

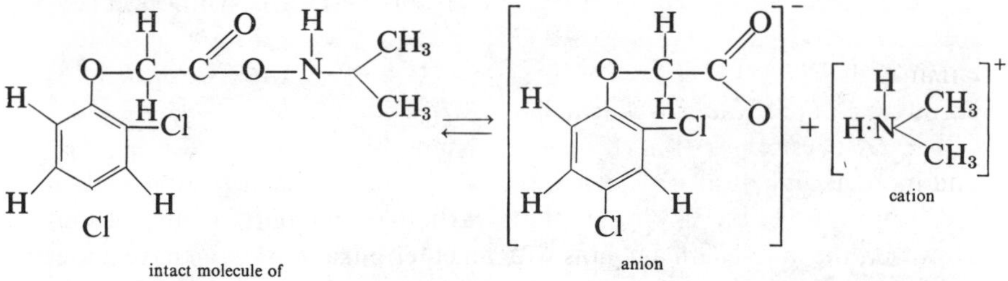

intact molecule of
dimethylamine salt
of 2,4-D

anion

cation

anion and cation of ionized
molecule of dimethylamine salt
of 2,4-D

Amine salts of 2,4-D are the most commonly used form of 2,4-D. The dimethylamine salt and its ionization in water is shown above. A mixture of alkanolamine salts (of the ethanol and isopropanol series) is the most widely used, but other amine salts are also available. The dimethylamine salt of 2,4-D is a white, crystalline solid. The amine salts of 2,4-D are usually liquids, and represents no volatility hazard to sensitive plants.

Amines are soluble in all proportions in water and therefore are well adapted to low-gallonage spray equipment, even at high rates of application. Most amine salts of 2,4-D dissolved in water form a clear solution, though it may be colored. Most amine salts are not soluble in petroleum oils.

An exception, however, is *N*-oleyl-1,2-propylenediamine salt of 2,4-D; it is oil-soluble and is essentially insoluble in water.

2,4-D Esters

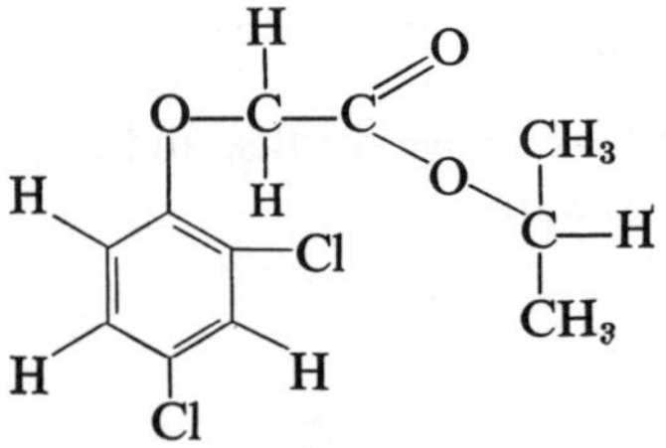

isopropyl ester of 2,4-D
(a volatile form)

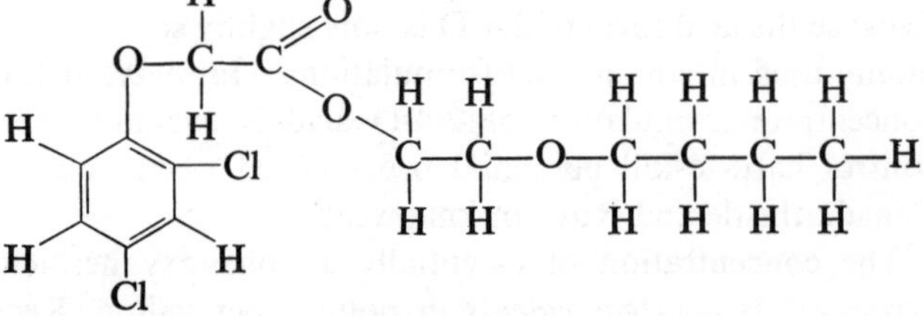

butoxyethyl ester of 2,4-D
(a low volatile form)

Esters of 2,4-D are colorless liquids. In contrast to amine formulations, esters are essentially insoluble in water. Esters of 2,4-D are synthesized by a reaction between 2,4-D acid and an alcohol. One molecule of water is eliminated. The alkyl group of the alcohol replaces the hydrogen of the carboxyl group of the 2,4-D acid.

The 2,4-D ester is identified by the name of the alcohol used. The cheaper and more abundant alcohols are commonly used, like methyl (one carbon), ethyl (two carbons), isopropyl (three carbons), and butyl (four carbons). However, the long-chain alcohols with an ether linkage (—O—) have a lower volatility hazard than the short-chain alcohols when used to formulate 2,4-D. Therefore, long-chain alcohols, even though more expensive, have assumed increasing importance (see the section on "Volatility" in this chapter).

2,4-D ester compounds are only slightly soluble in water, but they are soluble in some petroleum oils. The ester is usually slightly diluted in oil, after which an emulsifying agent is added. Esters are usually sold as a liquid. When mixed with water, the emulsifying agent keeps the tiny, oillike droplets suspended for a time, much the way butterfat is suspended in milk. Water is in the continuous phase and oil droplets are dispersed; thus, it is an oil-in-water (O/W) type of emulsion.

When mixed with water the 2,4-D emulsion appears milky. If properly emulsified, the ester can be mixed in all proportions with water, but if allowed to stand, the oil droplets may separate. Mixing will reform the emulsion. Once applied, the oil is not easily washed from plants by rain. The oillike characteristics make it difficult to remove ester forms from spray equipment.

So far we have considered an O/W emulsion. But if the oil is in the continuous phase and the water in the dispersed phase, the emulsion is a water-in-oil (W/O) type. This is often referred to as an invert emulsion. (For further details, see Chapter 6.)

Ester formulations of 2,4-D are generally considered the most toxic to plants. There are at least three possible explanations: (1) volatility permits absorption of the gases through the stomates; (2) wetting action of the oillike ester and the oil carrier may actually aid penetration of the stomates; and (3) ester forms, with their low polarity, are compatible with the cuticle and aid penetration directly through the cuticle.

Because of its greater toxicity, the ester form is often more effective on resistant species, especially on woody plants. It is also more likely to injure crop plants where excessive amounts are applied. Lower rates of 2,4-D esters—about two-thirds as much—than of salts are usually suggested for post-emergence application to crop plants.

2,4-D Mineral Salts

sodium salt of 2,4-D

If sodium replaces the hydrogen atom of the carboxyl group of 2,4-D acid, the sodium salt of 2,4-D is produced. It is a white, crystalline solid. Lithium, potassium, and ammonium salts of 2,4-D have been manufactured but are no longer used, and the sodium salt has only a limited use,

The sodium salt is sold as a water-soluble powder. Some stirring or mixing is usually required to get it into solution. It is not soluble enough in water to be applied in low-gallonage equipment at high rates of application. In most other respects it resembles the amine salts. The amine salts have gradually replaced other salts because the amines are more easily dissolved in water.

Precipitate Formation

When a salt of 2,4-D is dissolved in water, some of the molecules are dissociated into ions which are then free to combine with other ions in the solution. This creates a problem when the water is "hard," having a high calcium or magnesium content. The calcium and magnesium salts of 2,4-D are only slightly soluble in water—the calcium salt at 2.5 g/liter and the magnesium salt at 17.4 g, both at a temperature of 20°C. Thus if calcium or magnesium ions are present in the solution, a precipitate may be formed that will clog the filters and nozzles (28).

Esters cause little or no clogging in hard water. The 2,4-D ester is not in solution nor is it ionized. It is emulsified in the water with the oillike droplets dispersed or suspended in the water. 2,4-D ester droplets are "insulated" from the water by an emulsifying agent; therefore, there is little opportunity for a reaction to take place and a precipitate to form. (For further details, see Chapter 6.)

Mixtures of chemicals may form complex molecules resulting in a precipitate. Before placing in the sprayer, in a small container make such a mixture in the exact proportion to be used, and observe it over a period of time. If no precipate forms then, it is likely that none will form in the sprayer. This precaution may save you the job of removing a congealed mass from the sprayer or the difficult task of cleaning clogged filters and nozzles.

Volatility

Volatility means the tendency of a chemical to vaporize or give off fumes. Vapors of 2,4-D esters may kill or injure susceptible plants, and where the vapors are confined, they may inhibit the germination of seed (see Figure 6-9). The amount of fumes or vapors given off is related to the vapor pressure of the chemical (14); 2,4-D acid, amine salts, and sodium salts have very low volatility characteristics and cause little or no volatility hazard (see Figure 6-8 and Table 16-1).

As previously discussed under the section "2,4-D esters," high molecular-weight alcohols are used to make 2,4-D esters with reduced volatility. The butoxyethyl ester and the isooctyl ester (eight carbons) are both low-volatile esters. The high molecular-weight alcohols are relatively expensive; therefore, their 2,4-D esters are more expensive than the volatile formulations.

It should be emphasized that the above esters are *low-volatile* and not *nonvolatile*. Under hot, humid conditions they will volatilize enough to injure susceptible plants.

Uses

2,4-D is used to control annual and perennial broadleaf weeds on noncrop lands as well as in tolerant crops. It is applied to the foliage or stem (or both) of the plant to be controlled. At rates above 1 lb/acre it may serve as a soil-applied preemergence herbicide, with the effects lasting for 30 days or less. The usual rate is $\frac{1}{2}$–2 lb/acre; however, for woody-plant control, rates as high as 6 lb/acre are used. At high rates, it is likely that plants will be controlled through both root and foliar absorption.

The highly versatile 2,4-D is used on barley, corn, oats, pastures, rangeland, rice, rye, sorghum, sugarcane, and wheat, as well as for aquatic-weed control. In low concentrations it is used as a growth regulator to reduce fruit drop, increase fruit size, and increase storage life in certain citrus crops. For woody plant control it is often combined with 2,4,5-T.

The numerous details of timing, method of application, and rates will be discussed in later chapters on specific crops.

Soil Influence

Soil type and formulation of 2,4-D influences its leaching in soil. Water-soluble forms leach more readily than those that are slightly soluble. It is adsorbed by soil colloids and less leaching occurs in clay and organic soils than on sandy soils.

Table 16-1. General Characteristics of Different Forms of 2,4-D

Form of 2,4-D	Soluble in Water	Soluble in Oil	Appearance When Mixed with Water	Precipitates Formed in Hard Water	Volatility[1]	General Remarks
2,4-D acid, pure	No	No	Milky	Yes	No hazard	Only for specialized use.
Amine salt	In all proportions	Not soluble, usually	Clear	Yes	No hazard	Good for general farm use, lawns, turf, some woody plants
Sodium salt	Medium solubility	Not soluble	Clear	Yes	No hazard	Only for specialized use
Esters						
Volatile forms	No, but can be emulsified	Yes	Milky	No	Volatile	Dangerous to use near susceptible crops due to volatility
Low-volatile forms	No, but can be emulsified	Yes	Milky	No	Medium to low volatility	Some danger from volatility if used near susceptible plants, especially with high temperature

[1] The tendency to form volatile fumes or gases which can injure plants.

Chapter 4 covers factors affecting the length of time that 2,4-D may remain toxic in the soil. Of major importance is the decomposition by micro-organisms, the growth and activity of which is affected principally by food supply, soil temperature, soil moisture, and soil aeration. Low rates of 2,4-D will normally be decomposed in 1–4 weeks in a warm, moist loam soil. There is no risk of the chemical accumulating in the soil from one year to the next under such conditions. In very dry soils or in frozen soils, the rate of decom-position may be inhibited considerably.

At normal dosages, 2,4-D does not generally reduce the total number of microorganisms in the soil. When heavy rates of treatment are used, however, some may be inhibited and others stimulated. Aerobic microorganisms are more sensitive and 2,4-D may seriously inhibit their growth. Anaerobic organisms may not be significantly affected and some facultative anaerobic organisms may even be stimulated (26). Therefore, those organisms requiring free oxygen for respiration may be hindered by 2,4-D, while those not requir-ing free oxygen may actually be stimulated.

The effect of 2,4-D on legume nodulation is of vital importance to the vigor of the legume plant. Minute rates of 2,4-D slow down the formation of nodules on the bean plant (15), but these plants are far more sensitive to 2,4-D than are the Rhizobia bacteria (2). In the soil solution, 0.21 lb of 2,4-D/acre seriously restricted the germination and growth of beans, peas, red clover, and alfalfa. In the latter study as well, Rhizobia bacteria were grown in-dependently on culture media to learn the effects of 2,4-D on the organisms. The bacteria grew almost normally until equivalent test rates of 2,4-D reached 200 lb/acre. Hence the lessened nodulation associated with 2,4-D is primarily a plant response.

Mode of Action

Plant Structure

Twisting and curvature (epinasty) are among the most obvious effects of 2,4-D treatment on broadleaf plants. These plants generally develop grotesque and malformed leaves, stems and roots when treated with 2,4-D (see Figures 16-2, 16-3, and 16-4). The chemical appears to concentrate in young embryonic or meristematic tissues that are growing rapidly. It affects these tissues more than more mature or relatively inactive young tissues.

Histological studies with the kidney bean showed that the cambium, endodermis, embryonic pericycle, phloem parenchyma, and phloem rays all showed active cell division; the cortex and xylem parenchyma exhibited little response to 2,4-D treatment; and the epidermis, pith, mature xylem, sieve

Figure 16-2. A common burdock plant twisted and curled following treatment with 2,4-D.

Figure 16-3. Bean leaves showing the effect of 2,4-D. The leaf that is second from the left is normal.

Figure 16-4. Abnormal corn brace roots as a result of an excess of 2,4-D during a susceptible stage of growth. Corn yields were not reduced.

tubes, and the differentiated pericycle gave no response (21). The types of tissue affected in field bindweed and sow thistle (22) were much the same as those affected in the bean.

In further studies on the bean leaf, the malformed leaves resulted from continued growth of the apical meristem and failure of the lateral meristem derivatives to develop normally. Such growth reduced expansion of the areas between the veins so the veins become elongated and grew abnormally close together (see Figure 16-3). Buds that were undergoing rapid differentiation, and thus were in a state of great physiological activity when treated, showed the greatest effects (25). Similar effects have been noted in grapes and cotton (7). (See Figure 3-7.)

Most adventitious roots develop from the pericycle, and thus, if the pericycle of the root is developing rapidly when treated with 2,4-D, the number and structure of the new roots may be changed. Proliferation of the brace roots of corn occurred when 1–1½ lb of 2,4-D/acre were applied in the eight-leaf stage (1 ft high). When the corn root had outgrown the meristematic stage, similar applications did not produce malformed brace roots (19) (see Figure 16-4).

Absorption

Plant roots readily absorb 2,4-D, probably absorbing polar forms (salts) most readily. The leaves most readily absorb nonpolar forms of 2,4-D (the acid and ester forms), while the salt formulations are absorbed more slowly. The use of surfactants increases the rate of foliar absorption. Chapters 3 and 6 give more details.

The length of time that 2,4-D must be on a plant prior to a rain varies with the formulation of 2,4-D, the rate of 2,4-D application, and the temperature, humidity, and the susceptibility of the plants. In most cases, however, a rain-free period of 6–12 hr is adequate for effective weed control. The esters, being oillike, have a tendency to resist washing from the plant even though they are not absorbed.

Translocation

Translocation of a herbicide applied to the foliage is essential if plant roots are to be killed. This is especially important for control of perennial weeds. (The general principles of translocation are the same as those discussed in Chapters 3 and 5.) After 2,4-D migrates through the leaf cuticle, it moves to the phloem; it is then moved down through the phloem with the photosynthate. Little translocation from the leaves takes place if the readily available food supply has been diminished by continued darkness or reduced light.

Because 2,4-D moves through the phloem with the photosynthate the treatment of rapidly developing leaves of a perennial weed in the early spring causes little or no translocation of 2,4-D to the roots. Also, excessive rates of application would kill the living phloem cells, stopping translocation to the roots.

Perennial weeds are treated most effectively when large amounts of foods are being translocated to the roots, such as late spring or early fall. Furthermore, low rates of chemical application when applied repeatedly may give better perennial-weed control than a single heavy application. The above assumptions have proven to be generally valid, as tested through research and practical applications. The rates of chemical application and timing of treatments are different for different species.

Translocation of 2,4-D from the soil upward follows the transpiration stream, with the movement of water and soil nutrients through the xylem. Translocation up or down is favored by sufficient soil moisture to favor rapid plant growth. Dry soils may slow translocation.

Many techniques are used to trace translocation through the plant. Radioactive elements synthesized as a part of the herbicide molecule have proven especially valuable in translocation studies (1, 4, 5, 6, 11, 26).

Maturity of the Plant

Maturity of the plant directly influences its susceptibility to 2,4-D. In general, all plants are most susceptible during the time of seed germination. Even the grasses are affected by low concentrations of 2,4-D at this time. The plant gains tolerance with age; some are tolerant while still small, and others never gain more than a slight tolerance for 2,4-D.

Some plants may develop a second period of susceptibility. This usually coincides with a period of rapid growth. At this time the meristems are metabolically active and very susceptible to 2,4-D. Small grains, therefore, may be very susceptible to 2,4-D in the germinating and small-seedling stages. The plant becomes tolerant in the fully tillered stage, becomes susceptible again in the jointing and heading stage, and is very tolerant as the grain reaches the "soft-dough" stage (see Figure 21-3).

Molecular Fate

The degradation of the phenoxy herbicides by higher plants has been reviewed by Loos (12). 2,4-D is degraded to nonphytotoxic forms in higher plants, undergoing decarboxylation and demethylation of the side chain as well as dechlorination and hydroxylation of the ring. Ring cleavage ultimately occurs. The resistance of certain species to 2,4-D has been in part attributed to their ability to rapidly degrade 2,4-D to nontoxic molecules.

Mechanism of Action

The biochemical mechanism (or mechanisms) by which 2,4-D acts as a herbicide have proven very elusive. In 1966, Penner and Ashton (16) cited approximately 400 papers which discuss the numerous biochemical and metabolic changes in plants induced by chlorophenoxy herbicides, and many more papers have been published since then. The theories of Hanson and Slife (9) are perhaps the most accurate with regard to the activity of 2,4-D. These theories are presented in the following two paragraphs.

2,4-D does not appear to act as a simple inhibitor. Although certain enzymes can be inhibited *in vitro* by 2,4-D, there is no firm evidence that it acts *in vivo* by directly interfering with intermediary metabolism, respiration, or photosynthesis. It appears to be acting as an auxin, but accumulates to higher concentrations than the native auxin, indoleacetic acid, because it is degraded more slowly. Both inhibition and promotion of growth are involved when susceptible plants respond to 2,4-D, depending on the organ and tissue examined, and the amount of 2,4-D in them.

Hanson and Slife propose that the immediate cause of death is physiological disfunction of the plant brought about by abnormal growth. In turn, the abnormal growth is believed to be based on an abnormal nucleic acid metabolism.

MCPA

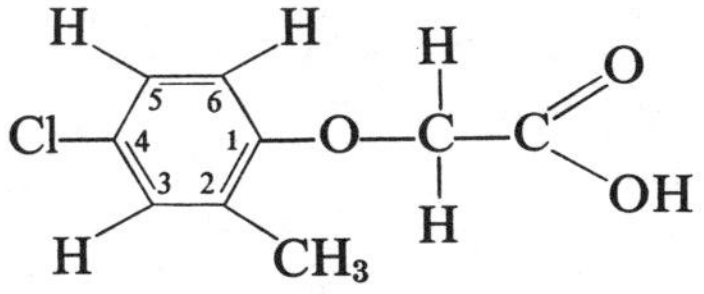

MCPA

MCPA is the common name for [(4-chloro-*o*-tolyl)oxy]acetic acid. Another name which has been used is (2-methyl-4-chlorophenoxy)acetic acid. It has several trade names. MCPA is identical to 2,4-D except it has a methyl (—CH_3) group on the number 2 position of the ring instead of a chlorine atom. It is a light-brown solid, essentially insoluble in water. The acute oral LD_{50} is about 700 mg/kg.

Its herbicidal properties, length of residual toxicity in the soil, and toxicity to man and animals are similar to 2,4-D. It can be formulated as the acid, salt, and ester forms much the same as 2,4-D.

MCPA was one of the first hormonelike herbicides tested in England. It still is an important herbicide of England, Sweden, and other Northern European countries. It remains a specialized chemical in the United States, used on a relatively small scale.

MCPA is less injurious to crops such as oats and rice, and it is more effective on some weed species. The reverse is also true. In the Northern European countries, MCPA is used for weed control in small grains. It has also been used for weed control in small-seeded legumes, small grains, pastures, flax, and peas. In the United States, MCPA has remained somewhat more expensive to use than 2,4-D, with a higher cost per pound of active ingredient, and larger quantities are usually needed for effective weed control.

2,4-DB

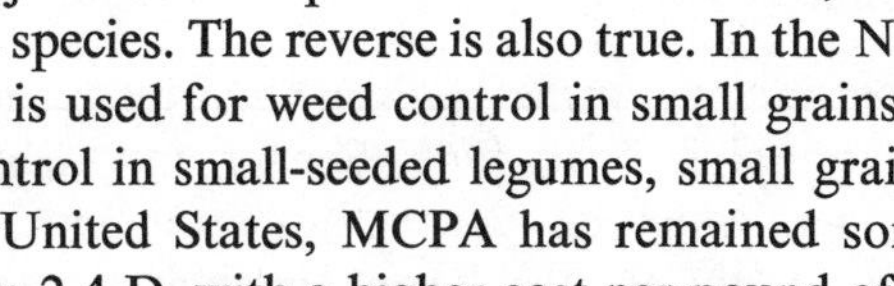

2,4-DB

2,4-DB is the common name for 4-(2,4-dichlorophenoxy)butyric acid. The trade names are Butoxone® and Butyrac®. It is a white, crystalline solid, practically insoluble in water. It is formulated as amine salts and low-volatile esters. The amine salts are water-soluble and nonvolatile; the esters are oil-soluble and usually emulsifiable in water.

2,4-DB has been particularly effective (1) as a postemergence treatment to seedlings of small-seeded legumes to control small broadleaf weeds or (2) in established legumes before flowering. It has also been used to control cocklebur (less than 3 in. high) in soybeans when the beans are 8–12 in. high.

2,4-DB is not highly phytotoxic *per se*; however, it undergoes beta-oxidation in plants and soils to form 2,4-D. Some plants are able to make this conversion rapidly while in others (e.g., small-seeded legumes) this reaction takes place slowly. Thus most broadleaf weeds are controlled by 2,4-DB, whereas the small-seeded legumes are not injured.

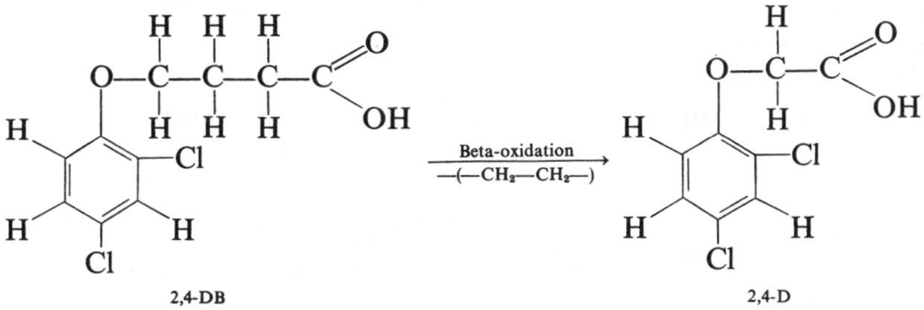

2,4-DB   Beta-oxidation   2,4-D
                —(—CH₂—CH₂—)

DICHLORPROP

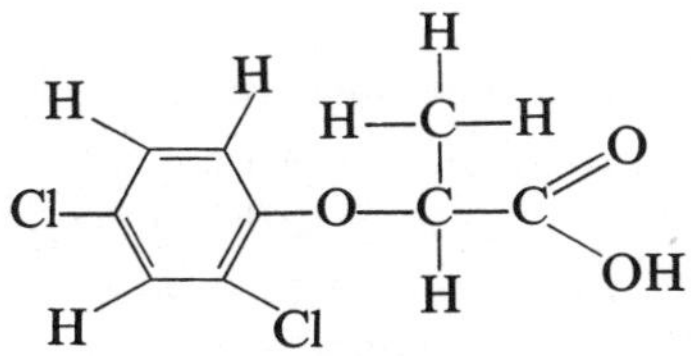

dichlorprop

Dichlorprop is the common name for 2-(2,4-dichlorophenoxy) propionic acid. It has also been referred to as 2,4-DP. It has several trade names. Dichlorprop varies from a white to tan color, and is a crystalline solid with a water solubility of 710 ppm. Although it could be formulated in various salts and esters common to 2,4-D, the available commercial formulations are usually low-volatile esters. The acute oral LD_{50} is 800 mg/kg.

In the United States, the use of dichlorprop is limited to noncrop land. It has been primarily used for woody plant control. It may be used in combination with 2,4-D to control 2,4-D-resistant species.

2,4,5-T

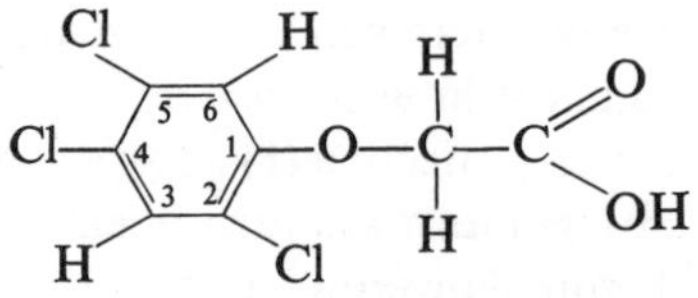

(2,4,5-trichlorophenoxy)acetic acid

2,4,5-T is the common name for (2,4,5-trichlorophenoxy)acetic acid. It has several trade names. It is identical with 2,4-D except for an additional chlorine atom on the number 5 position of the ring. A white solid with a water solubility of 238 ppm, it is usually formulated as the amine salts and esters much the same as 2,4-D. Amine salts of 2,4,5-T are water-soluble and non-volatile. Esters are oil-soluble and usually emulsifiable in water; esters are classified as either volatile or low-volatile.

2,4,5-T is effective on many woody species that are resistant to 2,4-D, but less effective on many other plants. Therefore, it is prudent to consider relative susceptibility and cost in selecting between the two herbicides.

Mixtures of 2,4-D and 2,4,5-T are frequently sold as "brush killers." These are especially effective on a mixture of brush species. 2,4-D may be more effective on some species and 2,4,5-T on others. 2,4,5-T has also been used as a plant-growth regulator for preharvest fruit-drop control in apples.

If temperatures rise too high during synthesis of 2,4,5-T, a dioxin material is formed. This dioxin is 2,3,7,8-tetrachlorodibenzoparadioxin. It is regarded as a potential carcinogenic and teratogenic toxicant. The use of 2,4,5-T (containing as much as 28 ppm of dioxin) as a defoliant during the war in Vietnam aroused differences of opinion over the continued use of 2,4,5-T.

With effective manufacturing control procedures, 2,4,5-T can be manufactured essentially free of dioxin (less than 0.1 ppm). Free of the dioxin, and used according to labeled instructions approved by the Environmental Protection Agency, 2,4,5-T is not considered a human health hazard.

SILVEX

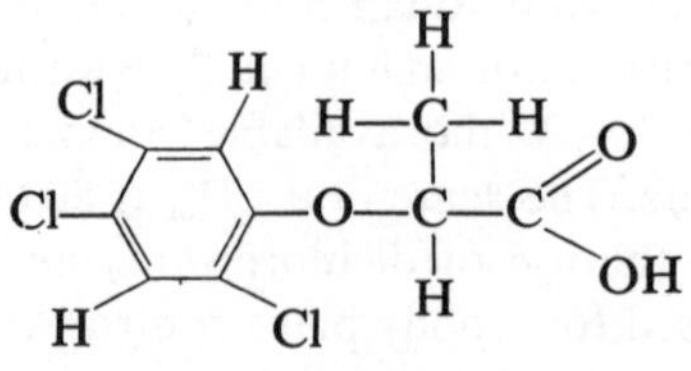

silvex

Silvex is the common name for 2-(2,4,5-trichlorophenoxy)propionic acid. It has also been referred to as 2,4,5-TP. There are several trade names. It is a white solid with a water solubility of about 180 ppm. It is usually formulated into various amine salts and esters similar to 2,4-D. These forms will have similar physical and chemical characteristics to their 2,4-D equivalents; thus, salts of silvex are water-soluble and have very low-volatility characteristics. Esters are oil-soluble and are usually emulsifiable in water; esters also are classed as either volatile or low-volatile.

The chemical controls some plant species that are resistant to both 2,4-D and 2,4,5-T. Some species especially susceptible to silvex are chickweed, henbit, wild strawberry, and a number of the oaks and maples. 2,4,5-TP has shown considerable promise against some aquatic weeds, and it has also been used as a plant-growth regulator for preharvest-drop control in apples and pears.

LITERATURE CITED

1. Barrier, G. E., and W. E. Loomis, *Plant Physiol.*, *32*, 225 (1957).
2. Carrol, R. B., *Contribs. Boyce Thompson Inst.*, *16*, 409 (1952).
3. Crafts, A. S., *Hilgardia*, *26*, 287 (1956).
4. Crafts, A. S., and S. Yamaguchi, *Hilgardia*, *27*, 421 (1958).
5. Crlyle, R. E., and J. D. Thorpe, *J. Amer. Soc. Agron.*, *39*, 929 (1947).
6. Davis, D. E., H. H. Funderburk, Jr., and N. G. Sansing, *Weeds*, *7*, 300 (1959).
7. Dunlap, A. A., *Phytopathology*, *38*, 638 (1948).
8. Hamner, C. L., and H. B. Tukey, *Botan. Gaz.*, *106*, 232 (1944).
9. Hanson, J. B., and F. W. Slife, *Residue Rev.*, *25*, 59 (1969).
10. King, L. J., J. A. Lambrech, and T. P. Finn, *Contribs. Boyce Thompson Inst.*, *16*, 191 (1950).
11. Leonard, O. A., and A. S. Crafts, *Hilgardia*, *26*, 366 (1956).
12. Loos, M. A., "Phenoxyalkanoid Acids," p. 1 in *Degradation of Herbicides*, P. C. Kearney and D. D. Kaufman, Eds., Marcel Dekker, New York, 1969.
13. Marth, P. C., and J. W. Mitchell, *Botan. Gaz.*, *106*, 224 (1944).
14. Mullison, W. R., and R. W. Humner, *Botan. Gaz.*, *111*, 77 (1949).
15. Payne, M. G., and J. L. Fults, *J. Amer. Soc. Agron.*, *39*, 52 (1947).
16. Penner, D., and F. M. Ashton, *Residue Rev.*, *14*, 39 (1966).
17. Peterson, G. E., *Agric. History*, *41*, 244 (1967).
18. Pokorny, R., *J. Amer. Chem. Soc.*, *63*, 1768 (1941).
19. Rodgers, E. G., *Plant Physiol.*, *27*, 153 (1952).
20. Slade, R. E., W. G. Templeman, and W. A. Sexton, *Nature*, *155*, 497 (1945).
21. Swanson, C. P., *Botan. Gaz.*, *107*, 522 (1946).
22. Tukey, H. B., C. L. Hamner, and B. Imkoffe, *Botan. Gaz.*, *107*, 62 (1946).

23. Vlitos, A. J., *Contribs. Boyce Thompson Inst.*, *16*, 435 (1952).
24. Vlitos, A. J., and L. J. King, *Nature, 171*, 523 (1953).
25. Watson, D. P., *Amer. J. Bot.*, *35*, 543 (1948).
26. Yamaguchi, S., and A. S. Crafts, *Hilgardia*, *28*, 161 (1958).
27. Zimmerman, P. W., and A. E. Hitchcock, *Contribs. Boyce Thompson Inst.*, *12*, 321 (1942).
28. Zussman, H. W., *Agr. Chem.*, *IV*, 27–29, 73 (1949).

FOR CHEMICAL USE, SEE THE MANUFACTURER'S LABEL AND FOLLOW THE DIRECTIONS. ALSO SEE THE PREFACE.

17 Triazines

In 1952, J. R. Geigy, S.A., of Basle, Switzerland, started investigations with triazine derivatives as potential herbicides (12). Herbicidal properties of chlorazine were reported in 1955 by Gast et al. (9) and Antognini and Day (1). Since then, numerous triazine derivatives have been synthesized and screened for their herbicidal properties. At least nine triazine herbicides are now being used in the United States and probably more are employed worldwide. Only these nine, however, will be covered here (plus metribuzin with slightly different structure).

Chemically, the triazines are heterocyclic nitrogen derivatives. The name heterocyclic is used to designate a ring structure composed of atoms of different kinds. In this case the ring is composed of nitrogen and carbon atoms. Most triazine herbicides are symmetrical, that is, they have alternating carbon and nitrogen atoms in the ring. However, one exception is metribuzin, which is asymmetrical.

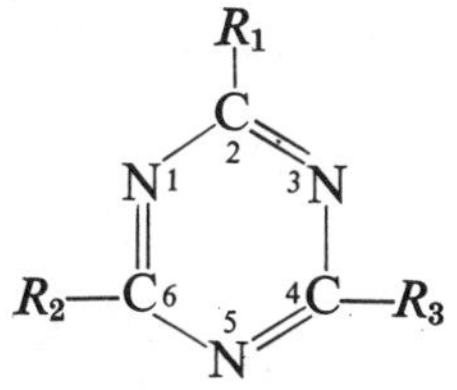

symmetrical triazine 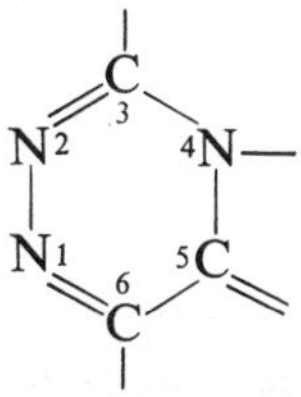   asymmetrical triazine

Structures of symmetrical triazine herbicides are given in Table 17–1. The substitution on the R_1 position of the ring determines the ending of the common name: -*azine* = chlorine atom; -*tryn* = methylthio group ($-SCH_3$); and -*ton* = methoxy group ($-OCH_3$). Water solubility of a series of triazines is markedly influenced by the substitution of R_1: prometon ($-OCH_3$), 750 ppm; prometryn ($-SCH_3$), 48 ppm; and propazine ($-Cl$), 8.6 ppm.

The greatest use of the triazine herbicides has been as a selective herbicide on cropland. Several excellent reviews have been written on various aspects of

Figure 17-1. Weed-free plots were treated with simazine, and applied preemergence to both weeds and corn. The corn was not injured. (Ciba-Geigy Corporation.)

triazine herbicides: their history (12), uses (8), degradation (15), molecular structure or function (11, 12), mode of action (2), and comprehensive (13). Volume 32 of *Residue Reviews* contains 15 chapters dealing with triazine–soil interactions.

AMETRYN

Ametryn is the common name for 2-(ethylamino)-4-(isopropylamino)-6-methylthio-*s*-triazine. The trade name is EVIK®. It is a white, crystalline solid with a water solubility of 185 ppm. It is formulated as a wettable powder. The acute oral LD_{50} is 1110 mg/kg.

Ametryn is a selective herbicide for control of annual grass and broadleaf weeds in pineapple, sugarcane, and bananas. It is effective when applied preemergence to annual weeds. It also has foliar activity, thus can also be applied postemergence to weeds. In pineapple it can be applied immediately after planting or after harvest is completed and before weeds emerge. Further applications may be made at 1- to 2-month intervals prior to differentiation. In sugarcane, ametryn can be applied at planting or after ratooning. Directed

Table 17-1. Common Name, Chemical Name, and Chemical Structure of the Symmetrical Triazine Herbicides

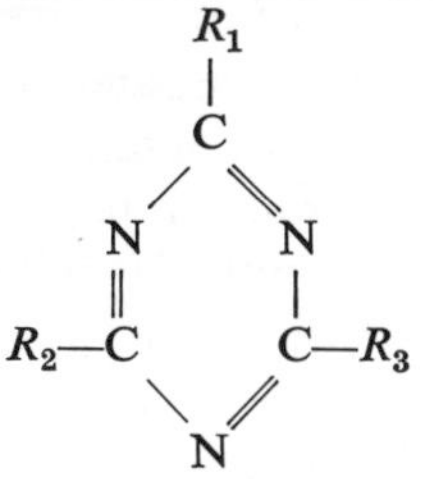

$$R_1$$
$$|$$
$$C$$

$$R_2-C \quad C-R_3$$

Common Name	R_1	R_2	R_3
Ametryn	—SCH$_3$	—NH·iso—C$_3$H$_7$	—NHC$_2$H$_5$
Atrazine	—Cl	—NH·iso—C$_3$H$_7$	—NHC$_2$H$_5$
Cyanazine	—Cl	—NHC$_2$H$_5$	H CH$_3$ | | —N—C—C≡N | CH$_3$
Cyprazine	—Cl	—NH·iso—C$_3$H$_7$	—NH—◁ 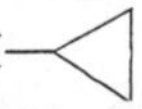
Prometon	—OCH$_3$	—NH·iso—C$_3$H$_7$	—NH·iso C$_3$H$_7$
Prometryn	—SCH$_3$	—NH·iso—C$_3$H$_7$	—NH·iso—C$_3$H$_7$
Propazine	—Cl	—NH·iso—C$_3$H$_7$	—NH·iso—C$_3$H$_7$
Simazine	—Cl	—NHC$_2$H$_5$	—NHC$_2$H$_5$
Terbutryn	—SCH$_3$	—NH·tert—C$_4$H$_9$	—NHC$_2$H$_5$

Note: With a symmetrical molecule, the numbering can start from any one of the nitrogen atoms in the ring.

postemergence application may also be used in Florida. In bananas it is used both preemergence and postemergence to annual weeds.

Ametryn is also used as a directed, postemergence application in corn when the crop is at least 12 in. high. And it can be used as a potato-vine desiccant.

ATRAZINE

Atrazine is the common name for 2-chloro-4-(ethylamino)-6-isopropylamino)-*s*-triazine. The trade name is AAtrex®. It is a white, crystalline solid with a water solubility of 33 ppm. It is formulated as a wettable powder, a flowable liquid suspension, and in a granular form. The LD$_{50}$ of atrazine is 3080 mg/kg.

Atrazine is widely used to control annual grass and broadleaf weeds in corn, macadamia orchards, pineapples, perennial ryegrass, sorghum, and sugarcane. It is effective when applied preemergence to annual weeds. When applied with an emulsifiable oil it controls many emerged annual weeds, but in this case crop selectivity may be reduced. Granular forms of atrazine combined with alachlor, propachlor, or butylate are available for annual weed control in corn.

Atrazine is used in some areas for selective weed control on conifer reforestation, Christmas-tree plantations, and grass-seed fields.

Atrazine is also widely used as a nonselective herbicide on noncrop land. In addition, it is available in granular form in combination with sodium chlorate, sodium metaborate, or both, for this same purpose.

CYANAZINE

Cyanazine is the common name for 2-[[4-chloro-6-(ethylamino)-s-triazin-2-yl]amino]-2-methylpropionitrile. The trade name is Bladex®. It is a white, crystalline solid with a water solubility of 171 ppm. It is formulated both as a wettable powder and in granular form. The acute oral LD_{50} is 334 mg/kg.

Cyanazine controls annual grasses and broadleaf weeds in corn. It is usually applied as a preemergence treatment. Under prolonged dry-soil conditions it should be incorporated into the top 2 in. of soil. Postemergence applications may be made through the four-leaf stage. Cyanazine is a relatively new herbicide and more uses may be developed.

CYPRAZINE

Cyprazine is the common name for 2-chloro-4-(cyclopropylamino)-6-(isopropylamino)-s-triazine. The trade name is Outfox®. It is a white, crystalline solid with a water solubility of 195 ppm. It is formulated as an emulsifiable liquid. The acute oral LD_{50} of cyprazine is about 1200 mg/kg.

Cyprazine is used as a postemergence treatment to control annual grass and broadleaf weeds in corn. It should be applied on rapidly growing, emerged weeds before they are 2 in. high. Cyprazine is a relatively new herbicide and other uses may be developed.

METRIBUZIN

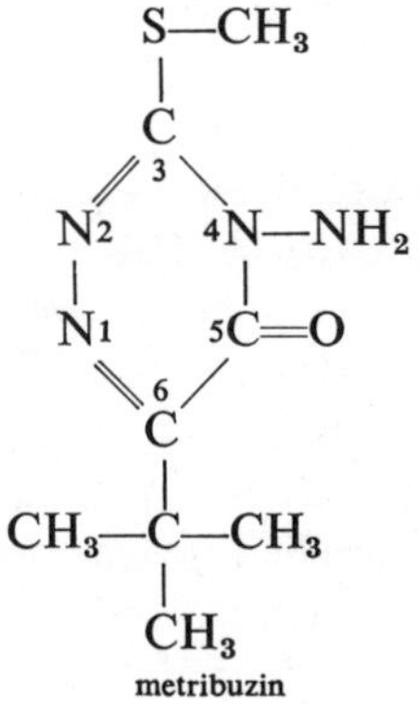

metribuzin

Because metribuzin is an asymmetrical triazine, its structural formula is given here rather than in Table 17-1. Metribuzin is the common name for 4-amino-6-*tert*-butyl-3-(methylthio)-*as*-triazine-5(4*H*)one. It has two trade names— Sencor® and Lexone®. It is a white, crystalline solid with a water solubility of 1200 ppm. It is formulated as a wettable powder and as granules. The acute oral LD_{50} of metribuzin is 1937–1986 mg/kg.

Metribuzin is registered for use in soybeans for control of some annual grasses and certain broadleaf weeds. It is a relatively new herbicide and additional uses may be developed.

PROMETON

Prometon is the common name for 2,4-bis(isopropylamino)-6-methoxy-*s*-triazine. The trade name is Pramitol®. It is a white, crystalline solid with a water solubility of 750 ppm. It is formulated as a liquid that can be applied in water or oil, and also applied as granules in combination with simazine, sodium chlorate, sodium metaborate, as well as with PCP. Prometon has an acute oral LD_{50} of 2980 mg/kg which is slightly toxic.

Prometon is a nonselective preemergence and postemergence herbicide used to control most annual and broadleaf weeds and certain perennial weeds on noncrop land. When prometon is combined with simazine, sodium chlorate, or sodium metaborate, a greater variety of perennial weeds is controlled and the period of weed control lasts longer. When prometon is combined with PCP, diesel oil, fuel oil, or weed oil, foliar contact activity is increased.

PROMETRYN

Prometryn is the common name for 2,4-bis(isopropylamino)-6-(methylthio)-*s*-triazine. The trade name is Caparol®. It is a white, crystalline solid with a water solubility of 48 ppm, and is formulated either as a wettable powder or in liquid form in combination with MSMA. Prometryn has an acute oral LD_{50} of 3750 mg/kg.

Prometryn, a selective herbicide, controls annual grass and broadleaf weeds in celery and cotton. In celery it is used as a postemergence treatment to seedbeds and transplants, and in cotton as a preplant, preemergence, or directed postemergence application. When combined with MSMA in cotton as a directed postemergence application, the mixture controls a wider range of weeds, especially nutsedge.

PROPAZINE

Propazine is the common name for 2-chloro-4,6-bis(isopropylamino)-*s*-triazine. The trade name is Milogard®. It is a colorless, crystalline solid with a water solubility of 8.6 ppm; formulated as a wettable powder. The acute oral LD_{50} of propazine is greater than 5000 mg/kg.

Propazine is widely used to control annual grass and broadleaf weeds in sorghum. It can be applied both before and after planting, but before weeds emerge. Shallow incorporation, not more than 2 in. deep, after application will generally give better weed control, particularly under dry or minimum soil-moisture conditions.

SIMAZINE

Simazine is the common name for 2-chloro-4,6-bis(ethylamino)-s-triazine. The trade name is Princep®. It is a white, crystalline solid with a water solubility of 5 ppm. The herbicide is formulated as a wettable powder and as granules, and its acute oral LD_{50} is greater than 5000 mg/kg.

Simazine was the first widely used triazine herbicide. Its largest crop use was in corn, but atrazine has largely replaced it for this purpose. It is primarily used to control annual grass and broadleaf weeds as a preemergence or preplant soil incorporation treatment; it should be applied before the weeds emerge.

Simazine is registered for use on more crops than any other triazine herbicide. These include alfalfa, artichokes, asparagus, several caneberries, corn, cranberries, currants, pineapples, numerous orchard-tree crops, and sugarcane. It is also used selectively on forage, bermudagrass, many species of woody plants in nurseries, Christmas tree plantings, shelter belts, and certain turf grasses for sod. It is also widely used as a nonselective herbicide on noncrop land.

TERBUTRYN

Terbutryn is the common name for 2-(tert-butylamino)-4-(ethylamino)-6-(methylthio)-s-triazine. The trade name is Igran®. It is a white, crystalline solid with a water solubility of 58 ppm formulated as a wettable powder. Its acute oral LD_{50} is 2400–2980 mg/kg.

Terbutryn is a selective herbicide used to control annual grass and broadleaf weeds in winter wheat and winter barley in Washington, Oregon, and Idaho, and it can also be used in sorghum. It can be applied either before or after the stem emerges. Postemergence treatments should be applied before weed rosettes reach 3 in. in diameter or 4 in. in height.

SOIL INFLUENCE ON THE TRIAZINES

Triazine herbicides are reversibly adsorbed by clay and organic colloids. They are not subject to excessive leaching in most soil types. In a study of five

Table 17-2. Relative Persistences of Several Triazine Herbicides in Soils

Persistence Order	Buchanan and Rogers, 1963 (5)		Sheets et al., 1962 (19)	Sheets and Shaw, 1963 (20)		Switzer and Rauser, 1960 (21)
	Laboratory	Field	Greenhouse	Greenhouse[1]	Greenhouse[2]	Field
Persistence decreasing →	Atraton	Atraton	Simeton	Atraton	Atraton	Prometon
	Simazine	Ipazine	Simazine	Prometon	Prometon	Simazine
	Atrazine	Simazine	Atrazine	Simeton	Simeton	Atrazine
	Ipazine	Atrazine	Propazine	Simazine	Ametryn	Propazine
	—	—	Ipazine	Atrazine	Prometryn	Ipazine
	—	—	Trietazine	Simetryn	Propazine	Trietazine
	—	—	Chlorazine	Ametryn	Atrazine	Chlorazine
	—	—	—	Propazine	Simetryn	—
	—	—	—	Prometryn	Simazine	

[1] Bosket sandy loam.
[2] Cecil sandy loam.

triazine herbicides on 25 soil types, adsorption almost always increased in the following order: propazine > atrazine > simazine > prometon > prometryn (22). Correlation analysis indicated that adsorption of the methylthio-(prometryn) and methoxy-(prometon) triazines was more highly related to clay content, whereas adsorption of chloro-triazines (simazine, atrazine, propazine) was more highly related to organic matter. The following relative leachability in Lakeland fine sand has been reported: atraton > propazine > atrazine > simazine > ipazine > ametryn > prometryn (17).

Note that the order of triazine herbicides common to these two studies is the same, indicating that the leachability of triazine herbicides is directly related to their adsorption to soil colloids. These two studies also indicate that adsorption and leachability have little or no relationship to water solubility of the compounds. A reduction in phytotoxicity of triazine herbicides is associated with increasing amounts of clay and organic matter in soil (23).

Triazine herbicides are relatively persistent in most soils. (See Table 4-1.) Some are persistent enough to injure sensitive plants in the season after application in some soils and under some environmental conditions. However, in noncrop land, prolonged persistence is very desirable. Table 17-2 shows relative persistence of several triazine herbicides in soils (18). Methoxytriazines are generally more persistent than methylthio- or chloro-triazines (18). Prometon is one of our most persistent herbicides.

A monograph has been published (10) containing several review papers on the interactions of triazine herbicides and soil.

MODE OF ACTION

The triazine herbicides inhibit plant growth, but this is considered to be a secondary effect caused by an inhibition of photosynthesis (3). At herbicidal concentration, triazine herbicides cause foliar chlorosis followed by death of the leaf. Other leaf effects include loss of membrane integrity and chloroplast destruction. At sublethal levels, however, increased greening of leaves may occur.

Triazine herbicides are absorbed by leaves, but translocation from them is essentially nil (3). The amount of foliar absorption varies for various compounds. Propazine and simazine are poorly absorbed by leaves, while ametryn and prometryn are readily absorbed. The others appear to be intermediate. All triazine herbidices are rapidly absorbed by roots and readily translocated throughout the plant by the transpiration stream. They are considered to be translocated almost exclusively in the apoplast system.

The rate of degradation of triazine herbicides in higher plants varies greatly with different species (3). In resistant species they are rapidly degraded,

Figure 17-2. Radioactive simazine, simazine-degradation products, or both. The pictures show marginal accumulation in (a) susceptible cucumber, (b) localized distribution in moderately susceptible cotton, and (c) general distribution in resistant corn (6).

Figure 17-3. Leaf chlorosis induced by the triazine herbicides. Upper: chlorotriazines (e.g., simazine) almost always show interveinal chlorosis. Lower: methylthiotriazines (e.g., ametryne) usually show veinal chlorosis. (C. L. Elmore, University of California, Davis.)

while in susceptible species, the herbicides degraded slowly; thus, the rate of degradation appears to be the primary basis of selectivity. This process occurs by hydroxylation, dechlorination, demethoxylation, or demethy-thiolation, depending on the parent substitution. Dealkylation of the alkyl side chains also occurs.

The mechanism of action of triazine herbicides involves a severe inhibition of the Hill reaction of photosynthesis (7, 16). Total herbicidal effect, however, must be more complex than this because the plants do not merely starve to death. It has been postulated that the action involves the interaction of light, chlorophyll, and triazine to produce a secondary phytotoxic substance (4).

LITERATURE CITED

1. Antognini, J., and B. E. Day, *Proceedings of the Eighth Southern Weed Conference,* 1955, pp. 92–98.
2. Ashton, F. M., *Proceedings of the Eighteenth Southern Weed Control Conference,* 1965, p. 596.
3. Ashton, F. M., and A. S. Crafts, *Mode of Action of Herbicides,* Wiley, New York, 1973.
4. Ashton, F. M., E. M. Gifford, and T. Bisalputra, *Bot. Gaz., 124,* 329 (1963).
5. Buchanan, G. A., and E. G. Rogers, *Proceedings of the Southern Weed Conference, 16,* 393 (1963).
6. Davis, D. E., H. H. Funderburk, Jr., and N. G. Sansing, *Weeds, 7,* 300 (1959).
7. Gast, A., *Experientia, 13,* 134 (1958).
8. Gast, A., *Residue Rev., 32,* 11 (1970).
9. Gast, A., E. Knüsli, and H. Gysin, *Experientia, 11,* 107 (1955).
10. Gunther, F. A., *Residue Rev., 32,* 413 (1970).
11. Gysin, H., *Weeds, 4,* 541 (1960).
12. Gysin, H., *Weed Sci., 22, 523 (1974)*
13. Gysin H., and E. Knüsli, *Adv. Pest Control Res.* 1960, p. 289.
14. Knüsli, E., *Residue Rev., 32,* 1 (1970).
15. Knüsli, E., D. Berrer, G. Dupuis, and H. Esser, *"s-Triazines,"* in *Degradation of Herbicides,* P. C. Kearney and D. D. Kaufman, Eds., Marcel Dekker, New York, 1969, p. 51.
16. Moreland, D. E., W. A. Gentner, J. L. Hilton, and K. L. Hill, *Plant Physiol., 34,* 432 (1959).
17. Rodgers, E. G., *Weed Sci., 16,* 117 (1968).
18. Sheets, T. J., *Residue Rev., 32,* 287 (1970).
19. Sheets, T. J., A. S. Crafts, and H. R. Drever, *J. Agr. Food Chem., 10,* 458 (1962).
20. Sheets, T. J., and W. C. Shaw, *Weeds, 11,* 15 (1963).
21. Switzer, C. M., and W. E. Rauser, *Proceedings of the Northeastern Weed Control Conference, 14,* 329 (1960).
22. Talbert, R. E., and O. H. Fletchall, *Weeds, 13,* 46 (1965).
23. Weber, J. B., *Residue Rev., 32,* 93 (1970).

SEE THE MANUFACTURER'S LABEL FOR CHEMICAL USE, AND FOLLOW THE DIRECTIONS. ALSO SEE THE PREFACE.

18 Ureas and Uracils

Substituted urea and uracil-type herbicides are included in the same chapter because they have several similar uses and their modes of action have many features in common. These herbicides have been arranged alphabetically, and the substituted ureas discussed first. The pioneering development of both classes of herbicides was carried out by E. I. duPont de Nemours and Company.

UREAS

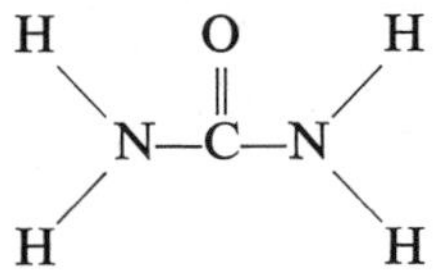

$$
\begin{array}{ccc}
H & O & H \\
\diagdown & \| & \diagup \\
& N-C-N & \\
\diagup & & \diagdown \\
H & & H
\end{array}
$$

urea

Urea is a common nitrogen fertilizer. By substituting three of the hydrogen atoms of urea with other chemical groups, effective herbicides are produced. Common groups include a phenyl group, a methyl group, and a methoxy group. The phenyl group often has one or two chlorine substitutions and occasionally a bromine substitution (see Table 18-1).

Most urea herbicides are relatively nonselective at high rates of usage, and they are usually applied to the soil. Certain ones, however, have foliar activity, which may be increased by the addition of surfactants. At low rates of application, they may be selective. Selectivity is obtained by both depth protection provided by little or no leaching characteristics of the herbicide, and differences in plant tolerance provided by inherent tolerance. The various substituted ureas are alphabetically arranged by their common names.

Buturon

Buturon is the common name for 3-(p-chlorophenyl)-1-methyl-1-(1-methyl-2-propynyl)urea. The trade name is Etapur®. It is a white solid with a water

238

solubility of 30 ppm; formulated as a wettable powder. The acute oral LD_{50} is 3000–5800 mg/kg.

Buturon is a pre- and postemergence herbicide and has been suggested for use in cereals and corn. It is not used in the United States.

Chlorbromuron

Chlorbromuron is the common name for 3-(4-bromo-3-chlorophenyl)-1-methoxy-1-methylurea. The trade names are Maloran® and Bromex®. It is a white, crystalline solid having a water solubility of 50 ppm; and formulated as a wettable powder. The acute oral LD_{50} is 2150 mg/kg.

Chlorbromuron is primarily used on soybeans and potatoes as a pre-emergence treatment to control many annual grass and broadleaf weeds. A wider range of annual weeds may be controlled by applying chlorbromuron in combination with alachlor as a tank mix for use in soybeans.

Chloroxuron

Chloroxuron is the common name for 3-[p-(p-chlorophenoxy)phenyl]-1,1-dimethylurea. The trade names are Tenoran® and Norex®. It is a white, crystalline solid with a water solubility of 2.7 ppm. It is formulated as a wettable powder. Its acute oral LD_{50} is 3700 mg/kg.

Chloroxuron is used to control many annual grasses and broadleaf weeds in carrots, celery, onions, soybeans, and strawberries. It is most effective on emerged, young weeds after the cotyledons have opened, but before they have reached a height of 2 in. Chloroxuron is usually applied after the crop has emerged or the transplants (strawberries and celery) are established.

Diuron

Diuron is the common name for 3-(3,4-dichlorophenyl)-1,1-dimethylurea. Trade names are Karmex® and Marmer®. It is a white, crystalline solid with a water solubility of 42 ppm. It is formulated as a wettable powder and as a flowable liquid suspension. Diuron has an acute oral LD_{50} of 3400 mg/kg.

Herbicidal uses of diuron are numerous on both crops and noncrop land, and it is combined with other herbicides. Diuron is used primarily to control annual grass and broadleaf weeds before emergence in at least 19 crops: alfalfa, artichokes, asparagus, barley, bermudagrass pastures, birdsfoot trefoil, blueberries, caneberries, gooseberries, corn, cotton, grapes, grass-seed

Table 18-1. Common Name, Chemical Name and Chemical Structure of the Substituted Urea Herbicides

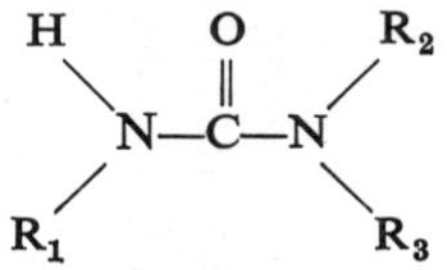

Common Name	Chemical Name	R_1	R_2	R_3
Buturon	3-(p-chlorophenyl)-1-methyl-1-(1-methyl-2-propynyl)urea	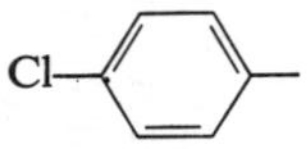	CH_3-	CH_3-CH- $C\equiv CH$
Chlorbromuron	3-(4-bromo-3-chlorophenyl)-1-methoxy-1-methylurea	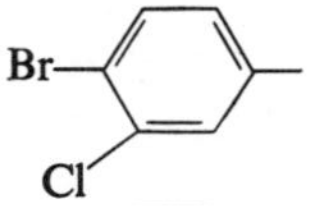	CH_3-	CH_3O-
Chloroxuron	3-[p-(p-chlorophenoxy)phenyl]-1,1-dimethylurea	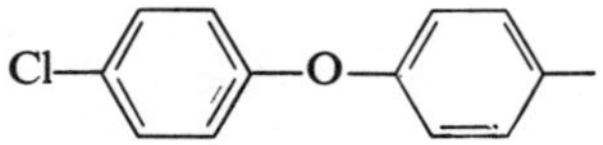	CH_3-	CH_3-
Diuron[1]	3-(3,4-dichlorophenyl)-1,1-dimethylurea	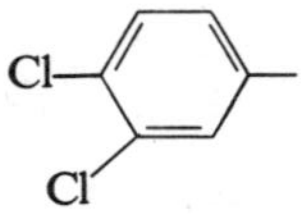	CH_3-	CH_3-
Fenuron	1,1-dimethyl-3-phenylurea		CH_3-	CH_3-
Fluometuron	1,1-dimethyl-3-(α,α,α-trifluoro-m-tolyl)urea	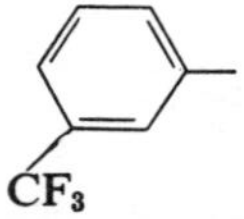	CH_3-	CH_3-

Name	Structure		
Karbutilate	*tert*-butylcarbamic acid ester with 3-(*m*-hydroxyphenyl)-1,1-dimethylurea	CH_3—	CH_3—
Linuron	3-(3,4-dichlorophenyl)-1-methoxy-1-methylurea	CH_3—	CH_3O—
Metobromuron	3-(*p*-bromophenyl)-1-methoxy-1-methylurea	CH_3—	CH_3O—
Monolinuron	3-(*p*-chlorophenyl)-1-methoxy-1-methylurea	CH_3—	CH_3O—
Monuron[1]	3-(*p*-chlorophenyl)-1,1-dimethylurea	CH_3—	CH_3—
Neburon	1-butyl-3-(3,4-dichlorophenyl)-1-methylurea	CH_3—	C_4H_9—
Norea	3-(hexahydro-4,7-methanoindan-5-yl)-1,1-dimethylurea	CH_3—	CH_3—
Siduron	1-(2-methylcyclohexyl)-3-phenylurea		H
Tebuthiuron	*N*-[5-(1,1-dimethylethyl)-1,3,4-thiadiazol-2-yl]-*N,N′*-dimethylurea	CH_3—	CH_3—

[1] Monuron and diuron were previously designated as CMU and DCMU, respectively; the photosynthetic biologists continue to use the older terminology.

crops, peppermint, pineapple, sorghum, sugarcane, winter wheat, and several fruit- and nut-tree crops. For these selective uses the amount of diuron applied is relatively low, usually 1–4 lb/acre. The rate varies for different crops and different soil types.

Diuron may also be used nonselectively on noncrop land for total vegetation control. For this use the rates are 4–16 lb/acre. When perennial weeds are to be controlled with diuron, rates as high as 16 to 40 lb/acre may be required.

Diuron alone has very little foliar activity on most plants. However, by adding certain surfactants to the spray solution, considerable foliar toxicity is obtained. In this way, emerged annual weeds as well as germinating seedlings may be controlled in several crops by use of directed sprays.

Diuron is also combined with other herbicides to control a wider variety of weeds. Some of these combinations are: (1) preplant soil-incorporation treatment of trifluralin followed by a preemergence application of diuron in cotton, (2) diuron and surfactant in combination with DSMA for postemergence weed control in Western irrigated cotton, (3) diuron plus bromoxynil in winter wheat in Washington, Oregon, and Idaho, and (4) diuron plus bromacil in citrus and noncrop land.

Fenuron

Fenuron is the common name for 1,1-dimethyl-3-phenylurea. The trademark name is Dybar®. It is a white, crystalline powder with water solubility of about 3850 ppm, and usually formulated as a pellet. The acute oral LD_{50} of fenuron is about 6400 mg/kg.

Because of its low adsorption by soils and high water solubility fenuron is readily leached. It is primarily used to control woody plants. Pellets are scattered on the soil at the base and within the dripline of the plant to be killed; rainfall then carries the herbicide into the rooting zone.

Fenuron-TCA is a combination used for woody plant control as well as for complete vegetation control on noncrop land. It has a molecule of TCA "loosely" bound to the fenuron molecule. The trade name is Urab®.

Fluometuron

Fluometuron is the common name for 1,1-dimethyl-3-(α,α,α-trifluoro-m-tolyl)urea. The trade names are Cotoran® and Lanex®. It is a white, crystalline solid with a water solubility of 90 ppm. It is formulated as a wettable powder, and the acute oral LD_{50} is in the range of 7900–8900 mg/kg.

Fluometuron is used to control annual grass and broadleaf weeds in cotton and sugarcane. It can be applied preemergence to the crop or as a directed spray after the crop and weeds have emerged. Sequential application may also be used, a preemergence application followed by one or two postemergence-directed applications. DSMA or MSMA may be tank-mixed with fluometuron for a directed postemergence application in cotton to broaden the variety of the weeds controlled.

Karbutilate

Karbutilate is the common name for *tert*-butylcarbamic acid ester with 3-(*m*-hydroxyphenyl)-1,1-dimethylurea. It is apparent from the chemical name and the molecular structure that karbutilate could be considered either a carbamate- or urea-type herbicide, but because its herbicidal properties are more similar to the urea herbicides, it is included here. The trade name is Tandex®. Karbutilate is a white, crystalline solid with a water solubility of 325 ppm. It is formulated as a wettable powder, as a flowable liquid suspension, and in granular form. The acute oral LD_{50} is about 3000 mg/kg.

Karbutilate is a soil-applied, nonselective herbicide used to control annual and perennial broadleaf weeds and grasses, brush, and vines on noncrop land.

Linuron

Linuron is the common name for 3-(3,4-dichlorophenyl)-1-methoxy-1-methylurea. The trade name is Lorox®. It is a white, crystalline solid with a water solubility of 75 ppm; formulated as a wettable powder. The acute oral LD_{50} of linuron is about 1500 mg/kg.

Linuron is applied to soil to control germinating annual seedlings, but has limited contact effect when applied to foliage. Best results from foliar applications are obtained when the weeds are young and succulent, temperatures are 70°F or higher, and humidity is high.

As a preemergence herbicide it is used in carrots, corn, parsnips, potatoes, sorghum, soybeans, and winter wheat; postemergence, it is used on carrots, corn, cotton, soybeans, and winter wheat. Postemergence applications in corn, cotton, and soybeans should be *directed* to minimize the amount of herbicide received by crop plants. Linuron is usually combined with another herbicide as a tank mix when used preemergence in corn (atrazine or propachlor) or sorghum (propazine or propachlor).

Metobromuron

Metobromuron is the common name for 3-(p-bromophenyl)-1-methoxy-1-methylurea. The trade name is Patoran®. It is a colorless, crystalline solid with a water solubility of 330 ppm, formulated as a wettable powder. The acute oral LD_{50} is 3000 mg/kg.

Metobromuron may be used as a preemergence herbicide to control many annual grass and broadleaf weeds in potatoes.

Monolinuron

Monolinuron is the common name for 3-(p-chlorophenyl)-1-methoxy-1-methylurea. The trade name is Aresin®. It is a white, crystalline solid with a water solubility of 735 ppm, and it is formulated as a wettable powder. The acute oral LD_{50} is 2250 mg/kg.

Monolinuron is effective both preemergence and postemergence. It has been used in asparagus, beans, caneberries, cereals, grapes, potatoes and certain other crops; however, it has not been registered for use in the United States.

Monuron

Monuron is the common name for 3-(p-chlorophenyl)-1,1-dimethylurea. The trade name is Telvar®. It is a white, crystalline solid with a water solubility of 230 ppm. It is formulated as a wettable powder. The acute oral LD_{50} of monuron is about 3600 mg/kg.

The herbicidal potential of monuron was first described by Bucha and Todd (3) in 1951. It was the first urea-type herbicide widely used. Monuron has been used to control annual weeds in certain crops but at the present time it is only used on noncrop land.

Monuron-TCA is used as a nonselective herbicide on noncrop land. It has a molecule of TCA "loosely" bound to the monuron molecule. The trade name is Urox®.

Neburon

Neburon is the common name for 1-butyl-3-(3,4-dichlorophenyl)-1-methyl-urea. The trade name is Kloben®. It is a white, crystalline solid with a water solubility of 4.8 ppm; formulated as a wettable powder. Its acute oral LD_{50} is greater than 11,000 mg/kg.

Neburon is a specialty herbicide used to control annual weeds in nursery plantings of certain ornamentals and dichondra.

Norea

Norea is the common name for 3-(hexahydro-4,7-methanoindan-5-yl)-1,1-dimethylurea. The trade name is Herban®. It is a white, crystalline solid with a water solubility of 150 ppm; formulated as a wettable powder. The acute oral LD_{50} ranges from 1476 to 6830 mg/kg.

Norea has been used to control annual weeds in cotton, sorghum, soybeans, spinach, and sugarcane.

Siduron

Siduron is the common name for 1-(2-methylcyclohexyl)-3-phenylurea. The trade name is Tupersan®. It is a white, crystalline solid with a water solubility of 18 ppm. It is formulated as a wettable powder. A variety of granular formulations including fertilizers, insecticides, or both, is available from formulators. The acute oral LD_{50} of siduron is greater than 5000 mg/kg.

Siduron is a speciality herbicide used almost exclusively to control certain annual grasses in turf (most turf grasses are tolerant even when germinating from seed) and is particularly effective on smooth and hairy crabgrass, foxtail, and barnyardgrass. It will not control annual bluegrass, clover, or most broadleaf weeds. Although most turf grasses are resistant to injury, a limited number of bentgrass strains and bermudagrass turfs may be injured by siduron.

Tebuthiuron

Tebuthiuron is the common name for N-[5-(1,1-dimethylethyl)-1,3,4-thiadiazol-2-yl]-N,N'-dimethylurea. The trade name is Spike® and Perflan®. Pure, it is a white, odorless crystalline powder, formulated both as a wettable powder and as pellets. It is light-stable with essentially no volatility.

When tebuthiuron is fed to rats, rabbits, dogs, mallard ducks, and fish, the chemical is rapidly absorbed, metabolized, and excreted through the kidneys.

Feeding 1000 ppm of tebuthiuron to both the rat and dog for 3 months caused no effect.

Herbicidal uses of tebuthiuron include total vegetation control at high rates of application, and at relatively low rates of usage for selective weed control in sugar cane, for woody plant control in pastures and ranges, and removal of certain undesirable species in reforestation programs.

At rates required for total vegetation control the chemical may persist for more than 1 year, providing effective year-around weed control on railroads and industrial sites. The chemical kills many woody species; therefore, *application to the rooting area of desirable trees and shrubbry must be avoided.*

SOIL INTERACTION OF UREA HERBICIDES

As a class, urea-type herbicides are relatively persistent in soils. Under favorable moisture and temperature conditions with little or no leaching, most of them can be expected to persist 3–6 months at selective rates and 24 months or more at higher nonselective rates. However, norea is less persistent than other urea herbicides, about 1 month; and siduron and linuron intermediate, about 3 and 4 months, respectively. (See Table 4-1.)

Principle factors affecting persistence of substituted ureas in the soil are microorganism decomposition, leaching, adsorption on soil colloids, and photo-decomposition. The latter is important only when the herbicide remains on the soil surface for an extended period of time. Researchers believe volatility and chemical decomposition are of minor importance.

Microorganism decomposition is probably the most important factor. Bacteria such as Pseudomonas, Xanthomonas, Sarcina, and Bacillus, and fungi such as Penicillium and Aspergillus can use some substituted ureas as a direct source of energy (4). Conditions such as moderate moisture and temperature with adequate aeration, favoring such organisms, would also favor decomposition. Therefore, under dry, cold, or very wet soil conditions (poor aeration), the chemicals would normally persist for a long time.

The adsorptive forces between the chemical and the soil colloids directly affect the chemical's rate of leaching; its solubility is a less-important factor.

In a study using four urea-type herbicides, leachability was correlated with adsorption and water solubility (1, 5). (See Table 18-2.) Fenuron was leached the most, followed in order by monuron, diuron, and neburon. Comparing these four herbicides, fenuron would be used to kill deep-rooted perennials, whereas if you wanted very little leaching, then neburon would be preferable.

As to be expected from the above facts, neburon is less toxic to many plants, giving less weed kill per unit of chemical than diuron and monuron. Therefore, you might increase weed control by using diuron or monuron

Table 18-2. Water Solubility and Adsorption on Soil of Urea and Uracil-Type Herbicides (1, 5)

Compound	Solubility in Water[1] (ppm)	Adsorption on Keyport Silt Loam[2]
Fenuron	3850	0.3
Bromacil	815	1.5
Terbacil	710	1.7
Monuron	230	2.6
Diuron	42	4.0–5.2
Neburon	4.8	16.0

[1] At 25°C.
[2] ppm (active ingredient) present on soil in equilibrium with 1 ppm in soil solution at about 22°C

where slight leaching is permissible. Plants absorb the ureas mainly through the roots. The chemical must reach the rooting zone to be effective.

Substituted ureas applied to a dry soil surface, without enough rain to leach the chemical into the soil, will give poor weed control. Under such conditions, you will likely find little or no crop injury, even to susceptible crops.

If you apply fenuron or monuron as a preemergence treatment on a sandy soil (low-adsorptive capacity) and considerable rain falls soon after planting, the crop may be seriously injured. This is especially true if the crop has only moderate tolerance to the herbicide. For this reason, fenuron is seldom, if ever, recommended for preemergence treatment, and monuron is less favored than diuron.

MODE OF ACTION OF UREAS

Phytotoxic symptoms of urea-type herbicides are primarily observable in the leaves. They may show merely a slight chlorosis which develops slowly with low-rate applications of the herbicide or a water-soaked appearance that becomes necrotic in a few days at higher rates (2).

Most of the urea herbicides are readily absorbed by roots and rapidly translocated to upper plant parts via the apoplastic system. Applications to the leaves are also translocated apoplastically with little if any being translocated from the treated leaf. However, the actual amount absorbed and translocated from roots to shoot varies greatly with various compounds. Furthermore, differences in absorption and translocation between species

Figure 18-1.   Diuron induced chlorosis in peaches. Usually veinal chlorosis but sometimes interveinal. Left to right: untreated to increasing rates. (C. L. Elmore, University of California, Davis.)

have been sufficient with different urea herbicides to allow their selective use in certain crops (2).

Inhibition of the Hill reaction (basic to photosynthesis) is generally acknowledged to indicate the primary site of action of the urea herbicides. However, many workers do not believe that this explains the light-dependent phytotoxic symptoms of the urea herbicides. In other words, plants do not merely starve from lack of photosynthate; rather, it has been postulated that a secondary phytotoxic substance is formed in the oxygen-liberating pathway of photosynthesis. The nature of this secondary phytotoxic substance has not been determined (2) (see Figure 18-1).

Urea herbicides are subject to degradation by higher plants, but a given herbicide may be degraded at different rates by various species. This is the basis of certain selective uses of some urea herbicides. Demethylation is the primary detoxification mechanism for the urea herbicides. Demethoxylation also occurs in those compounds containing a methoxy group. Following dealkylation, dealkoxylation, or both, the molecule may be subject to hydrolysis. This would involve a deaminization and decarboxylation yielding the corresponding aniline. This aniline may be subject to oxidation to yield the corresponding nitrite or to conjugation with normal cellular constituents (2).

URACILS

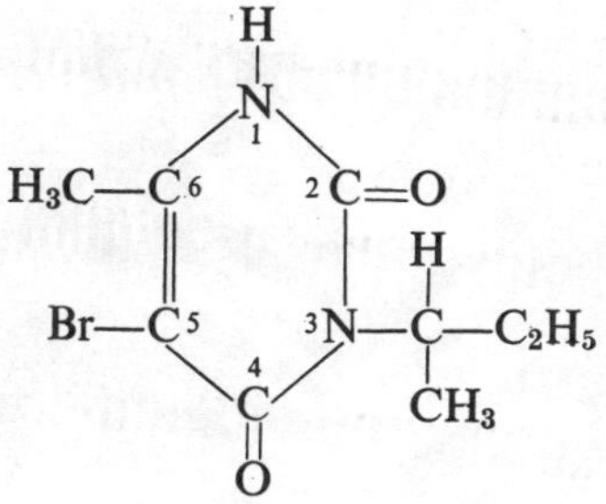

bromacil 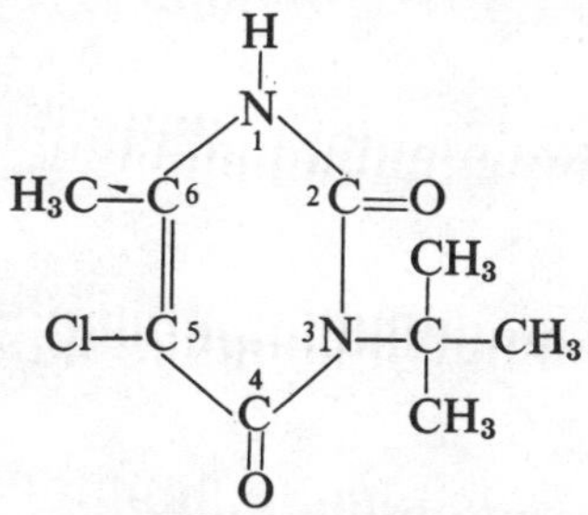   terbacil

Bromacil and terbacil are the two uracil-type herbicides used in the United States. Their chemical differentiation lies in the fact that bromacil has a bromine atom in position 5 and a *sec*-butyl group in position 3. However, terbacil has a chlorine atom in position 5 and a *tert*-butyl group in the position 3. They are both applied to the soil. Lenacil is another uracil used primarily in Europe for weed control in sugar beets and strawberries.

Bromacil

Bromacil is the common name for 5-bromo-3-*sec*-butyl-6-methyluracil. The trade name is Hyvar®. It is a white, crystalline solid with a water solubility of 815 ppm. It is formulated as a wettable powder (Hyvar X®), a pellet (Hyvar X-P®), and as the lithium salt in a water-soluble liquid form (Hyvar X-L®). The acute oral LD_{50} is 5200 mg/kg.

Bromacil is used for selective weed control in citrus and pineapple crops, and for general vegetation control on noncrop land. Many annual grass and broadleaf weeds are controlled at rates of 3–5 lb/acre. Some perennial weeds and several brush species are controlled with 6–10 lb/acre, while johnsongrass and other somewhat resistant perennial weeds may require 12–24 lb/acre. These higher rates are not selective to citrus or pineapple (see Figure 18-2).

Terbacil

Terbacil is the common name for 3-*tert*-butyl-5-chloro-6-methyluracil. The trade name is Sinbar®. It is a white, crystalline solid with a water solubility of 710 ppm. It is formulated as a wettable powder. The acute oral LD_{50} is greater than 5000 mg/kg, but less than 7500 mg/kg.

Figure 18-2.   Interveinal chlorosis in walnuts induced by bromacil. (C. L. Elmore, University of California, Davis.)

Terbacil is used at relatively low rates to control many annual grass and broadleaf weeds in apples, citrus, mint, peaches, and sugarcane. At higher rates, 5–8 lb/acre, bermudagrass and certain other perennial weeds can be controlled in citrus trees 2 or more years old.

SOIL INTERACTIONS OF URACILS

Bromacil and terbacil are adsorbed less on soil colloids than the urea-type herbicides monuron, diuron, or neburon, but more tightly than fenuron (see Table 18-2). Therefore, they are leached more readily than monuron, diuron, or neburon, but less readily than fenuron.

Bromacil and terbacil have a half-life of about 5–6 months when applied at 4 lb/acre, but at sterilant rates they persist for more than 1 season (see Table 4-1). This loss is apparently a result of microbiological degradation, because volatilization and photo-decomposition losses are negligible and leachability is limited. Soil diphtheroids, Pseudomonas and Penicillium species, have been shown to be able to degrade bromacil (1).

MODE OF ACTIONS OF URACILS

The uracil-type herbicides are similar to the urea-type herbicides in their mode of action (2). They are readily absorbed by roots and translocated apoplastically to the leaves, and block photosynthesis via the Hill reaction. These herbicides cause chlorosis and necrosis in leaves, and bromacil has been shown to inhibit root growth. The structure of leaf chloroplasts is grossly altered and cell-wall development of roots is modified by bromacil.

LITERATURE CITED

1. *Herbicide Handbook WSSA*, 1970.
2. Ashton, F. M., and A. S. Crafts, *Mode of Action of Herbicides*, Wiley, New York, 1973.
3. Bucha, H. C., and C. W. Todd, *Science*, *114*, 493 (1951).
4. Hill, G. D., and J. W. McGalen, *Proceedings of the Southern Weed Conference*, *8*, 284 (1955).
5. Wolf, D. E., R. S. Johnson, G. D. Hill, and R. W. Varner, *Proceedings of the North Central Weed Conference*, *15*, 7 (1958).

FOR CHEMICAL USE, SEE THE MANUFACTURER'S LABEL AND FOLLOW THE DIRECTIONS. ALSO SEE THE PREFACE.

19   Other Organic Herbicides

This chapter covers those organic herbicides that do not fall into any chemical class previously discussed, and by themselves do not provide sufficient information to justify a separate chapter.

AMITROLE

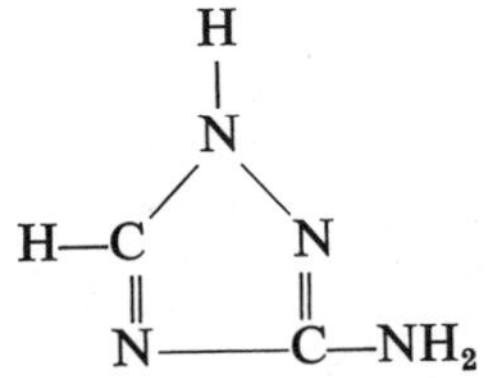

$$
\begin{array}{c}
H \\
| \\
N \\
H-C \qquad N \\
\parallel \qquad \parallel \\
N----C-NH_2
\end{array}
$$

amitrole

Amitrole is the common name for 3-amino-*s*-triazole. The trade names are Weedazol® and Amino Triazole Weed Killer 90®. It is a white, crystalline solid with a water solubility of about 28%. It is formulated as a water-soluble powder. It is also formulated with ammonium thiocyanate as a liquid (Amitrole-T®, and Cytrol®) as well as with simazine as powder (Amizine®). The acute oral LD_{50} is 24,600 mg/kg. However, 2-year, lifelong rat studies at high feeding levels indicate that amitrole is a goitrogen (enlarges the thyroid gland).

Uses

Amitrole is applied as a foliar spray to control essentially all emerged annual weeds and several perennial weeds on noncrop land. Rate of application is usually ½–1 oz/gal with thorough wetting of foliage. Perennial weeds may require the treatment of regrowth to gain adequate control. With certain perennial weeds, the time of year when the initial application is made is critical.

Addition of ammonium thiocyanate to the formulation appears to increase its activity when applied to foliage. When simazine is applied with amitrole, amitrole controls emerged annual plants and simazine controls annual plants that germinate later.

Mode of Action

The most striking symptom of amitrole phytotoxicity is the albino appearance of leaves and shoots that develop after application. Amitrole is one of our most readily translocated herbicides. It is translocated in both the symplastic and apoplastic systems, and is therefore classified as a systemic herbicide.

Although amitrole is rapidly degraded in soil (less than a 1-month persistence), it appears to be considerably more stable in plants. In plants it forms conjugates with sugars and amino acids which then may be degraded to yield free amitrole. Symptoms of amitrole may be observed in perennial plants a year or more after application.

Amitrole reportedly blocks development of chloroplast ribosomes (3). Such a blockage would interfere with formation of chloroplast membranes, enzymes and pigments. Accumulation of chloroplast DNA is also inhibited by amitrole (2). These effects on chloroplasts would explain the chlorosis one observes following amitrole treatments. However, it is possible the inhibition of histidine biosynthesis also contributes to the phytotoxicity of amitrole in higher plants (49, 50).

BENSULIDE

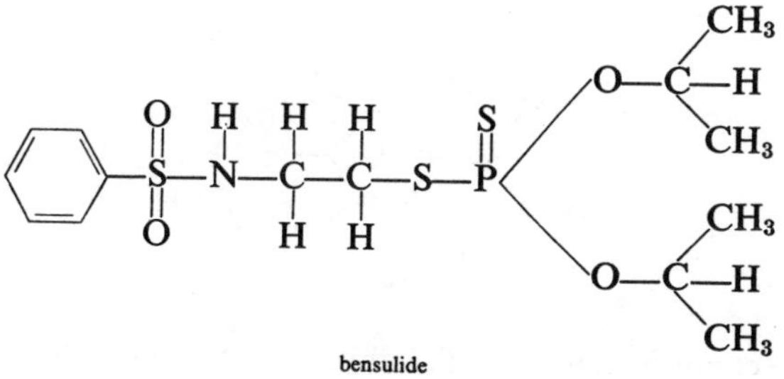

bensulide

Bensulide is the common name for *O,O*-diisopropyl phosphorodithioate *S*-ester with *N*-(2-mercaptoethyl)benzenesulfonamide. The trade names are Betasan® when the product is to be used on turf, ornamentals, or ground covers, and Prefar® when used in crops. Bensulide has a relatively low melting point (34.4°C) and may therefore be a liquid or crystalline solid when pure. It has a water solubility of 25 ppm, and it is formulated as an emulsifiable concentrate and in granular form. The acute oral LD_{50} is 770 mg/kg.

Uses

Bensulide controls several annual grasses including crabgrass, annual blue-grass, and goosegrass, as well as certain broadleaf weeds in established grass and dichondra lawns. It can also be used to control weeds in many established flowers, ornamentals and ground covers. Bensulide can also be used in carrots, cole crops, cotton, lettuce, cucurbits, onions, peppers, tomatoes, and grass-seed crops. It is applied as a preemergence or preplant, soil-incorpora-tion treatment.

Mode of Action

Bensulide inhibits the growth of roots and partially inhibits cell division (12). It is adsorbed on root surfaces and a small amount is absorbed by the root. However little, if any, is translocated upwards to the leaves. It appears to be degraded by higher plants.

Bensulide is relatively persistent in soils, for 8–12 months. Therefore, it is suggested that crops that are not listed above should not be planted until 18 months after the last application. Bensulide is inactivated in soils con-taining high amounts of organic matter. The leaching of bensulide is restricted in all soil types.

DCPA

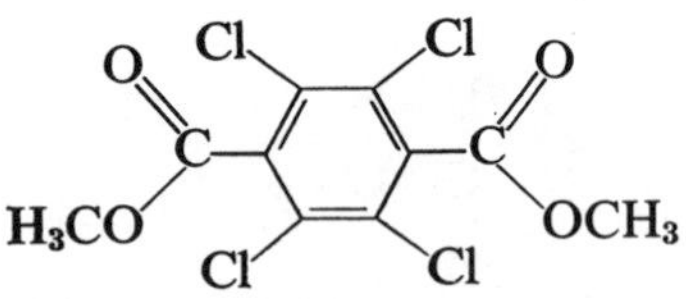

DCPA

DCPA is the common name for dimethyl tetrachloroterephthalate. The trade name is Dacthal®. It is a white, crystalline solid with a water solubility of 0.5 ppm. It is formulated as a wettable powder and in granular form. The acute oral LD_{50} is greater than 3000 mg/kg.

Uses

DCPA is applied as a preemergence or preplant, soil-incorporation treatment to control most annual grasses and many broadleaf weeds in beans, cole crops, collards, corn, cotton, cucurbits, eggplant, garlic, kale, lettuce, mustard greens, onions, peppers, potatoes, southern peas, soybeans, strawberries, sweet potatoes, tomatoes, turnips, and yams. It is also used in grass turf, nursery stock, and established ornamentals.

Mode of Action

DCPA is absorbed by roots, but it is thought not to be absorbed by leaves. Two studies reported that DCPA is absorbed by the coleoptiles of grass seedlings (44) and is readily absorbed by the hypocotyl of cucumber and translocated into the foliage (35). It is translocated from roots slightly, if at all. DCPA appears to be a general growth inhibitor and especially inhibits germinating seeds and roots. It does not appear to be metabolized by higher plants.

When used to control dodder in alfalfa, DCPA prevented the parasite from attaching itself to alfalfa (4).

This herbicide is adsorbed to organic matter in the soil and thus, is not subject to leaching. It is slowly degraded in soils by microorganisms and chemical hydrolysis, persisting for 4 to 6 months in most soils.

ENDOTHALL

endothall

Endothall is the common name for 7-oxabicyclo[2,2,1]heptane-2,3-dicarboxylic acid. There are several trade names. It is a white, crystalline solid with a water solubility of about 10%, formulated as water-soluble liquids and in granular forms. Various endothall salts are available such as disodium, dipotassium, or amine. Endothall is also formulated in combination with other herbicides such as silvex, TCA, or ammonium sulfate. The acute oral LD_{50} of sodium or amine salts ranges from 182 to 206 mg/kg.

Uses

Endothall may be applied preemergence or preplant, soil-incorporated or postemergence to small, emerged, annual weeds in sugar beets. It is also used as a preharvest desiccant for several seed crops including alfalfa, clover, . soybeans, trefoil, and vetch. In cotton it is used as a preharvest defoliant. Endothall also controls aquatic weeds in irrigation and drainage canals, lakes, and ponds.

Mode of Action

Endothall is absorbed readily by leaves and roots. It is translocated to a limited extent from roots to shoots of plants via the xylem, but it is not phloem-mobile and thus not translocated from leaves to other plant parts. Its action appears to be contact in nature, causing rapid desiccation to germinating seedlings, browning of the foliage, or both.

One advantage of endothall over certain other aquatic herbicides is its low toxicity to fish (46, 51).

Endothall is subject to considerable leaching in soils. It is rapidly degraded in both soil and water.

FENAC

fenac

Fenac is the common name for (2,3,6-trichlorophenyl)acetic acid. The trade name is Fenac®; however, when combined with other herbicides, it has several trade names. It is a white, crystalline solid, only slightly soluble in water. As the sodium salt, fenac is formulated as an aqueous solution, and it is also formulated as a granule. In combination with other herbicides such as 2,4-D, amitrole, bromacil, and atrazine, various formulations are available. Fenac has an acute oral LD_{50} of 1780 mg/kg.

Figure 19-1.   Modification of leaf structure of cotton induced by fenac. (W. B. McHenry, University of California, Davis.)

Uses

Fenac is used as a preemergence herbicide in sugarcane to control many annual weeds, especially johnsongrass seedlings. It also controls aquatic weeds. On noncrop areas such as roadsides, fence rows, drainage ditchbanks, and industrial sites, it is often combined with one or more herbicides listed in the preceeding paragraph to control a wider variety of weeds. Fenac is not a general vegetation-control herbicide; it works primarily through root systems of specific plants (see Figure 19-1).

Mode of Action

Fenac has growth-regulating properties causing epinasty, bud inhibition, and bud necrosis. However, unlike auxin-type herbicides, its translocation appears to be primarily restricted to the apoplast; therefore, it is usually applied to the soil and absorbed by the roots.

Fenac is strongly adsorbed by soil colloids and resists leaching. It is degraded slowly in soil, remaining phytotoxic from 1 to 2 years.

MH OR MALEIC HYDRAZIDE

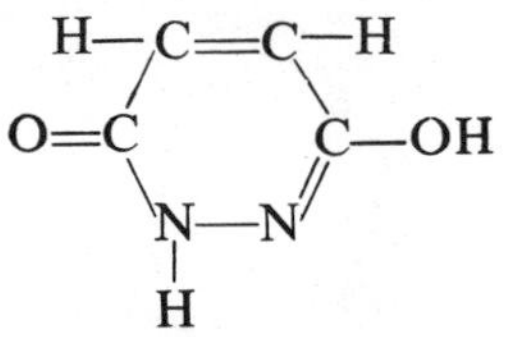

$$H-C=C-H$$
$$O=C \qquad C-OH$$
$$N-N$$
$$H$$

maleic hydrazide

MH, maleic hydrazide, is the common name for 1,2-dihydro-3,6-pyridazine-dione. There are several trade names. It is a white solid, very soluble in water—about 6000 ppm. It is usually formulated as a water-soluble solution in the diethanolamine or potassium salt form. The acute oral LD_{50} of the sodium salts is 6950 mg/kg.

Uses

MH is basically a growth inhibitor rather than a herbicide. It has been used to slow down growth of trees, shrubbery, and grass, thereby reducing pruning and cutting costs, to inhibit sucker development on tobacco, and induce dormancy in citrus. When applied before harvest, it will also inhibit sprouting of onions and potatoes in storage.

MH has also been used as a herbicide to control quackgrass, curly dock, wild onions, and wild garlic.

Mode of Action

MH is absorbed slowly through both the upper and lower surfaces of the leaf. It penetrates faster when the leaf cells are turgid under conditions of high humidity. Wilted plants absorb almost no MH. Spraying is more effective in late afternoon and evening than early morning. This may be caused by differences in humidity, since the relative humidity increases at this time and usually remains at a high percentage throughout the night.

Under favorable conditions a rain-free period of 6 hr was necessary for tobacco to absorb MH, and 24 hr for potatoes and quackgrass (41). However, clipping the tops of bermudagrass 2–4 days after spraying completely removed the effects of MH. With an 8-day interval, top growth was inhibited for 3 months, but the plants recovered growth in 5 months. With a 16-day interval, inhibition lasted for 5 months (9). Quackgrass 4–8 in. tall treated with MH was controlled best when thoroughly tilled 4–8 days after treatment (6).

MH is truly systemic, being readily translocated in both the symplast and apoplast.

MH inhibits cell division in the meristematic regions, but not cell enlargement (8). Therefore, the active meristematic region is replaced by abnormally large cells in a state of maturity (18).

MH persists in the soil a very short time, less than 1 month. It has been suggested that it is broken down by microorganisms.

DIPHENYL ETHERS

fluorodifen   nitrofen

Fluorodifen

Fluorodifen is the common name for p-nitrophenyl α,α,α-trifluoro-2-nitro-p-tolyl ether. The trade name is Preforan®. It is a yellow, crystalline solid, soluble in water at less than 2.0 ppm. It is formulated as an emulsifiable concentrate, wettable powder, or granules. The acute oral LD_{50} of fluorodifen is about 15,000 mg/kg.

Fluorodifen is applied as a preemergence herbicide to control annual weeds in soybeans, southern peas, and several green and dry beans. In the future, other selective uses may be found.

Nitrofen

Nitrofen is the common name for 2,4-dichlorophenyl-p-nitrophenyl ether. The trade name is TOK®. It is a dark-brown solid and essentially insoluble in water. It is formulated as an emulsifiable concentrate, wettable powder, or granules. The acute oral LD_{50} is about 2630 mg/kg.

As a preemergence herbicide nitrofen controls annual weeds in cole crops, carrots, celery, horseradish, onions, and parsley, as well as sugar beets in the Imperial Valley of California. In all these crops except sugar beets, it may also be applied postemergence, posttransplant, or both, to the crop to control emerged annual weeds and those that may emerge later.

Mode of Action

These two diphenyl ethers, fluorodifen and nitrofen, appear to cause loss of membrane integrity. However, their action when applied to foliage may be different than their action when applied to soil. When applied to foliage, the selectivity of nitrofen seems to be primarily related to differential wetting and absorption (24, 36).

The translocation of these herbicides is quite limited, with somewhat more transport from roots to shoots than vice versa. Translocation appears to be primarily restricted to the apoplast (13, 14, 17, 24, 36, 47).

The degradation of these compounds in higher plants is relatively rapid. Ester linkage is cleaved and ring substitutions may be modified. Conjugates of parent molecule, degradation products, or both, have been detected (14, 15, 16, 25). They are also rapidly degraded in soil.

PICLORAM

picloram

Picloram is the common name for 4-amino-3,5,6-trichloropicolinic acid. The trade name is Tordon®. It is a white solid with a water solubility of 430 ppm. It is available in three basic forms: potassium salt, triisopropanolamine salt, and isooctyl ester.

1. The potassium salt is formulated as a water-soluble liquid and as a pellet. A special water-soluble formulation of the potassium salt is available for mixing with certain water-soluble 2,4-D and 2,4,5-T amine formulations. A potassium salt-disodium tetraborate bead formulation is also made.

2. The triisopropanolamine form is used in combination with a similar 2,4-D salt as liquid formulations.

3. The isooctyl ester form is used in combination with 2,4,5-T esters as a liquid formulation.

The acute oral LD_{50} of picloram is about 8200 mg/kg.

Uses

Picloram is effective on most perennial, broadleaf, herbacious weeds and many woody species; however, most grasses are resistant. Although we think

Figure 19-2. Picloram induced injury to wheat. Left: untreated. Right: treated. (C. L. Elmore, University of California, Davis.)

of picloram in terms of perennial broadleaf or woody plant control, it will also control many annual weeds. Rates as low as $\frac{1}{50}$ lb/acre will control many annual broadleaf weeds. On some species it is 10 times as potent as 2,4-D. When applied to the foliage of woody plants and broadleaf perennial weeds as a liquid spray, rates from $\frac{1}{2}$ to $1\frac{1}{2}$lb/acre are used. When applied to thick stands of brush as pellets, rates of 6–8 lb/acre may be needed. In combination with 2,4-D or 2,4,5-T, it may also be applied as bark or cut-surface treatment to control woody plants. (See Chapter 27, and Figure 19–2.)

Mode of Action

Picloram has high phytotoxicity, it is easily absorbed by roots and foliage, truly systemic, and it is degraded slowly. It is absorbed by leaves, stems, and roots and is translocated both in the symplast and apoplast. Toxicity symptoms of picloram are quite similar to 2,4-D and other auxin-type herbicides—epinasty, cuplike leaves, and tissue proliferation. The mechanism of action is probably associated with modification of nucleic acid metabolism and certain species may be resistant because of their high levels of nucleases (32).

Picloram is adsorbed by soil colloids and the degree of leaching is inversely correlated to this adsorption and water-holding capacity of the soil. It is most easily leached through sandy, montmorillonitic soils low in organic matter

In contrast, it is leached with greatest difficulty through soils high in organic matter and lateritic clay soils. Picloram is very persistent in soils, but subject to slow degradation by microorganisms. Phytotoxicity may often be detected well over 1 year after application.

PYRAZON

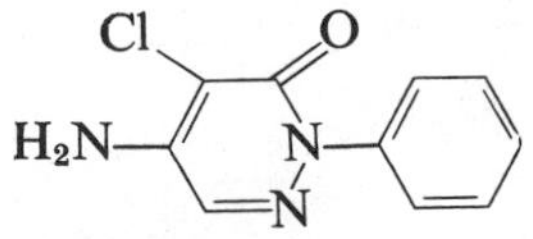

pyrazon

Pyrazon is the common name for 5-amino-4-chloro-2-phenyl-3(2H)-pyridazinone. The trade name is Pyramin®. It is a tan-to-brown solid with a water solubility of 30 ppm, formulated as a wettable powder and as granules. The chemical is also combined with dalapon as a wettable powder, Pyramin® Plus. Pyrazon has an acute oral LD$_{50}$ of 3000 mg/kg.

Uses

Pyrazon is used for annual, broadleaf-weed control in sugar beets and red beets as a preemergence or preplant, soil-incorporation application. It is also used as a postemergence treatment in combination with dalapon or phenmedipham for control of annual grasses and broadleaf weeds in sugar beets. These postemergence treatments should be applied after beets have at least two true leaves, but before weeds reach the four-leaf stage.

Mode of Action

Pyrazon induces wilting, chlorosis, and necrosis in leaves as well as inhibiting growth in susceptible species (20, 38). Microscopic studies have shown that chloroplast structure is also altered (1). Although pyrazon is absorbed by leaves, it is not translocated from them to a significant degree. It is readily absorbed by roots and distributed throughout the plant; thus, pyrazon is primarily translocated in the apoplast. Although pyrazon is degraded in plants, the primary basis of the resistance of sugar beets appears to be associated with the conjugation of pyrazon with glucose to form a nontoxic molecule. Pyrazon has been shown to inhibit photosynthesis (19, 26).

Soil type seems to have considerable influence on effectiveness and selectivity of pyrazon. Soil applications are not recommended on sands or loamy sands because of leaching and possible crop injury. On the other hand, adsorption on soils containing more than 5% organic matter precludes adequate weed control. Pyrazon is degraded fairly rapidly in warm, moist soils, persisting from 1 to 3 months.

PETROLEUM OILS

Petroleum oils are hydrocarbons. Oils have gained commercial use as weed killers mostly since about 1940. They act as contact herbicides and may be used either as selective or as general weed killers. As selective weed killers, the varsol or Stoddard solvent-type oils are used to control both grass and broadleaf weeds in carrots, celery, parsnips, parsley, and in certain conifer seedlings. As a directed spray they effectively control weeds in cotton and woody ornamental plantings. General or nonselective oils are used to kill all plant life. The oils have been popular on railway roadbeds, on canal banks, roadsides, barnyards, and in orchards where nontillage is practiced.

Most oils used as herbicides are by-products of petroleum, but some are produced during destructive distillation of other organic materials such as coal, wood, and peat. Certain specific chemical products appear to be the same regardless of the source. In general, products obtained by destructive distillation are highly unsaturated hydrocarbons. These products, highly important to the chemical industry, are very toxic to plants but are too costly to be widely used as herbicides.

Size and structure of the oil molecule influences toxicity of oils to plants. Range in size and structure of molecules is partially reflected by the number of saturated and unsaturated compounds, boiling point (closely related to volatility and flash point), viscosity, and specific gravity. Most oils wet vegetation more rapidly and more thoroughly than water, because of their low surface tension.

Classification

Petroleum products can be classified into *saturated* and *unsaturated* hydrocarbons.

Saturated Hydrocarbons

Saturated hydrocarbons have *single valence bonds between the carbon atoms*. Chemically, they are relatively inert: at cool temperatures they resist the action of most acids and bases; they are affected only slowly by bromine, chlorine, and iodine; and they are oxidized only slowly in the presence of air.

Saturated hydrocarbons are the main components of low-octane gasolines, kerosene, and lubricating oils.

Saturated hydrocarbons can be further divided into straight-chain or the normal paraffin hydrocarbons, branched-chain paraffins, and ring or cyclic saturated hydrocarbons also known as naphthenes and as cyclo-paraffins.

Straight-Chain—Saturated   Branched-Chain—Saturated

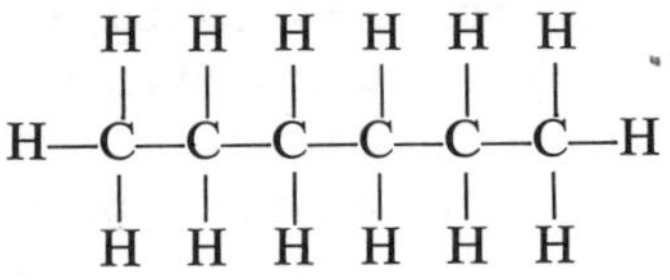

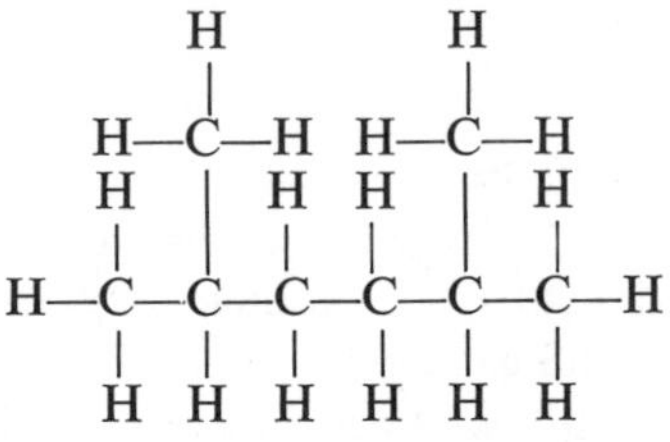

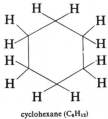

n-hexane (C_6H_{14})   2,5-dimethyl hexane (C_8H_{18})

Ring Structure—Saturated

cyclohexane (C_6H_{12})

These structural formulas should help to distinguish the three types of saturated hydrocarbons. Note that the chemical names end in "ane," and also that the chemist uses the six-sided figure to indicate a ring structure with six carbons arranged in a ring. In structural formulas, the carbon atoms in each corner of the ring are not shown.

Unsaturated Hydrocarbons

Unsaturated hydrocarbons (*deficient in hydrogen*) have *double* or *triple valence bonds between the carbon atoms*. Each molecule may have from one to many such bonds. Unsaturated hydrocarbons can be further classified as olefins (ethylene series) which include both straight-chain and branched-chain unsaturated hydrocarbons, and ring or cyclic unsaturated hydrocarbons, also referred to as aromatic hydrocarbons. The basic structure of this group is the unsaturated benzene ring. Two or more rings may condense; thus, the naphthalene molecule may be represented as two benzene rings joined together, as illustrated. Naphthalene is a solid commonly used as "moth balls."

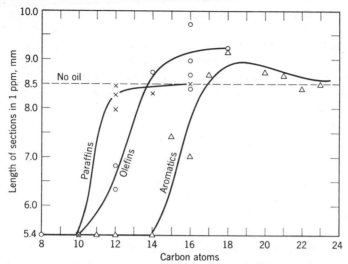

Figure 19-3. The effect of a 2-min exposure of various hydrocarbons on the growth of corn coleoptiles. The smaller the molecule, the more toxic the oil was to the plants. Also, as a class, the aromatics were most toxic, the olefins intermediate, and the paraffins least toxic to the plant. (J. van Overbeek and R. Blondeau, Shell Development Company, Modesto, Calif.)

Among olefins the multiple-valence bonds of this group are easily broken to unite with other substances such as sulfur dioxide, sulfuric acid, nitric acid, bromine, and chlorine. Olefin compounds are easily oxidized; they can also be reduced to the saturated hydrocarbons by the catalytic addition of hydrogen. For some unknown reason the aromatics exhibit this characteristic only partially (see Figure 19-3).

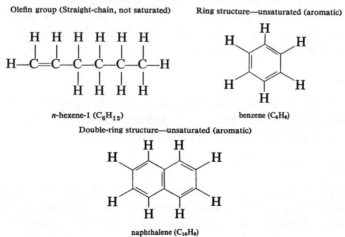

The aromatic molecule appears to be more toxic to plants than the olefin; and the saturated, straight-chain (paraffin) is the least toxic of the three (23). The aromatic oil content of a given product is expressed as a percentage based on weight or volume. Unsaturation may be indicated by the percentage of "maximum sulfuric acid absorption," or as the "unsulfonated residue."

The aromatic or ring molecule may combine with a paraffin molecule to produce a structure like this:

ethylbenzene

As one or more side chains are added to the aromatic group, toxicity to plant tissue increases; it increases until the molecular weight of the side chain equals that of the aromatic portion. Above this point the molecule becomes more like its paraffinic component (5).

Forms Sold on Market

Stoddard Solvents or Varsol

Stoddard solvents or varsol, petroleum spirits, and mineral spirits are petroleum distillates of low flammability often used for dry-cleaning clothes or as paint thinners. They are normally clear and smell like refined naphtha. Most Stoddard solvents have a gravity API of 42° or above, a minimum flash point of 100°F, and an aromatic content of 8–20%.

These products are valuable for selective weed control in vegetable crops such as carrots, parsnips, celery, and parsley (42). They are also used effectively in conifer-tree nurseries.

Gasoline

Gasoline kills some weeds but it is not used as a herbicide because of its fire and explosion hazard. Gasoline may kill susceptible plants very rapidly. Its effects disappear quickly if the plant is not killed shortly after application.

Kerosene

Kerosene, if highly refined, contains a low percentage (less than 4%) of unsaturated compounds, with a high percentage of straight-chain compounds. Its toxicity to plants therefore is usually low. Most kerosenes have a boiling point range between 350 and 500°F, a gravity API of 38–48°, and a flash point of 120–180°F. With less refinement, the percentage of unsaturated and ring-structure compounds increases. Toxicity increases accordingly. Kerosene is used as a diluent or a solvent in formulation of oil-soluble herbicides.

Fuel Oil

Fuel oil is sold by grades, depending on the degree of refining, flash point, and distillation temperatures. Number 1 is the most highly refined; it is a relatively volatile, light oil, generally with a low content of unsaturated compounds. It is similar to, and may be the same as, kerosene. Number 6 is usually the lowest grade, having considerable quantities of impurities, a high viscosity, and usually many unsaturated and ring-structure compounds. Usually the lower the grade, the greater the contact toxicity to plants. Stove oils, a term most commonly used in western states, are high-grade fuel oils.

Diesel Oils

Diesel oils are usually moderately heavy, low-grade oils designed for burning in diesel motors. The gravity API commonly ranges from 35 to 43° and the flash point from 150 to 180°F. The product appears to be quite variable, with the aromatic and unsaturated content differing with the source and degree of refinement. In general, diesel fuel oils are too phytotoxic (toxic to plants) to use for selective weed control. They are effective as general contact herbicides and are often made even more toxic by fortification with PCP or Dinoseb.

Factors Affecting Toxicity

Storage

Some oils apparently increase in toxicity with storage (10). This effect is largely a result of acid formation (27). Therefore, any oil that has been stored for any appreciable time should be tested on a small area before using it for field-scale weed control.

Boiling Point and Range of Distillation

The boiling point of oil is related to molecular weight, volatility, and toxicity to plants. Crude petroleum oils are composed of fractions with differing boiling points. Therefore, an oil may have 20% of its total weight distilled at at 350°F, 50% at 450°F, and 90% at 550°F. If the oil has a very low boiling point, being volatile, it may evaporate before it affects the plant (22). High-boiling materials are viscous liquids or solids with molecules so large that they cannot penetrate the tissues (5). In general, hydrocarbons within the boiling range of 150–275°C (302–527°F) are most toxic to plants (22).

Flash Point

Flash point is of little importance in determining herbicidal toxicity, but it is especially important in determining explosion and fire hazards. Flash point is the temperature at which an oil vaporizes so rapidly that it forms an ignitable mixture with air in a container of given dimensions. Flash point is determined by exposing the vapors periodically to a flame. When the temperature is hot enough a "flash" results, but the temperature is not high enough to continue burning.

Most contact oils have flash points above 180°F; most selective oils (Stoddard solvents) have flash points above 100°F.

Viscosity

Viscosity describes the flowing quality of an oil: it is the resistance of the fluid to motion. Light oils flow more rapidly, or pump more easily than heavy oils. Viscosity of oils is usually determined experimentally as the time required for a given volume of the oil at a given temperature to drain through a specified opening. Viscosity influences the rate at which an oil will spread over and penetrate into a plant.

Specific Gravity

Specific gravity indicates the density or weight of an oil. Specific gravity of oil is of special importance when used in connection with aquatic-weed control. The oil may be directly toxic to aquatic weeds, or it may act as a carrier for some other herbicide. Whether the oil-chemical rises to the surface, remains suspended in the water, or sinks to the bottom, can be controlled by specific gravity. Specific gravity of most oils lies between 0.73 and 0.95 as compared with 1 for water. High-molecular-weight aromatics are usually heavier than water. Also, the specific gravity of aromatics can be increased through such chemical reactions as halogenation, nitration, and methylation.

Specific gravity is measured by a hydrometer; it operates on the same principle as the gauge generally used at filling stations to test antifreeze in auto radiators or to check the charge of the battery. The petroleum industry uses a Baumé scale (hydrometer) that has been adopted by the American Petroleum Institute (API). Therefore the "gravity API" will usually be given. The heavier the oil, the lower the degree reading. With other characteristics constant, the lower the reading, the higher the percentage of aromatics, and, therefore, the greater the toxicity.

General contact herbicidal oils usually have a gravity API maximum of 32° (heavy); selective oils have a gravity API of approximately 42° (light) (5).

How Oils Affect the Plant

Oils vary considerably in their toxicity to plants. As with most herbicides, the basic cause for herbicidal action of oils is only partially understood, although the effects of such an action are much better understood.

Plants sprayed with oil usually first show a darkening of the youngest leaf tips, presumably a result of a leakage of cell sap into intercellular spaces. This gives the plant a water-soaked appearance. There is loss of turgor and drooping of stems and leaves. The plants have an odor of macerated tissue or freshly mown hay (11).

Theory of Oil Action

Scientists have depicted the cell membrane or plasma membrane as a double layer of lipoids (fat-like substances) held in place by one layer of protein on each side.

Other researchers have proposed that once oil reaches the inside of the leaf, it solubilizes the lipoids of the cell membrane. This process makes the semipermeable membrane more permeable, cell sap leaks into intercellular spaces, and the cell collapses (45) (see Figure 19-4). Because of cell-sap leakage, the plant appears water soaked.

Selective Action of Oils

The plasma membrane of some plants seems to resist the solubilizing action of Stoddard solvent types of oils. Thus carrots, celery, parsnips, parsley, and some conifer tree seedlings are not seriously injured by treatments that kill most annual weeds.

Penetration into Plants

Oils may penetrate into plants through the stomates, thin cuticle, epidermis, bark, and even through injured roots. Plant leaves with a heavy cuticle and a small number of stomates permit little penetration of the oil. This explains why some desert plants with their heavy cuticle and few stomates absorb oils slowly. All oils exhibit low interfacial tension with plant cuticle and, if viscosity permits, the oil rapidly spreads over the waxy surface.

Penetration through stomates has been demonstrated on citrus leaves (29, 43). Oil may also be absorbed directly through the cuticle (21, 28, 29).

Stomatal penetration can be demonstrated readily. If an emulsion of a light oil is sprayed on young plants in light when the stomates are open, the plant is killed. However, when the same emulsion is applied during the night when the stomates are closed, the plant is not harmed (45).

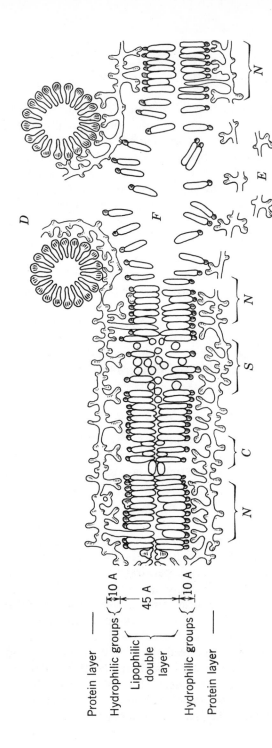

Figure 19-4. The plasma membrane as affected by various toxic molecules. N—Normal membrane consists of a double layer of fatty molecules stabilized by protein layers. Cell sap is kept inside the cell by fatty portion of the plasma membrane. C—Fatty molecules being pushed apart by polycyclic hydrocarbons. The large molecules penetrate slowly. S—Xylene solubilized into the fatty layer; penetration is rapid. D—Detergent micells are pulling away the protein layer, rendering the fatty layer F unstable. E—Disruption of the protein layer is brought about by agents which liquify the protein. Solubilization of the plasma membrane causes leaks in the membrane, permitting the cell sap to leak into the intercellular spaces. [J. van Overbeek and R. Blondeau, Shell Development Company, Modesto, Calif. (45).]

Oils penetrate the cuticle, especially if it is thin. However, unfortified oils almost never penetrate through a tough, continuous, waxy cuticle such as the upper-leaf surface of the apricot (45).

Oils wet plant surfaces readily and tend to spread as a thin film. Oils containing a dye were placed on 1-year-old apple twigs. Lower-viscosity oils easily penetrated the bark and the oil was slowly translocated through the twigs. The rate of penetration is related to viscosity of the oil (21) and to its surface tension (30). Light oils have lower viscosity and lower surface tension; they enter plant tissue more readily than heavy oils.

Translocation

After penetrating the leaf, the oil moves into the intercellular spaces (28, 40). Although oils move from one part of the plant to other parts, scientists still do not fully know how they move. Most research workers believe that oils move principally through the intercellular spaces (33, 39, 52, 53) with little or no movement through the vascular system. Intercellular translocation greatly lessens the amount, rate, and distance of movement. Oil may move in any direction—up, down, radial, or tangential.

Scientists have studied movement of kerosene-like oils in dandelions, carrots, and parsnips. Oil applied to cut roots moved up into the leaves, and oil applied to leaves moved down to the roots. Movement was confined to intercellular spaces. In a large, turgid dandelion root oil moved 4–5 cm (2 in.)/hr (33). When oils were applied to dandelion tops, respiration rates increased in the roots (37). Kerosene sprayed on bluegrass at proper rates will effectively remove dandelions; the dandelion is killed and the bluegrass is injured little or not at all (31).

LITERATURE CITED

1. Anderson, J. L., and J. P. Schaelling, *Weed Sci.*, *18*, 455 (1970).
2. Bartels, P. G., and A. Hyde, *Plant Physiol.*, *46*, 825 (1970).
3. Bartels, P. G., K. Matsuda, A. Siegel, and T. E. Weier, *Plant Physiol.*, *42*, 736 (1967).
4. Bayer, D. E., E. C. Hoffman, and C. L. Foy, *Weeds*, *13*, 92 (1965).
5. Bell, J. M., and W. L. Norem, *Agr. Chem.*, *5*, 31 (1950).
6. Buchholtz, K. P., *Proceedings of the North Central Weed Control Conference*, *11*, 62 (1954).
7. Busbey, R. L., *U.S.D.A.*, *Bur. Entomol. Plant Quarantine*, E610. 1944.
8. Compton, W., *Torrey Bot. Club Bull.*, *79*, 205 (1952).
9. Crafts, A. S., H. B. Currier, and H. R. Drever, *Hilgardia*, *27*, 723 (1958).
10. Crafts, A. S., and H. G. Reiber, *Hilgardia*, *18*, 77 (1948).
11. Currier, H. B., *Hilgardia*, *20*, 383 (1951).

12. Cutter, E. G., F. M. Ashton, and D. Huffstutter, *Weed Res.*, *8*, 346 (1968).

13. Eastin, E. F., *Plant Physiol.*, *44*, 1397 (1969).

14. Eastin, E. F., *Weed Sci.*, *19*, 261 (1971).

15. Eastin, E. F., *Weed Res.*, *11*, 63 (1971).

16. Eastin, E. F., *Weed Res.*, *11*, 120 (1971).

17. Ebner, L., D. H. Green, and P. Pande, *Proceedings of the Ninth British Weed Control Conference*, 1968.

18. Esau, K., *Hilgardia*, *27*, 15 (1957).

19. Eshel, Y., *Weed Res.*, *10*, 196 (1969).

20. Frank, R., and C. M. Switzer, *Weed Sci.*, *17*, 344 (1969).

21. Ginsbury, J. M., *J. Agr. Res.* *43*, 469 (1931).

22. Havis, J. R., *Cornell Agricultural Experimental Station Memoria No. 298*, 1950.

23. Havis, J. R., *Proc., Amer. Soc. Hort. Sci.*, *51*, 545 (1948).

24. Hawton, D., and E. H. Stobbe, *Weed Sci.*, *19*, 42 (1971).

25. Hawton, D., and E. H. Stobbe, *Weed Sci.*, *19*, 555 (1971).

26. Hilton, J. L., A. L. Scharen, J. B. St. John, D. E. Moreland, and K. H. Norris, *Weed Sci.*, *17*, 541 (1969).

27. Johnson, C. M., and W. M. Hoskins, *Plant Physiol.*, *25*, 507 (1952).

28. Kendall, J. C., *New Hampshire Agricultural Experiment Station Bulletin*, 262, 1932.

29. Knight, H., J. C. Chamberlin, and C. D. Samuels, *Plant Physiol.*, *4*, 299 (1929).

30. Knight, H., and C. R. Cleveland, *J. Econ. Entomol.*, *27*, 269 (1934).

31. Loomis, W. E., *J. Agr. Res.*, *56*, 855 (1938).

32. Molhotra, S. S., and J. B. Hanson, *Weed Sci.*, *18*, 1 (1970).

33. Minshall, W. H., and V. A. Helson, *Proceedings of the Northeastern States Weed Control Conference*, 1949, p. 8.

34. Montheith, J., Jr., and A. E. Rabbitt, *Turf Culture*, *1*, 63 (1939).

35. Nishimoto, R. K., and G. F. Warren, *Weed Sci.*, *19*, 156 (1971).

36. Pereira, J. F., Ph.D. Dissertation, Univ. of Illinois, Urbana, 1970.

37. Rasmussen, L. W., *Plant Physiol.*, *22*, 377 (1947).

38. Rodebush, J. E., and J. L. Anderson, *Weed Sci.*, *18*, 443 (1970).

39. Rohrbaugh, P. W., *Plant Physiol.*, *9*, 699 (1934).

40. Rohrbaugh, L. M., and E. L. Rice, *Botan. Gaz.*, *11*, 85 (1949).

41. Smith, A. E., *Bethany Information Sheet No. 88*, Naugatuck Chemical Division, U.S. Rubber Co., November 1, 1955.

42. Sweet, R. D., R. Kunkel, and G. J. Raleigh, *Proc. Amer. Soc. Hort. Sci.*, *48*, 475 (1946).

43. Turrell, F. M., *Botan. Gaz.*, *108*, 476 (1947).

44. Utter, G., Private Communication, 1960.

45. VanOverbeek, J., and R. Blondeau, *Weeds*, *3*, 55 (1954).

46. Walker, C. R., *Weeds*, *11*, 226 (1963).

47. Walter, J. P., E. F. Eastin, and M. G. Merkle, *Weed Res.*, *10*, 165 (1970).

48. Welton, F. A., and J. C. Carrol, *Ohio Agricultural Experiment Station Bulletin*, 1941, p. 619.

49. Wiater, A., T. Klopotonski, and G. Bagdasarian, *Acta Biochim. Pol.*, *18*, 309 (1971).

50. Wiater, A., K. Krajewska-Grynkiewicz, and T. Klopotonski, *Acta Biochim. Pol.*, *18*, 299 (1971).

51. Yeo, R. R., *Weed Sci.*, *18*, 283 (1970).

52. Young, P. A., *J. Agr. Res.*, *49*, 559 (1934).

53. Young, P. A., *J. Agr. Res.*, *51*, 925 (1935).

FOR CHEMICAL USE, SEE THE MANUFACTURER'S LABEL AND FOLLOW THE DIRECTIONS. ALSO SEE THE PREFACE.

20 Inorganic Herbicides

Inorganic herbicides are those weed-control chemicals which contain no carbon atom in their molecules. The principal ones are the arsenicals, borates, and chlorates, but the cyanates, calcium cyanamide, and ammonium sulfamate (AMS) also fall into this group.

Most inorganic herbicides were used before the modern era of organic herbicides began with 2,4-D in the mid-1940s. Although various organic herbicides have replaced these inorganic herbicides for many uses, they are still used.

SODIUM CHLORATE

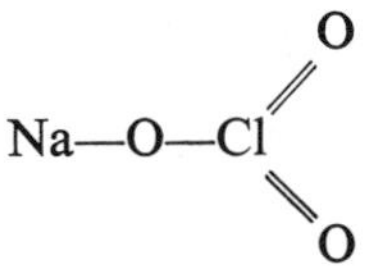

$$Na—O—Cl \begin{smallmatrix} O \\ // \\ \\ \backslash\backslash \\ O \end{smallmatrix}$$

sodium chlorate

Sodium chlorate ($NaClO_3$) is a white, crystalline salt that looks like common table salt (sodium chloride). Weight for weight, sodium chlorate is 30–50 times more toxic to plants than sodium chloride. Sodium chlorate is very soluble in water; 100 ml of water at 0°C will dissolve 75 g. The acute oral LD_{50} is about 5000 mg/kg.

Sodium chlorate has a salty taste. "Salt-hungry" animals may eat enough to be poisoned; 1 lb of this chemical/1000 lb of animal weight is considered lethal. Also, after spraying, some poisonous plants ordinarily avoided by livestock become palatable.

Fire Danger

Sodium chlorate has three atoms of oxygen per molecule. The oxygen is easily released, making sodium chlorate a strong oxidizing agent. It is therefore highly flammable when mixed with organic materials such as clothing,

274

wood, leather, or plant materials, and allowed to dry. It has been known to be ignited by the heat of the sun, clothing friction, or shoes scraping a rock. The fire cannot be smothered, since the chemical provides the needed oxygen. Moist sodium chlorate will not burn, so water will quickly extinguish the flame.

To summarize, three things are necessary to produce chlorate fire:

$$\text{Chlorate} + \text{organic matter (dry)} + \text{spark} \rightarrow \text{fire}$$

In its pure form, sodium chlorate is safe to store and handle. Also, it presents no fire hazard in a thoroughly moist condition. It is commonly mixed with borates to reduce the fire hazard.

Sodium chlorate can be applied either as a spray or as dry crystals. Spraying gives more uniform coverage and a more rapid kill, but the fire hazard may be serious when the foliage dries. If applied as dry crystals to *dry* vegetation there is usually no serious fire hazard. The chemical is leached into the soil with the first rain. However, if the crystals are applied to *damp* vegetation, they will adhere. When the vegetation dries it will be flammable, much the same as if it has been sprayed.

The person applying sodium chlorate must follow several precautionary measures. He can avoid contamination of his leather shoes by wearing rubber boots or overshoes; the crystals have a tendency to shed from rubber. He ought to wear trousers without cuffs. He should also change contaminated clothing before it dries, and *wash it immediately*. If allowed to dry, the clothes will be a serious fire hazard.

Uses

Sodium chlorate is generally used as a sterilant to kill all vegetation. It is widely used on cropland for spot treatment to control the spread of serious perennial weeds, and is broadcast on rights-of-way of highways and railroads. Most rates of treatment vary from 1 to 3 lb/100 ft^2. Other chemicals have proved more effective for selective weed control. (See Figure 20-1).

In practice, sodium chlorate is often combined with sodium borates with or without certain organic herbicides. These are covered in a later section in this chapter entitled "Borates."

Persistence in the Soil

Leaching quickly removes sodium chlorate from the soil (16). Also, soil microorganisms decompose chlorates to chlorides. Decomposition is usually

Figure 20-1. Single plants or small patches of johnsongrass, bermudagrass, nutsedge, and many other serious perennial weeds can be destroyed by spot treating the soil with soil-sterilizing chemicals. Here, sodium chlorate is being broadcast in the stubble of tall johnson-grass. (Kentucky Agricultural Experiment Station.)

most rapid in moist soils above 70°F. As would be expected, the effects of rainfall, soil texture and structure, organic-matter content, and temperature are very important. With low rainfall, chlorate may remain toxic for 5 years or longer. In the humid Southeastern states, toxicity may disappear in 12 months on heavy soils and in 6 months on sandy soils.

Ease of leaching may be a disadvantage—heavy rains or irrigation soon after application may remove the chemical from the upper 2–3 in. of soil. Shallow-rooted weeds such as bermudagrass may escape the toxic effects of the chemical and continue to grow.

Mode of Action

Absorption

The plant absorbs sodium chlorate rapidly through both roots and leaves. Dormant seeds in the soil usually survive the rates commonly used.

When sodium chlorate is sprayed on leaves, chlorate ions penetrate the cuticle and come into direct contact with living cells (12). The stomates need not be open for the chemical to enter the leaf (13).

Translocation

Chlorate moves rapidly from the roots upward through the xylem tissues. Since xylem tissue is composed principally of dead cells, chlorate moves freely regardless of its toxicity.

Since the chlorate kills living cells rapidly, it probably moves downward through the phloem very slowly, if at all. This is because only *living* phloem cells are active in translocation.

Effect of Metabolism

Sodium chlorate does three things—it depletes the plant's food reserves (2, 4, 11), temporarily increases rate of respiration, and decreases catalase activity (14). Catalase is an enzyme associated with the oxidation mechanism in plants; it is directly concerned with decomposition of hydrogen peroxide to water and oxygen. If hydrogen peroxide accumulates to more than very dilute concentrations, it is highly toxic to the plant.

The basic cause for the chlorate toxicity to plant tissues is unknown. The chemical disturbs metabolic processes so much that they no longer function properly. Scientists believe its high toxicity is a result of one or more of these factors: high oxidizing capacity of the chlorate ion, presence of the pentavalent chlorine (15), possible complete oxidation of respiratory chromogens (7), and possible toxic effects resulting from decomposition products such as chlorite and hypochlorite.

Plants increase in susceptibility to frost (or freezing) injury after treatment with chlorates (11). This is probably caused by depleted food reserves.

BORATES

Boron is an essential minor element for plant growth. In some areas, approximately 30 lb/acre are applied to improve plant growth. However, in large quantities boron is toxic to plants; it acts as a soil sterilant.

Borates used as herbicides are boron salts containing sodium and oxygen plus water. Herbicidal salts of boron are sodium metaborate tetrahydrate ($Na_2B_2O_4 \cdot 4H_2O$), sodium metaborate hexahydrate ($Na_2B_2O_4 \cdot 6H_2O$), sodium tetraborate pentahydrate ($Na_2B_4O_7 \cdot 10H_2O$), and disodium octaborate tetrahydrate ($Na_2B_8O_{13} \cdot 4H_2O$).

Boron, a nonselective herbicide, has been used effectively by railroads, farms, government agencies, and industry to destroy fire-prone and unsightly vegetation. The paving industry uses it under asphalt to prevent weeds from growing up through the asphalt.

These sodium borates are nonflammable; noncorrosive to ferrous metals, rubber, and plastics; nonvolatile; only slowly leached; and can be stored with little or no risk to fungicides, insecticides, and fertilizers. They are not deactivated by light. When used in sufficiently large quantities to be an effective herbicide, they are toxic to most soil microorganisms.

The toxicity of sodium borate to man and animals is low when normally handled and applied. Oral toxicity to mice was 2–3 g/kg of body weight. For 15-month-old rainbow trout, 1800 ppm were required for a 48-hr exposure to give an LD_{50}.

Uses

In the past, borates have been used alone as herbicides; however, now they are usually used in combination with other herbicides such as sodium chlorate, bromacil, diuron, and monuron. Combinations can all be applied with granular application equipment, and several can also be dissolved or suspended in water and applied with standard spray equipment.

The combination of borates and sodium chlorate reduces fire hazards of sodium chlorate discussed previously in this chapter. This combination also gives more effective, long-term, nonselective weed control, because sodium chlorate is leached more readily than the borates. Diuron or monuron is added to some formulations to give better control of certain annual weeds and seedlings of most perennial weeds. Bromacil is added to some formulations to control certain perennial grasses more effectively, and because boron is toxic to most soil microorganisms, it greatly retards microbial degradation of diuron, monuron, and bromacil.

Mode of Action

Sodium borates are absorbed principally by roots, translocated through the xylem to all parts of the plant, and accumulated in the leaves. In herbicidal quantities borates cause plant desiccation, beginning with burning and necrosis of leaf margins.

The chemical is usually carried into the soil by rainfall. The herbicide is most effective on young and tender plants. Therefore, time your treatment so that the borate will have reached the root-absorption zone by the time plant growth is just starting. The quantity required will vary with soil type, rainfall, and the weed species to be controlled. Recommended rates vary between 3 and 12 lb/100 ft².

In warm, moist soils the chemical usually remains effective as a soil sterilant for about 1 yr, but applications can usually be reduced in succeeding years. In dry soils or in frozen soils, sterilant effects would continue for several years.

ARSENICALS

Compounds of arsenic have been known since the early days of chemistry. They occur in nature as highly colored sulfides (As_2S_2 and As_2S_3), as several minerals (FeAsS and CoAsS), and occasionally as the oxide As_2O_3. Most arsenic used in the United States is the As_2O_3 form obtained as a by-product of smelting copper or lead ores. The organic arsenicals are discussed in Chapter 8.

Arsenic is used in minute quantities as a medicine, in alloying metals, in insecticides, for debarking trees, and as a herbicide. As a herbicide it controls aquatic weeds, crabgrass in lawns and turf, and acts as a soil sterilant. However, inorganic arsenical herbicides are rarely used as herbicides today because of their high toxicity to man. In addition, other much less toxic compounds have been developed.

Arsenic compounds have long been known as poisons to both animals and plants. Arsenic, often as As_2O_3, was a favorite tool of poisoners in the Middle Ages and it remained in favor with criminals until recently. Modern medical and chemical tests now make detection of such crimes rather easy. The fatal dose for a man is about 0.2 g.

In some countries mountaineers take arsenic to increase their endurance. They gradually become accustomed to daily portions greater than a fatal dose for an ordinary person. In the seventeenth century, women took small doses to heighten their color.

Minute quantities of arsenic may act as a stimulant for man, animals, or plants, whereas larger quantities act as strong inhibitors. The arsenic ion may react with proteins, having a denaturing or precipitating effect on the protoplasm. This same process may inactivate the enzymes. In both animals and plants toxic amounts of arsenic slow down the rate of respiration; oxygen uptake is reduced. This is true of both aerobic and anaerobic organisms.

Plants absorb arsenic compounds from foliage-applied arsenic sprays or from arsenic in the soil through the roots. If toxic quantities accumulate in the plant, it begins to die. In this condition the plant is usually less palatable to livestock. Most arsenic poisoning of livestock is caused by spray residue on plants when animals feed on them, or the chemical itself if animals gain access to it. The organic arsenicals are much less toxic to animals than are the inorganic arsenicals (see page 147).

Arsenic is the active ingredient of arsenical herbicides. Because their composition varies, the arsenic content is usually given as the As_2O_3 equivalent.

Rates of application of arsenic vary widely for different soils and climates. For equal weed control, sandy soils require much less arsenic than clay soils or soils high in organic matter. Also, these last two soils have a tendency to resist leaching of arsenic, whereas the chemical is easily leached through sandy soils.

In low rainfall areas, one heavy application of arsenic may sterilize the soil for 5–8 years; in high-rainfall areas, the period of toxicity is much shorter, provided the soil is porous enough to permit leaching. Rates of application, based on As_2O_3 content, vary from 1.0 to 8.0 lb/100 ft².

AMS (AMMONIUM SULFAMATE)

Ammonium sulfamate, sold as "Ammate," is an effective woody-plant killer. It is nonflammable, nonvolatile, and as normally used, nonpoisonous to man or livestock. You can apply it safely where 2,4-D and related products are hazardous to apply. That is why it is used principally on rights-of-way, roadsides, and ditches which adjoin crops such as cotton, tomatoes, grapes, and tobacco (see Figure 20-2).

It is also applied to stumps to prevent sprouting and in frills or notches to kill undesirable hardwood trees without cutting. Undesirable woody plants mixed in with a desirable woody hedge can be killed by the jar method. This involves placing and leaving the tops of undesirable plants in a jar or bucket of ammonium sulfamate solution.

The chemical acts as both a contact and translocated herbicide. It will also give temporary soil sterility. Because of its nitrogen and sulfur content, it may have a fertilizing effect after breakdown in the soil.

Ammonium sulfamate as sold commercially is a yellow, crystalline substance. The pure chemical is a colorless, crystalline substance. It is very soluble in water and absorbs moisture when exposed to the air.

$$\underset{\text{ammonium sulfamate}}{H_2N-\overset{\displaystyle O}{\underset{\displaystyle O}{\overset{\|}{\underset{\|}{S}}}}-O-NH_4}$$

Figure 20-2.   Ammonium sulfamate used as a spray to control the brush on the left side of the road. The tobacco on the right side of the road was not injured. (E. I. du Pont de Nemours and Company.)

In solution the chemical corrodes some metals, especially brass and copper; it also affects steel surfaces exposed to air such as the pump, the outside of tanks, and truck or tractor parts. Stainless steel, aluminum, and bronze are resistant. Metal parts covered by the spray solution inside the tank, pump, and lines corrode much more slowly.

Coat exposed areas with protective paints or cover them with oil. Rinsing exposed surfaces with water after each day's use is important. Thorough cleaning inside and out at the end of the season, followed by a coating of oil, will preserve the equipment. Corrosive effects are considered negligible on fences, guy wires, and telephone wires.

Ammonium sulfamate has a very low order of toxicity to humans and livestock. When the chemical was included in the daily feed of rats, and even when reasonably large quantities were injected into their bloodstreams, no serious ill effects were noted (1). Sheep have been fed up to 0.5 lb/day without apparent injury.

Ammonium sulfamate was tested on 194 human subjects using the procedure described by the Office of Dermatoses Investigation, U.S. Public Health Service. No contact irritation or evidence of sensitization was found.

OTHER SALTS

Many salts are toxic to plant tissues if applied in high concentrations. This is especially true of many fertilizer salts such as ammonium nitrate, urea, and potassium chloride. In sufficient concentration these provide a contact burning effect. Dissolved in water and added as a wetting spray, many annual weeds, especially broadleaf weeds, are easily killed. Addition of a wetting agent may double the contact killing effect (6, 10) (see Figure 6-6).

LITERATURE CITED

1. Ambrose, A. M., *J. Ind. Hyg. Toxicol.*, *25*, 26 (1943).
2. Bakke, A. L., W. G. Gaessler, and W. E. Loomis, *Iowa Agricultural Experiment Station Research Bulletin*, *254*, 1939.
3. Crafts, A. S., *Hilgardia*, *7*, 361 (1933).
4. Crafts, A. A., *Plant Physiol.*, *10*, 699 (1935).
5. Crafts, A. S., *J. Agr. Res.*, *58*, 637 (1939).
6. Davis, J. C., and G. C. Klingman, *Southern Weed Conference Proceedings*, *7*, 174 (1954).
7. Harvey, R. B., *J. Amer. Soc. Agron.*, *23*, 481 (1931).
8. Helgeson, E. A., *North Dakota Agricultural Experiment Bimonthly Bulletin*, *3*, 9 (1940).
9. Kennedy, P. B., and A. S. Crafts, *Plant Physiol.*, *2*, 503 (1927).
10. Klingman, G. C., and J. C. Davis, *Southern Weed Conference Proceedings*, *7*, 167 (1954).
11. Latshaw, W. L., and J. W. Zahnley, *J. Agr. Res.*, *35*, 757 (1927).
12. Loomis, W. E., E. V. Smith, R. Bissey, and L. E. Arnold, *J. Amer. Soc. Agron.*, *25*, 724 (1933).
13. Meadly, G. R. W., *J. Dept. Agr. W. Australia*, *10*, 481 (1933).
14. Neeler, J. R., *J. Agr. Res.*, *43*, 183 (1931).
15. Offord, H. R., *U.S.D.A. Technological Bulletin No. 240*, 1931.
16. Seely, C. E., K. H. Klages, and E. G. Schafer, *Washington Agricultural Experiment Station Bulletin*, 505, 1948.

FOR CHEMICAL USE, SEE THE MANUFACTURER'S LABEL AND FOLLOW THE DIRECTIONS. ALSO SEE THE PREFACE.

21 Small Grains and Flax

Small grains discussed here include wheat, oats, barley, rye, and rice. Flax is included because its cultural practices and weed problems are similar to those of small grains.

Winter varieties are planted in the fall, live through winter, and are harvested the following summer. Spring varieties are planted in early spring and harvested in mid to late summer. As an average, tolerance to cold is in this order: rye, wheat, barley, oats, and rice. Therefore, winter rye is grown in far northern areas. Rice in the United States is normally planted in the spring.

In general, winter varieties are most often infested with winter-annual weeds and to a lesser extent by summer annuals that germinate in early spring. Spring varieties are primarily infested by summer annuals that germinate in the early spring. Perennial weeds are also troublesome in certain areas.

Several review papers on weed control in these crops were prepared for the Food and Agriculture Organization of the United Nations (FAO) International Conference on Weed Control. These included four for small grains (9, 12, 15, 16), two for rice (13, 22), and one for flax (20).

EFFECT OF WEEDS ON YIELD

Weeds compete directly with the grain crop for light, moisture, carbon dioxide, and soil nutrients. Grain yield reductions range from total loss to losses so slight that they are not measurable (see Figure 21-1).

Fifty fields were selected at random in 1956 and another 50 in 1957 near Winnipeg, Canada, to clearly establish actual losses in small grain and flax resulting from weeds. These fields included a complete range of weed infestations from those with satisfactory weed-control programs to those badly infested with weeds. In each field 10 paired plots were staked out. One plot of each pair was kept weed free by hand weeding. Principal weed species

Figure 21-1. Oats sprayed with 2,4-D to remove wild mustard. (New Jersey Agricultural Experiment Station.)

included wild mustard, wild oats, Canada thistle, sow thistle, and green foxtail. Average losses in yields caused by weeds are shown in Table 21-1 (7).

Lowest losses were recorded where the crop-management program in previous years had provided effective weed control. Highest losses occurred when there was neither herbicide-spraying program nor crop rotation. Farmers with no weed-control programs experienced much heavier losses than indicated in Table 21-1.

Table 21-1. Yield Reductions Expressed as a Percentage from Weed Competition (7)

Year	Wheat (%)	Barley (%)	Oats (%)	Flax (%)	Total Crop Average (%)
1956	19.5	—	10.6	26.6	15.9
1957	14.3	18.6	6.5	27.6	17.3

Weed competition early in the season reduces yields more than late-season competition. Although yields are not greatly reduced, late-season weeds may cause difficulty in harvesting. As well as causing smaller yields and harvesting difficulties, weeds lower crop quality and may reduce the protein content of grain.

WEED CONTROL METHODS

Weed-control methods in small grains and flax include clean seed, crop rotation, good seedbed preparation, full use of competition, and application of herbicides.

Clean Seed

The importance of clean seed can hardly be overemphasized. To eliminate repetition, however, see Chapter 2.

Crop Rotation

With continuous production of small grains or flax, weeds that grow during the same season and are favored by the crop-management programs build up rapidly.

Farmers learned that small grain or flax could usually be grown for several years on "new land" before weeds became a serious problem. After this time, they were forced to use more effective methods of weed control. Crop rotation proved effective against many species.

Crop rotation is a strong link in the chain of improved weed-control practices. Herbicides can be combined effectively into a crop-rotation program. In some areas, the rotation of cotton, corn, and small grain with lespedeza illustrates this point. As a result of variations in the time of seedbed preparation, time of cultivations, and period of growth, plus the use of different herbicides in each of the four crops, the buildup of many weed species is prevented. Here "crop rotation" and "chemical rotation" are combined for maximum weed control.

Seedbed Preparation

Weed control is one of the principal purposes of seedbed preparation. Because small grains cannot be effectively cultivated after sowing, the importance of controlling weeds before sowing is obvious. In some areas where weed seedlings appear before sowing, weed competition can be reduced by a presowing cultivation. Where practical, summer fallow is especially effective because weed-seed production can be prevented during an entire summer season, and the growth of perennial weeds checked.

Competition

Thickly planted, fast-growing small grains offer considerable competition to weeds. In such grains few or no weeds may be found at harvest time. Yet had the grain not crowded them out, the area would have been solidly covered with weeds. (See Figure 21-2.)

Figure 21-2.   Competition from the crop crowds out many weeds. (North Carolina State University.)

Table 21-2. Effect of Stand of Flax on Yields of Seed, Straw, and Weeds (9)

Stand of Flax on July 25 (%)	Flax Seed (bushels/acre)	Flax Straw (lb/acre)	Weed Plants (lb/acre)
130	15.8	2068	291
100	17.3	2073	312
55	14.2	1537	576
17	11.1	1083	1329
LSD (0.05)	3.5	329	350

Well adapted, disease-resistant varieties, planted at the proper time, with adequate soil fertility and moisture, may be able to grow slightly faster than weeds. If the crop emerges ahead of the weeds, it may compete effectively with many troublesome weed species.

Flax is only partially effective in competing with weeds. With its slow early growth and small leaf surface, the flax plant offers little competition to the weeds. Nevertheless, a thick stand of flax is important for weed control, as shown in Table 21-2. It should be noted that even with good stands of flax, the amount of weeds was still large.

Chemical Control

Weeds are often a serious problem in small grains and flax, even with good cultural practices. Some weeds are favored by the same management programs that favor small grains.

Chemicals can often be used in small grains in such a way that crop competition, as discussed above, is increased. The herbicide may stunt weeds with little or no reduction in the growth of crop plants. With this slight competitive advantage, small grain may "crowd out" the weeds (see Figure 21-2).

Information given here and in following chapters on chemical weed control should be considered only as a guide. The performance of herbicides can and does vary under different climatic and soil conditions as well as the stage and rate of plant growth (see Chapters 3 and 4). Therefore, consult your state or county extension agents and local company field men. It is essential to follow directions on the product label. These may vary for different geographical areas.

BARLEY, OATS, AND WHEAT

The principal chemicals used for weed control in barley and wheat are 2,4-D, MCPA, dicamba, bromoxynil, barban, and triallate. About nine other herbicides are also used for special weed problems in small grains including trifluralin, linuron, and terbutryn.

2,4-D

More acres of small grain are treated with 2,4-D than any other crop or group of crops. The use of 2,4-D on small grains is a widely accepted herbicide program.

Time of Treatment

Good weed control without crop injury usually depends on proper timing in the application of herbicides. This is particularly true for 2,4-D in small grains. For our discussion, growth of small grains is divided into four stages:

1. Zero-to four-leaf stage.
2. Four-leaf to boot stage.
3. Boot stage to flowering.
4. Soft-dough-grain stage to maturity.

It should be pointed out that periods of greatest susceptibility to 2,4-D are during the periods of rapid growth. It appears that the rate of meristematic development is closely related to the plant's susceptibility to 2,4-D.

The four-leaf up to just before the boot stage is the recommended stage for the application of 2,4-D for reasons discussed below.

Zero to four-leaf stage. Small grains, including rice, are very sensitive to 2,4-D during germination and seedling stages. Therefore, treatment during this time will usually cause many malformations of the head, onion leaves, general stunting of the plant, and reduced yields (see Figure 21-3).

Four-leaf to boot stage. Four-leaf up to just before the boot stage includes the fully tillered stage and is considered to be the *most desirable time* to apply 2,4-D (see Figure 21-3). In addition to the grain's tolerance at this time, weeds are usually small and easily killed. Also, weeds have not caused serious

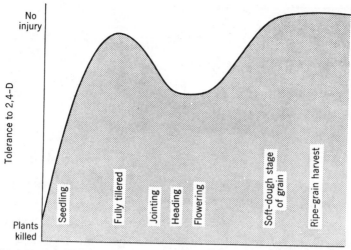

Figure 21-3.   Stages of small-grain growth and the degree of tolerance to excessive rates of
2,4-D, *with most tolerant stage listed first.* 1. Soft dough of grain stage to maturity. 2. Fully
tillered; four leaves or more per plant; 5–8 in. tall. 3. Jointing stage through flowering.
This includes the boot stage to the flowering stage. 4. Germination to four-leaf stage.
(North Carolina State University.)

damage as far as competition is concerned and ground spray equipment
causes but slight mechanical injury to the grain.

Most varieties will withstand up to $\frac{1}{2}$ lb of 2,4-D (acid equivalent)/acre at
this time without yield reductions; tolerant varieties will withstand 2 lb/acre
(9). In general, wheat varieties are the most tolerant, barley is intermediate,
and oats are the least tolerant. Rates over $\frac{1}{2}$ lb/acre are suggested only where
resistant weeds require this amount for control. The above high rates are for
amine forms of 2,4-D. Ester forms are used at about half the above rates
($\frac{1}{4}$–$\frac{1}{2}$ lb/acre).

Boot stage through flowering.   During the boot stage the internodes elon-
gate rapidly. Also, this stage is known as the jointing stage. From the start
of the boot stage through flowering, small grains are susceptible to injury.
2,4-D sprays at this time may reduce yields seriously because injury to the
heads may occur as they develop.

Soft-dough-grain stage to maturity.   Small grains are very tolerant to
2,4-D at this time. However, this treatment is not usually recommended
because: ground-spray equipment will crush much of the grain, weed competition has already done its damage to crop yield, and there are possible
residues of 2,4-D in harvested grain (see Figure 21-4).

Figure 21-4. Thatcher wheat injured by premature treatment with 2,4-D. The wheat was about 3 in. tall when treated, and from 60 to 70% of the heads were abnormal. (J. B. Harrington, University of Saskatchewan.)

Effect of 2,4-D on Germination

Germination studies indicate little or no damage to germination of grain from plants treated with 2,4-D when the grain is treated at the proper time and rate (3, 5, 6, 19, 21). When treated during a susceptible stage (boot stage), wheat showed an 11% drop in germination (10).

Three barley varieties were treated with 2,4-D at rates from $\frac{3}{4}$ to 2 lb/acre when the barley was 6 in. tall. An additional treatment was made after heading, in which $1\frac{1}{4}$ lb/acre of 2,4-D ester were used. Research workers concluded that treatments did not lower germination of the barley.

Ten oat varieties were given the same treatments as barley. Eight oat varieties showed little or no loss in germination. Two varieties showed a significant drop in percentage of germination; the greatest reduction was 18.5% (8).

Effect of Soil Fertility on Crop Tolerance to 2,4-D

Oats were grown under three levels of soil fertility in New Jersey. All plots were treated at 1 lb/acre of 2,4-D. A 10-10-10 fertilizer was applied to provide medium and high fertility plots. Yields were not reduced on low-fertility plots as compared to low-fertility check plots, even though several 2,4-D sprays were applied at stages when grain is normally susceptible to the rate of 2,4-D used.

Yields were reduced on medium- and high-fertility plots when treated during a period of susceptibility. Treatment in the fully tillered stage (tolerant stage) did not reduce yields. Therefore, it is concluded that high soil fertility can be expected to increase the crop's susceptibility to injury.

If, as previously discussed, plants are most susceptible during periods of rapid growth or rapid development of meristematic tissues, the above facts are to be expected. Higher rates of fertility will have a tendency to stimulate growth, and thus increase the rate of meristematic tissue development.

Variety Tolerance to 2,4-D

Many studies have been made of variety tolerance to 2,4-D. Each study has shown considerable variation in tolerance among varieties.

Due to widespread use of the chemical, 2,4-D tolerance might well be one point of selection in an improved variety breeding program (17).

MCPA

MCPA is less injurious to small grains, flax, and some legumes than 2,4-D (this is especially so for flax and oats) and is more effective on a few broadleaf weeds; for example, hemp nettle. However, it is less effective than 2,4-D on most broadleaf weeds, and is usually more expensive in the United States.

MCPA is less injurious to most legumes, especially red clover, than is 2,4-D. There is no difference in alfalfa.

MCPA is especially suggested for weed control in flax, in those varieties of oats known to be susceptible to 2,4-D, in tolerant clovers such as red clover, and for control of weeds known to be susceptible.

2,4-DB

Less injurious to legumes than 2,4-D, this herbicide can be used to control many broadleaf weeds in wheat, barley and oats when they are used as nurse crops for small-seeded legumes (grain underseeded with legumes). However, 2,4-DB is usually less effective than 2,4-D as a broadleaf herbicide.

Dinoseb

Small grains underseeded with legumes can be effectively treated with amine salts of dinoseb for broadleaf annual-weed control. The chemical is applied as a wetting spray (20–40 gal/acre) at the rate of $1-1\frac{1}{2}$ lb/acre of active ingredient. It is usually applied when grain is 3–6 in. high and when legumes have four–six leaves. Temperatures of 60°F with no rain for at least 24 hr after treatment will make the treatment more effective.

Dicamba

Dicamba, an auxin-type herbicide, controls some broadleaf weeds in small grains not readily controlled with 2,4-D; for example, wild buckwheat, smartweeds, gromwell, and corn cockle. It is applied in the spring at $\frac{1}{8}$ lb/acre before the boot stage of fall-seeded grain; or at about $\frac{1}{8}$ lb/acre to spring-seeded grain at the two- to five-leaf stage.

Bromoxynil

Bromoxynil also controls certain 2,4-D resistant broadleaf weeds in wheat and barley; for example, fiddleneck, gromwell, wild buckwheat, and fumitory. It is usually applied at $\frac{1}{4}-\frac{1}{2}$ lb/acre alone or in combination with $\frac{1}{4}$ lb/acre of MCPA in the spring or fall when the grain is in the three-leaf to boot stage.

Barban

Barban controls wild oats in several crops, but is used mainly in barley and wheat. It is applied as a postemergence treatment when most wild oats are in the two-leaf stage. This stage lasts from the time the second leaf first appears until the third leaf first appears. Timing of the application is extremely critical for good control.

It is applied as a spray at $\frac{1}{8}-\frac{1}{5}$ lb/acre. Uniform and accurate coverage of wild oats with at least five droplets per plant is essential.

Triallate

Triallate is another wild-oat herbicide for use in wheat and barley. However, it is a soil-applied material. As a preplant, soil-incorporation treatment, the crop should be planted within 3 weeks after application. As a preemergence, soil-incorporation treatment, it should be applied before the shoots of the crop are $\frac{1}{2}$ in. long. It can also be incorporated into the soil in the fall and the crop planted the following spring. The rates used are $1\frac{1}{4}$ to $1\frac{1}{2}$ lb/acre.

Trifluralin

Trifluralin is used in cereal grains, especially wheat and barley for the control of trifluralin-susceptible weeds. This includes the annual grasses such as foxtails, windgrass, ryegrasses, Alopecurus, and annual bluegrass plus such broadleaved weeds as field poppy, pimpernel, fumitory, hempnettle, henbit, wild buckwheat, corn spurry and speedwell. Partial control of wild oats can be expected. The chemical is usually applied postplant as a surface-applied chemical or with very shallow incorporation. It is also applied in combination with other herbicides to broaden the spectrum of weed control.

Most of the above weeds are not cereal-grain problems in the United States. Thus, trifluralin has *not* been cleared for use in the cereal grains in the United States.

RICE

Rice responds to essentially the same weed-control practices as wheat, oats, and barley, and its response to herbicides is also similar (17). However, it may be grown under "paddy" conditions (flooded with water until just before harvest) or "upland" conditions (grown under dryland conditions). These cultural differences influence weed-control practices. For example, barnyardgrass can be partially controlled by maintaining 6 in. of water over the soil surface in paddy rice, but this condition does not exist in upland rice.

Principal weeds of upland rice are coffee weed, curly indigo, and grassy weeds, especially barnyardgrass. Main weeds of paddy rice are barnyardgrass and aquatic species, especially arrowhead, water plantain, sedges, and bulrush.

Phenoxy-Type Herbicides

2,4-D has been used widely to control many broadleaf weeds in rice at rates of $\frac{1}{2}$–1 lb/acre. It is applied late postemergence, after tillering, but before the boot stage. However, MCPA is more selective to rice and a safer material for control of some broadleaf weeds. It is used at $\frac{3}{4}$–$1\frac{1}{4}$ lb/acre.

Propanil

Propanil controls barnyardgrass and other annual grasses, sedges, and several broadleaf weeds. In upland rice it is applied as a postemergence spray when barnyardgrass is in the one–three-leaf stage, at 3–6 lb/acre in 10–15 gal of water. In flooded rice, 5 lb/acre is recommended when most barnyardgrass is 6–9 in. above the water. This is usually 35–45 days after planting.

Rice plants may be severely or fatally injured if carbaryl or any organic phosphate insecticide is applied within 14 days before or after the application of propanil.

Molinate

Molinate is specific for barnyardgrass control in rice, but some other weeds may also be controlled. It is applied before planting and incorporated into the soil by shallow discing, harrowing, or flooding within 6 hr. These types of application are for rice seeded by aircraft after flooding.

Molinate can also be applied postemergence to water-seeded or drilled rice when barnyardgrass measures from 2 to 5 in. tall and is at least two-thirds submerged by water. This water level (not less than 2 in.) must be maintained until the barnyardgrass dies (usually in 4–7 day). A rate of 3 lb/acre is usually used.

Benthiocarb

Benthiocarb is an effective herbicide for transplanted rice. Mixed with sime-tryn, the herbicide mixture effectively controls a broad spectrum of weeds. Research is examining the possibility of its use on direct-seeded rice.

FLAX

Weed control in flax is similar to that of small grains. As discussed earlier under "Competition", flax has a small leaf surface. It offers little competition to weeds for light, so naturally weeds are often a serious problem in flax.

Flax is usually sprayed with MCPA when it is 2–6 in. tall and before weed seedlings are 2 in. tall. (See Figure 21-5).

MCPA (amine form) is usually applied at the rate of $\frac{1}{4}$ lb/acre for weeds such as wild mustard, lambsquarters, ragweed, fanweed, or cocklebur; and up to $\frac{1}{2}$ lb/acre for smartweed.

Barban at $\frac{1}{4}$ lb/acre is used as a postemergence treatment to control wild oats in flax. Wild oats should be at the two-leaf stage, but do not treat them after flax reaches the 12-leaf stage or later than 14 days after crop emergence.

EPTC, although not widely used, can be used as a preplant soil-incorporation application in flax to control wild oats, other annual grasses, and many broadleaf weeds. It is applied at the rate of 3 lb/acre.

Figure 21-5. Flax weeds that should have been killed earlier in the season. The herbicide applied will be MCPA, up to 0.6 lb/acre; ¾ lb of dalapon/acre may be added to MCPA. (L. A. Derscheid, South Dakota State University.)

Low to medium rates of the above chemicals are not expected to affect germination of flax seeds, oil content, or iodine number.

LITERATURE CITED

1. Bell, A. R., and J. D. Nalewaja, *Weed Sci.*, *16*, 505 (1968).
2. Bowden, B. A., and G. Friesen, *Weed Res.*, *7*, 349 (1967).
3. Buchholtz, K. P., *North Central Weed Control Conference Research Report*, *5(11)*, 10 (1948).
4. Burrows, V. D., and P. J. Olson, *Can. J. Agr. Sci.*, *35*, 68 (1955).
5. Derscheid, L. A., L. M. Stahler, and D. E. Kratochvil, *Agron. J.*, *44(4)*, 182 (1952).
6. Derscheid, L. A., L. M. Stahler, and D. E. Kratochvil, *Agron. J.*, *45(1)*, 11 (1953).
7. Friesen, G., *North Central Weed Control Conference Proceedings*, *14*, 40 (1957).
8. Grigsby, B. H., and B. R. Churchill, *Michigan Agricultural Experiment Station, Quarterly Bulletin*, *30*, 448 (1948).
9. Hay, J. R., *FAO International Conference on Weed Control*, WSSA, Urbana, Illinois, 1970, p. 38.
10. Helgeson, E. A., K. L. Blanchard, and S. D. Sibbitt, *North Dakota Experiment Statin. Bimonthly Bulletin*, *10(5)*, 166 (1948).

11. Haemoeller, W. A., M.S. Thesis, North Dakota State University, 1967.

12. Jakobsons, P.. *FAO International Conference on Weed Control*, WSSA, Urbana, Illinois, 1970, p. 65.

13. Matsunaka, S., *FAO International Conference on Weed Control*, WSSA, Urbana, Illinois, 1970, p. 7.

14. Messersmith, C. M., and J. D. Messersmith, *Abstracts of the WSSA*, No. 15, 1969.

15. Mukula, J., *FAO International Conference on Weed Control*, WSSA, Urbana, Illinois, 1970.

16. Nalewaja, J. D., and W. E. Arnold, *FAO International Conference on Weed Control*, WSSA, Urbana, Illinois, 1970, p. 48.

17. Price, C. D., and G. C. Klingman, *Agron. J.*, *50*, 200 (1958).

18. Robinson, R. G., *Agron. J.*, *41*, 483 (1950).

19. Robinson, R. G., R. S. Dunham, and O. H. Shulstad, *North Central Weed Control Conference Research Report*, *7*, 78 (1950).

20. Santelmann, P. W., E. W. Hauser, and E. Knake, *FAO International Conference on Weed Control*, WSSA, Urbana, Illinois, 1970, p. 260.

21. Shafter, N. E., *North Central Weed Control Conference, Research Report*, *7*, 78 (1950).

22. Smith, R. J., Jr., *FAO International Conference on Weed Control*, WSSA, Urbana, Illinois, 1970, p. 24.

For chemical, see the manufacturer's label for method and time of application, rates to be used, weeds controlled, and special precautions. Label recommendations must be followed—regardless of statements in this book. Also see the preface.

22 Small Seeded Legumes

Weed control is a major problem in growing legumes. Alfalfa, white clover, and ladino clover are perennial crops usually grown for hay and pasture. Red clover, alsike clover, crimson clover, sweet clover, and lespedeza are hay or soil-improving crops. They are often interplanted in small grain. With either group there is little opportunity after seeding to control weeds by cultural methods, because of the random nature of crop-plant spacing.

Weed control of large-seeded legumes planted in rows, such as soybeans and peanuts, is discussed in Chapter 23.

Weeds in small-seeded legumes can decrease yields, lower quality, and increase disease and insect problems, as well as causing premature loss of stand, harvesting problems, and irritation in animals when a spiny weed is eaten.

Most weeds in small-seeded legumes can be controlled by one or more of the following methods:

1. Clean seed
2. Weed control before seeding
3. Proper date of seeding
4. Competitive nature of crop
5. Mowing
6. Flaming
7. Cultivation
8. Chemicals

CLEAN SEED

The use of certified seed is of prime importance to avoid sowing weed seeds along with the crop. This is especially true for dodder.

The seeds of many serious weeds are nearly the same size, shape, and weight as seeds of small-seeded legumes, and once contaminated can be only partially cleaned by present methods. In recent years several mechanical devices have been used that clean seed well enough to meet most state seed-law requirements. Many of these devices are quite ingenious and take advantage of small, physical differences between the seeds to be separated. Shape, size, weight, differences in seed coat, hairiness, and appendages have all proved useful in separations.

One example is the dodder cleaner. Seed coats of most legumes are smooth and waxy, but seed coats of dodder are rough and pitted and stick to felt cloth; legume seeds do not. Adjoining, felt-covered rollers rotate in opposite directions and seeds pass between them. Dodder seeds are carried to the outside while legume seeds remain between the rollers. Rollers are slanted so that legume seeds slide out at the lower end.

WEED CONTROL BEFORE SEEDING

Small-seeded legume seedlings are not vigorous growers and offer little competition to aggressive weeds. Where the area is infested with serious perennial weeds that will persist after the legume crop is seeded, control *before seeding* may be necessary. Control methods must be appropriate for the weed species concerned.

Annual weeds are often brought under control by two methods. The first is crop rotation before seeding. For example, a rotation of row crops and small grain *in which weeds are effectively controlled for* 2 *years or more* usually reduces weed-seed populations in the soil. The second method is preparation of the seedbed well before seeding. This timing permits killing one or more crops of weeds. Cultural implements are preferred that *do not* bring deeply buried seed to the surface. Deeply buried seed usually remains dormant, so there will likely be no problem if undisturbed.

DATE OF SEEDING

Date of seeding may determine weediness of the crop (2). Most small-seeded legume crops can be either fall planted or spring planted. Weediness will depend on whether winter- or summer-annual weeds are more serious. Summer-annual weeds are often more serious. Thus, alfalfa is usually fall planted. However, there are the hazards of fall drought and winter injury in some areas and excessive rainfall in other areas. With fall planting the legume crop is well established by spring and competes well with summer-annual weeds.

Spring planting may be preferred if winter-annual weeds are the major problem. Spring planting avoids the flush of winter-annual-weed growth. However, with spring planting, summer-annual weeds may crowd out legume seedlings before they become established. More effective herbicides may make it possible to take advantage of more ideal seeding conditions usually found in the spring.

If legume crops are seeded when soil temperature and moisture especially favor them, they may germinate and grow fast enough to crowd out most weeds.

COMPETITION

One effective control for many weeds is to maintain a thick stand of the legume. Therefore, conditions that make the legume more vigorous usually reduce weed growth. Conversely, less competition from the legume means more weeds. Proper fertilization, drainage where needed, moisture conservation where moisture is limited, disease and insect control, proper mowing time, use of adapted varieties, and other proper management practices help maintain a thick stand of legumes.

MOWING

Few weeds can survive competition with vigorously growing alfalfa plus being mowed two or more times per growing season. This double-barreled attack weakens and may kill most annual weeds and many serious upright growing perennial weeds. Prostrate weeds escape control by mowing. Mowing controls weeds especially well when you also maintain a thick legume stand.

If you consistently cut alfalfa when it is too immature, its vigor will be reduced, thus giving weeds a chance to grow (1). (See Table 22-1)

Table 22-1. Effect of Cutting on Yield and Stand of Alfalfa and on Number of Weeds (3). (2-yr treatment in Wisconsin)

Stage of Cutting Alfalfa	Number of Cuttings/ Season	Yield of Weed-Free Alfalfa (tons)	Number of Alfalfa Plants at End of Experiment (/ft²)	Weeds in Hay (%)
Full bloom	2	3.8	12	17.2
Early bud	3	2.1	3	45.1
Succulent	4	0.7	0	72.8

Mowing weeds that outgrow legume seedlings is an effective method of control.

FLAMING

Propane or diesel burners can be used to control weeds in established alfalfa. Winter-annual broadleaf weeds are controlled with flaming just prior to resumption of growth in the spring. This treatment also suppresses alfalfa weevil-larval populations.

Flaming just after a cutting has controlled established dodder plants. However, it usually results in a few day's suppression of alfalfa growth. This treatment is usually applied only in the dodder infested areas.

CULTIVATION

Tillage of alfalfa fields to control annual weeds has been frequently recommended and practiced, even though little experimental evidence supports the practice.

In Nebraska, cultivation of alfalfa did not increase yields; however, neither did it reduce yields. (See Table 22-2.)

Therefore, where a good stand of alfalfa persists, one cultivation may help control weeds subject to such cultural methods. The spring-tooth harrow kills many annual weeds without serious injury to alfalfa crowns. The disc harrow may cause considerable damage to alfalfa by cutting the crowns. The spike-tooth harrow is effective only on very small weeds.

Weed control is vital in legume-seed production. When grown mainly for seed, alfalfa and certain other legumes are occasionally planted in rows. Row planting permits weed control by cultivation. It also permits directed and shielded application of herbicides.

Table 22-2. Yield of Alfalfa Receiving Various Types of Cultivation (1). (4-Yr Average)

Treatment	Yields of Alfalfa (tons/acre)
No cultivation	3.39
Three disk-harrow treatments	3.22
Three spring-tooth harrow treatments	3.31
Spike-tooth harrow treatments	3.37

CHEMICAL WEED CONTROL

Small-seeded legumes vary widely in their tolerance to various herbicides. This tolerance plus susceptibility of weeds is important in choosing an appropriate herbicide treatment. We only have space here to discuss the most common crop-weed combinations.

Chemicals are applied before planting to eliminate troublesome species that cannot be controlled efficiently after planting. For example it may be desirable to treat bermudagrass, johnsongrass, horsenettle, or wild garlic to control them before seeding alfalfa. Length of residual herbicide toxicity in the soil and its effect on the newly seeded crop must also be considered (see Table 4-1).

Since seedlings of most species, especially small-seeded legumes, are much less tolerant to herbicides than mature plants, the following discussion is divided into weed control for seedling establishment and in the established crop.

Seedling Establishment

Chemical treatments used for seedling establishment can be classified as preplanting, preemergence, and postemergence

Preplanting

Benefin or EPTC is used as a preplant, soil-incorporation treatment to control most annual grasses and many annual broadleaf weeds in alfalfa, birdsfoot trefoil, and several clovers. With benefin, planting can be delayed—see the label. However, with EPTC, legumes should be planted soon after treatment. Benefin is applied at $1-1\frac{1}{2}$ lb/acre and EPTC at 3–4 lb/acre; use lower rate on light soils and higher rate on heavy soils.

Propham can be used as a preplant, surface application (do not incorporate) to control annual grasses and some annual broadleaf weeds in alfalfa, birdsfoot trefoil, and clovers. It is applied at 4 lb/acre. Rainfall or over-headirrigation is required for propham to be effective as a surface application.

Propham is degraded fairly rapidly in moist warm soil (less than 1 month), EPTC persists somewhat longer (1–3 months), and benefin still longer (3–8 months). Effective weed control would last corresponding periods of time.

Preemergence

Propham can be applied as a preemergence treatment when rainfall can be expected or sprinkle irrigation applied. It controls most annual grasses and some annual broadleaf weeds in alfalfa, birdsfoot trefoil, and clovers. It is applied at 5 lb/acre.

Postemergence

After small-seeded legume and annual weeds have emerged, two herbicides can be sprayed over the crop and weeds to control many annual, broadleaf weeds with minimal crop injury. These are dinoseb (ammonium or amine salts) and 2,4-DB (ester or amine salt).

Dinoseb is applied when the crop has two trifoliate leaves and broadleaf weeds are very small. The ammonium salt is applied at $\frac{3}{4}$–1 lb/acre, while the amine salt is applied at 1–1$\frac{1}{2}$ lb/acre. 2,4-DB ester is applied at $\frac{1}{2}$–$\frac{3}{4}$ lb/acre, while the amine salt is applied at 1 to 1$\frac{1}{2}$ lb/acre when the weeds are in the one- to three-leaf stage. It is essential that broadleaf weeds be quite small when all of these herbicides are applied. Grassy weeds are not controlled.

Established Small-Seeded Legumes

Most herbicides named above can also be used on established small-seeded legumes; however, they are usually used somewhat differently and at different rates. Several more phytotoxic herbicides may also be used.

Dinoseb

Dinoseb (nonselective-phenol form) has been used widely to kill all emerged annual-weed seedlings in established alfalfa. It is applied during the dormant season to control annual-winter weeds, or applied immediately after first cutting (before regrowth starts) to control dodder and annual weeds. The current Federal registration states that the area cannot be grazed or treated foliage used for feed or forage. This has tended to restrict its use.

It is usually prepared using 1$\frac{1}{4}$–2 lb in 20–50 gal of diesel oil and made up to 100 gal with water. The resulting emulsion is sprayed to thoroughly wet all weed foliage; this takes about 100 gal/acre.

This treatment also suppresses alfalfa weevil larval populations in the following cutting.

Chlorpropham

Chlorpropham is used preemergence to weeds in established alfalfa and clovers. It is particularly effective on annual grasses and dodder, but also controls some annual broadleaf weeds. When the crop is dormant, rates as high as 8 lb/acre may be used; in alfalfa, this rate may also be used after first cutting. A rate of 5 lb/acre may also be used in alfalfa after the seedlings have reached the trifoliate stage, or in clovers after seedling clovers have reached the four-leaf stage. Granular formulations are often used.

EPTC

In irrigated areas of the West, EPTC has been applied in irrigation water in established alfalfa, ladino clover, or both, before weeds emerge. Rates of 2–3 lb/acre are used. It is particularly effective on annual grasses, but some annual broadleaf weeds may also be controlled. Repeat treatments are usually necessary for seasonlong control.

2,4-DB

The amine or ester form of 2,4-DB may be used in established alfalfa, birdsfoot trefoil, and clovers for control of small, annual, broadleaf weeds as described above for seedling crops.

Diuron

Diuron is used in established alfalfa, at least 1 year old, when it is dormant or semidormant to control most annual weeds. It is applied at rates of $1\frac{1}{2}$–3 lb/acre. Diuron is relatively persistent and often gives seasonlong weed control. It is not suitable for use on light, sandy soils. Do not replant treated areas to any crop within 2 yr after application because of the broad spectrum of species controlled by diuron.

Nitralin and Trifluralin

Nitralin or trifluralin can be used to control most annual grasses and many annual broadleaf weeds in established alfalfa. Both require shallow incorporation into the soil to be effective and are usually applied to the dormant crop or after the last cutting. Both must be incorporated into the soil with care to avoid excessive physical damage to the crop. Depth of incorporation varies from 1 to $1\frac{1}{2}$ in. and can be done with a treader mulcher or a disc set to minimize crop injury and yet incorporate the herbicide into the top inch of soil.

Nitralin is applied at $\frac{1}{2}$–$1\frac{1}{2}$ lb/acre. Trifluralin is applied at $\frac{3}{4}$ to 1 lb/acre. Use lower rates on light soils and higher rates on heavy soils.

LITERATURE CITED

1. Kiesselback, T. A., and A. Anderson, *Nebraska Experiment Station Bulletin*, 1927, p. 222.
2. Klingman, D. L., *FAO International Conference in Weed Control*, WSSA, Urbana, Illinois, 1970, pp. 401–424.
3. Nelson, N. T., *J. Amer. Soc. Agr.*, *17*, 100–113 (1925).

For chemical, see the manufacturer's label for method and time of application, rates to be used, weeds controlled, and special precautions. Label recommendations must be followed—regardless of statements in this book. Also see the preface.

23 Field Crops Grown in Rows

Why are crops planted in rows? What determines the width between rows? Just why are row crops cultivated? What effect does cultivation have on crop yields? These are important questions to any farmer planning his future operations. Unfortunately we can only partially answer all of these questions.

While man depended on the hoe for weed control, undoubtedly he was also looking for easier ways to do the job. He found one great labor-saver in dragging a heavy hoe behind a horse. Early agricultural writings refer to "horse hoeing." Planting crops in rows made horsehoeing easy.

Thousands of research reports and articles have been written on weed control in various crops, especially in row crops. These reports are available in *Weed Science* and *Weed Research*, two scientific publications; Regional Weed Conference Proceedings and Research Reports; College of Agriculture bulletins, leaflets, and circulars; commercial company publications; and in various trade journals and farm magazines. These sources give detailed information, but no attempt will be made to review all this literature here. Only those practices that have proven their usefulness or that appear especially promising are discussed below.

Several review papers on weed control in these crops were prepared for the FAO International Conference on Weed Control. They include corn (6, 7), grain sorghum (7, 8), cotton (3), soybeans (5, 9), peanuts (4, 9), and sugar beets (10).

Weed control in row crops includes all methods that were discussed in Chapter 1—mechanical, competition, crop-rotation, fire, and chemical methods. It could include biological control through predators and diseases; however, this refinement has not been perfected for weed control in cultivated crops.

Top-yielding crops usually provide maximum competition to weeds. (See Figure 21-2.) Planning for and using such practices are the first steps in an effective weed control program.

Farmers cultivate principally *to control weeds* (see Chapter 1, subsection "Tillage"). Some crops may benefit from one early cultivation to loosen the soil if it becomes hard and packed when dry. However, on other soils, research data indicate that cultivation is of no value if weeds are controlled.

Cultivation has disadvantages. It requires a considerable amount of energy in the form of fuel. Also, large tractors are an expensive investment for the farmer. Wet weather or other work may prevent timely cultivation and weeds may get ahead of the crop. Some crops grow slowly and weeds may get a head start. Some weeds are very difficult to control in the row; cultivation may injure roots and reduce yields, and heavy weed growth may develop after the last cultivation even though the crop was clean when cultivated last. Also, repeated cultivation, especially in wet soils, injures the physical condition of the soil.

Weed control in most annual crops can be divided into *early-season* and *late-season* phases. Early-season weeds usually have a greater effect on crop yields than late-season weeds. Late-season weeds make harvesting difficult, may contaminate the harvested crop (grass in cotton, weed seeds in grain), and reinfest the soil with weed seeds.

WEED CONTROL IN SPECIFIC ROW CROPS

The following methods of weed control have proven effective in various parts of the United States, but have not necessarily proven themselves in all areas. Many differing factors such as soil type, rainfall, and temperature influence a herbicide's effectiveness. Therefore, local recommendations must be considered as final authority. These can usually be obtained from your state college of agriculture or from local commercial company representatives.

Corn

Modern weed control in corn often includes appropriate use of both cultivation and herbicides. Location of corn roots should determine depth and placement of cultivator shovels. Corn should be cultivated only deep enough to remove the weeds or deep enough to cover them, to minimize root pruning. When corn is $2\frac{1}{2}$ ft tall, cultivation within 6 in. of the stem to a depth 6 in. will cut off much of the root system (see Figure 23-1). In dry weather such plants may seriously wilt following deep cultivation. Deep, late-season cultivation will nearly always reduce crop yields. Therefore, if late cultivation

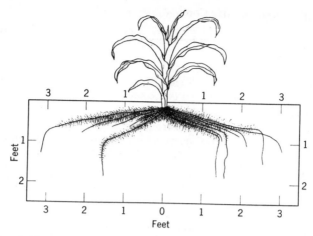

Figure 23-1. Cultivation should be shallow to avoid destroying corn roots, especially after the plant is 15 in. tall. (Adapted from the *Nebraska Research Bulletin No. 161*, 1949.)

is done, it should be with shallow sweeps. Roots dragging on the cultivator shank are a sign that you are cultivating too deeply.

Farmers have many methods of cultivation. Rotary weeders, finger weeders, and harrows are often used when the corn is small. By the time the corn is 3–4 in. tall, shovel or sweep cultivation can be initiated, throwing soil into the row to bury small weeds. Special shields may control the amount of soil thrown to the corn to avoid burying the corn. Cultivation is repeated as often as needed until the corn is 20–24 in. tall. The last cultivation is often called "lay-by" cultivation. Where cultivation is the only method of weed control, farmers usually do so 3–4 times.

Many farmers now combine a reduced-cultivation program with a herbicide program and get better weed control than by cultivation alone.

Very effective herbicides are available for weed control in corn. They can be applied during the entire life span of the corn plant. However, because of application difficulties, most herbicides are applied before the corn reaches 30 in. tall. After this, high-clearance sprayers are needed to prevent breaking the corn.

Chemical Weed Control

Since 2,4-D was introduced in the mid-1940s to control weeds in corn, many herbicides have been developed for this purpose (see Table 23-1). Space limitations, however, prevent covering each of these in detail. These herbicides are applied before the crop is planted (*preplant*); after the crop is planted, but before it emerges (*preemergence*); or after it emerges (*postemergence*). In some cases, a preplant or preemergence application is followed by a postemergence

Table 23-1.   Herbicides Used in Corn

Preplant	Preemergence	Postemergence
Alachlor	Alachlor	Ametryn
Atrazine	Atrazine	Atrazine
Butylate	CDEC	Cyanazine
Chlorbromuron	Chloramben	Cyprazine
Dalapon	Chlorbromuron	2,4-D
Diallate	Cyanazine	Dicamba
EPTC[1]	2,4-D	Dinoseb
Paraquat	Diallate	Linuron
Simazine	Dinoseb	Propachlor
	EPTC[1]	
	Paraquat	
	Propachlor	

[1] A special formulation of EPTC is usually used in corn. It contains a corn antidote.

application to give seasonlong weed control. Atrazine, alachlor, and 2,4-D are major corn herbicides.

Most of these herbicides are used to control annual weeds, both grasses and broadleaf weeds; however, a few may also give temporary control of certain perennial weeds. Combinations of two corn herbicides are often used to increase the spectrum of weed species controlled. Atrazine and alachlor are two important weed killers in corn, as measured by their widespread use. (See Figure 23-2).

Several preplant herbicides, but not all, perform better when physically incorporated into the soil. Most preplant herbicides are applied just before planting; however, with some, they may be applied several weeks ahead of normal planting date. Most preemergence herbicides are applied immediately after planting, but with some, application is delayed until just before emergence of the crop. This delay usually increases safety to the crop. Timing of application of the postemergence herbicides varies greatly for various herbicides used. Some are applied shortly after emergence, while others are delayed until lay-by time; others are intermediate. See manufacturer's label for method and time of applications, rates to be used, weeds controlled, and special precautions.

We are giving details here on the use of 2,4-D as a postemergence, directed spray to control broadleaf weeds in corn; this is because the treatment is relatively inexpensive and readily available. These factors are particularly important for developing countries where some of the herbicides listed in Table 23-1 may not be available.

Figure 23-2. Atrazine applied as a band treatment over the corn row at planting time. Usually, the weeds in the middle should be controlled by cultivation while the weeds are much smaller. The rows to the left were not treated. (Ciba-Géigy Corp.)

When corn is 3–4 in. tall, 2,4-D applied at the rate of $\frac{1}{6}$–$\frac{1}{4}$ lb/acre will control most broadleaved, above-ground annual weeds; corn that is 4–12 in. tall will withstand double this rate. Postemergence treatments with 2,4-D will not control established grasses.

Corn 12–30 in. tall will withstand $\frac{1}{3}$–$\frac{1}{2}$ lb of 2,4-D/acre for broadleaved weed control. The spray is preferably applied with nozzles dropped between the corn rows to avoid spraying most of the corn plant.

Usually 2,4-D esters are applied at the lower rates suggested above, and 2,4-D amines at the higher rate. If there are 2,4-D-susceptible crops in the area, use the amine form of 2,4-D to avoid the volatility hazard.

Corn becomes more susceptible to 2,4-D after excessively high temperatures with high soil-moisture content. After such periods the rate of 2,4-D should be lowered to reduce corn injury.

Corn varieties differ considerably in tolerance to 2,4-D. Some varieties will withstand rates of treatment higher than rates suggested above.

Milo (Grain Sorghum)

Weed control in milo is similar to corn, because several herbicides used in corn can also be used in milo. However, in general, milo is less tolerant than corn. With the development of suitable herbicides to give seasonlong weed control in milo, the rows can be planted close together.

Atrazine may be used preplant, preemergence, or postemergence for annual-weed control in milo (see Table 23-2). However, geographical limitations are

Table 23-2.   Herbicides Used in Milo

Preplant	Preemergence	Postemergence
Atrazine	Atrazine	Atrazine
Propazine	Propachlor	2,4-D
	Propazine	Dicamba
		MCPA

placed on these applications. Preplant and preemergence treatments of 2–3 lb/ acre control most annual broadleaf weeds and grasses, but postemergence treatment of about 1 lb/acre is primarily for broadleaf weeds; 1–1½ gal/acre of a nonphytotoxic oil added as a tank mix increases the postemergence activity of atrazine.

Propazine, less toxic than atrazine to milo, is used as a preplant or pre-emergence application. It is applied at rates of 2–3 lb/acre. Propazine controls most annual, broadleaf weeds and grasses. Both propazine and atrazine are fairly persistent in soils; crops other than corn or milo may be injured if planted within 18 months following their application.

Propachlor also controls most annual grasses and certain broadleaf weeds in milo. It is used as a preemergence treatment soon after planting and before weeds emerge. The rates of application are from 4 to 5 lb/acre.

Combinations of propachlor with atrazine or propazine are also used in milo as a preemergence treatment. These combinations give better overall weed control than either one alone. In general, atrazine and propazine are more effective on broadleaf weeds, whereas propachlor is more effective on grasses. When in combined use, the rate of each is reduced somewhat from rates when they are used alone.

Sorghum and corn have similar development stages of susceptibility and tolerance to 2,4-D. Grain sorghum, from 4 to 12 in. in height, is reasonably tolerant to 2,4-D. In the five to eight-leaf stage, most varieties will tolerate ⅓ lb/acre of a 2,4-D ester or ½ lb of an amine salt of 2,4-D. At about 12 in. in height, when the head starts to develop rapidly, the plant becomes susceptible to 2,4-D. When the grain is half formed, the plant again becomes tolerant to 2,4-D. (See Figure 23-3).

Soybeans

Soybeans are not planted until the soil is thoroughly warmed; thus, several crops of weed seedlings can be destroyed before planting. Management

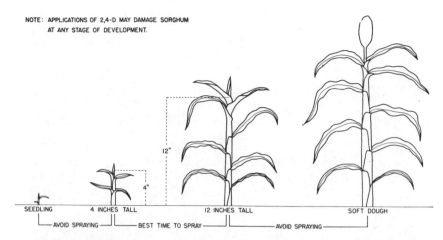

NOTE: APPLICATIONS OF 2,4-D MAY DAMAGE SORGHUM
AT ANY STAGE OF DEVELOPMENT.

Figure 23-3. Diagram indicating sorghum tolerance to 2,4-D at various stages of development. (W. M. Phillips, Kansas Agricultural Experiment Station and U.S. Department of Agriculture.

practices such as thorough seedbed preparation, adequate soil fertility, choice of a well-adapted variety, and use of good seed all contribute to a soybean crop that will compete with weeds.

The rotary hoe is an effective and economical weed killer in soybeans. For best results, use it when the ground is slightly crusted and when weeds are just emerging, and not more than $\frac{1}{4}$ in. high. This treatment is most effective when the ground is dry. However, if weeds are not more than $\frac{1}{4}$ in. tall, research shows that rotary hoeing is effective even though the soil is moist. (See Figures 23-4 and 23-5).

Figure 23-4. Left: Before rotary hoeing; a heavy infestation of green foxtail in soybeans. Fully 90% of the weeds have not emerged, but are germinated (in the white). This is the time to rotary hoe. Right: After timely rotary hoeing; note the lack of both weeds and soybean injury. (Iowa State University, Ames, Iowa.)

Figure 23-5.   Right: Soybeans with oryzalin plus metribuzin surface applied after planting, plus one early cultivation. Left: Cultivated plot, no herbicide treatment. (Elanco Products Company, a Division of Eli Lilly and Company.)

Chemical Weed Control

Almost as many herbicides have been developed to control annual weeds in soybeans as in corn (see Table 23-3). Trifluralin, alachlor, chloramben, and oryzalin are major soybean herbicides. Most farmers start their weed control program in soybeans with either a preplant, incorporated herbicide treatment or a surface-applied, preemergence treatment following planting. This may be followed with a rotary hoeing and one or two cultivations. The use of post-emergence treatment usually depends on the number of weeds that have survived earlier treatments. (see Figure 23-5).

Because space limitations do not permit a detailed discussion of each treatment in the table, see manufacturer's label, reference (2), or both, for method and time of application, rates, weeds controlled, and special precautions.

Table 23-3.   Herbicides Used in Soybeans

Preplant	Preemergence	Postemergence
Alachlor	Alachlor	Barban
	Bifenox	
Dalapon	Chloramben	Chloroxuron
Dinitramine	Chlorbromuron	2,4-DB
Metribuzin	DCPA	Linuron
Nitralin	Dinoseb	Paraquat
Paraquat	Diphenamid	TCA
Trifluralin	Fluorodifen	
Vernolate	Linuron	
	Naptalam	
	Metribuzin	
	Nitralin	
	Oryzalin	
	Paraquat	
	Propachlor	

Peanuts

The rotary weeder, flexible spike weeder, and cultivator remain as efficient tools for weed control in peanuts. The rotary weeder is often used just before and again just after peanuts emerge to control early, small, weed seedlings. The timely use of the rotary weeder may reduce cultivating time by 25% and hand hoeing up to 50%. Cultivating 2–4 times is common.

Chemical Weed Control

The peanut root grows very fast after planting, even though leaves emerge slowly. Under favorable conditions, the root grows downward in about 24 hr and may reach 2 in. long in 4 days, 6 in. long in 6 days, and 15 in. long in 12 days. In contrast, the first leaves usually do not appear above the soil until 7–12 days after planting.

Therefore when the leaves emerge, the root is deep in the soil. Root tips are well protected from preemergence herbicides applied about 6 days after planting. A delayed preemergence spray usually reduces the risk of injuring peanuts. A delayed application, however, may reduce the effectiveness of weed control because germinating weed seedlings may also have passed the optimal control period. Several preplant incorporated herbicides are selective enough to be used before planting. Preemergence herbicides are usually applied immediately after planting.

Table 23-4. Herbicides Used in Peanuts

| Preplant | Preemergence | |
	At Planting Time	At Cracking Time
Alachlor	Alachlor	Dinoseb
Dinitramine	Benefin	Diphenamid
Benefin	Chloramben	Naptalam-dinoseb
Nitralin	Dinoseb	
Trifluralin	Diphenamid	
Vernolate	Nitralin	
	Oryzalin	
	Sesone	
	Trifluralin	
	Vernolate	

Most herbicides used for annual-weed control in peanuts are applied preplant incorporated or applied preemergence (see Table 23-4). Benefin, alachlor, vernolate, and a mixture of naptalam plus dinoseb are major peanut herbicides.

Preplant-incorporated treatments are usually applied as part of seedbed preparation. Preemergence herbicides are applied either at planting time or at cracking time. Planting-time treatment is usually applied during actual planting with application equipment attached to the tractor doing the planting. Cracking-time treatment is a delayed, preemergence treatment applied when the emerging peanut shoot cracks the surface soil, but has not yet emerged. This treatment usually comes 6–10 days after planting; it may be postemergence to early germinating weeds and, if so, will kill those weeds hit by the spray.

Cotton

Weed control in cotton in previous years required considerable hand labor plus an equal amount of cultivation. Even so, cotton often was still weedy at harvest time. With mechanical harvesting, weed control became increasingly important because weedy trash gets mixed with cotton lint.

Production methods are needed that provide for the best possible growth and yields of cotton. Among these are a well-adapted variety, adequate fertility, and proper control of diseases, insects, and weeds. A vigorous crop of cotton helps control weeds through competition, especially late-season weeds.

Flaming

A flame cultivator may be used first when cotton stems are $\frac{1}{4}$ in. in diameter. Best results are obtained if the preceding weed-control program has kept weeds small. The flat-type burner which uses volatile gases, such as propane or butane, is suggested. The continued use of flaming depends on the price of fuel.

Chemical Weed Control

Cotton was once planted thick, and the stand thinned during the first hoeing. Herbicides eliminated much of the need for hand hoeing, and the desired cotton stand has been obtained by *planting to a stand.* Preparing the seedbed well, using high-quality seed, sideplacing fertilizer properly, and delaying planting until the soil is warmed to at least 60°F have provided the desired stand without chopping. In many areas growers want a stand of 45,000 plants/acre.

Herbicides used in cotton are given in Table 23-5. Trifluralin, cotoran, diuron, DSMA, and MSMA are major cotton herbicides. Certain preplant-incorporated treatments, such as trifluralin, can be done in the fall preceding spring planting, or at any time up to the date of planting. Usually the herbicide is incorporated as a part of seedbed preparation. (See Figure 23-6.)

Most preemergence herbicides are applied just after planting, often with application equipment attached to the tractor doing the planting.

Table 23-5. Herbicides Used in Cotton

Preplant	Preemergence	Postemergence
Bensulide	Bensulide	DCPA
Dalapon	DCPA	Diuron
Dinitramine	Diphenamid	DSMA
Diuron	Diuron	EPTC
Fluometuron	Nitralin	Fluometuron
Paraquat	Oryzalin	Linuron
Prometryn	Prometryn	Monuron
Nitralin	Trifluralin	MSMA
Trifluralin		Nitralin
		Oil (selective)
		Prometryn
		Trifluralin

Figure 23-6.   Diuron applied in a band over the row controlled weeds for 6 weeks after planting. Areas between rows have been cultivated. There is no need for weed chopping in the row here. (E. I. du Pont de Nemours and Company.)

Several postemergence herbicides used in cotton require directed sprays to minimize crop injury. These include diuron, DSMA, fluometuron, linuron, monuron, MSMA, nitralin, oil (selective), prometryn, and trifluralin.

Herbicidal oils like varsol or Stoddard solvent are effective postemergence sprays on cotton. Oils are applied when cotton is small; for effective control, oils must be applied while weed seedlings also are small. Oil is aimed into an 8–10-in. band in the cotton row so as not to hit the cotton leaves. The usual rate is 35–50 gal/acre; however, because it is applied in a band, only 7–10 gal are required/acre planted. Treatments are repeated every 5–7 days (see Figure 1-12). Usually two or three treatments are made, and the treatment is stopped when cracks appear in the bark of the cotton stalk; if not, the cotton may be injured.

Sugar Beets

The major weed problem in sugar beets is annual weeds, both grasses and broadleaf weeds. They are particularly troublesome at emergence through

Table 23-6. Herbicides Used in Sugar Beets

Preplant	Preemergence	Postemergence
Cycloate	Endothall	Barban
Dalapon	Nitrofen	Dalapon
Diallate	Paraquat	Endothall
EPTC	TCA	EPTC
Paraquat		Phenmedipham
Pebulate		Trifluralin
Propham		Propham
Pyrazon		Pyrazon

thinning time and then later after lay-by. Before selective preplant and pre-emergence herbicides were developed for sugar beets, these annual weeds were serious competitors with the crop. The use of herbicides now permits sugar beets to stay essentially weed free until thinning time. Moreover, they allow use of precision planting and mechanical thinning.

Herbicides now used preplant or preemergence in sugar beets are degraded fairly rapidly in the soil. Therefore, an additional herbicide is usually required later to give seasonlong weed control.

Herbicides used in sugar beets are shown in Table 23-6. Three of these are thiocarbamates—EPTC, pebulate, and cycloate. They were developed in the order listed. With each successive compound, the safety to sugar beets has increased with little, if any, loss of weed-killing ability. EPTC was still being used in the mid-1970s in only a few specific areas.

Sugarcane

Sugarcane is a perennial crop. The crop is started by planting pieces of the stalk in deep drills or furrows. New plants are vegetatively produced from buds located at the nodes of the stem; this crop is known as *plant cane*. After harvest, preparation is made for succeeding crops known as ratoon or stubble cane. Depending on weed control, disease, soil fertility and overall productivity, 2–4 ratoon crops are produced.

Sugarcane is normally grown on fertile soils, in areas with high rainfall, irrigation, or both, and usually in tropical and semitropical climates with a long growing season. These conditions, plus the perennial nature of the crop, are favorable to weeds of all kinds—annuals, biennials, and perennials.

The control of weeds in sugarcane is a serious and costly problem. Weeds must be effectively controlled until the cane "closes in" and provides sufficient shade to prevent weed growth.

Figure 23-7. Right: Tebuthiuron provides effective weed control in sugar cane. Shallow early cultivation increases effectiveness of treatment. Left: Not treated with herbicide. (Elanco Brasil.)

Historically, the crop was heavily cultivated, and methods included hand hoeing. Crop management techniques were developed to control weeds. With the advent of selective herbicides, however, sugar-cane producers quickly adapted herbicide usage to their program. See Figure 23-7.

The following herbicides are used on sugar cane: ametryn, atrazine, dalapon, 2,4-D, 2,4,5-TP, diquat, diuron, fenac, fluometuron, norea, tebuthiuron, terbacil, and trifluralin. Each of these controls a specific group of weeds and requires special instructions for application. See the herbicide label for these instructions.

Tobacco

Tobacco has very small seeds and, at first, the seedlings grow slowly; thus, the plants are started from seed in plant beds. These plants are subsequently transplanted to the field. These facts lead to a serious weed problem in the plant bed and also a weed problem in the field.

In the plant bed, weeds, soil-borne diseases, and insects and nematodes may constitute a problem, and a highly effective general-purpose soil fumigant is often used. Three materials are applied: methyl bromide, dazomet, and metham.

Methyl bromide is applied as a gas under a gas-proof cover at least 2 days before seeding. Metham is sodium methyldithiocarbamate. The seedbed is prepared and the chemical applied 3 or more weeks before planting. Covering the treated area requires half as much chemical than where no cover is used. Dazomet is tetrahydro-3,5-dimethyl-2H-thiadiazine-2-thione. The seedbed is prepared and the chemical applied 4 weeks or more before planting. The chemical is watered down and no gas-proof cover is needed.

In the field, benefin, isopropalin, diphenamid, and pebulate are applied as herbicides. Diphenamid can be applied up to 7 days before transplanting and the chemical incorporated into the soil. Also, diphenamid can be applied immediately after transplanting and the chemical applied as an over-the-top spray.

Isopropalin and pebulate are incorporated into the soil. Tobacco can be set directly into the treated soil. Also, beds can be formed (as is customary with flue-cured tobacco) after herbicide incorporation. Avoid throwing untreated soil onto the bed, or poor weed control may be the result.

Benefin is also incorporated into the soil. Tobacco plants can be set directly into the treated soil, providing that the tobacco is not made into beds. If beds are made following an original flat incorporation, the chemical is placed twice as deep in the soil. Tobacco plants set on top of the bed have difficulty in their roots penetrating through the deeply placed benefin. Therefore, benefin is recommended where the tobacco is planted essentially on flat land (as is the case with most burley and dark tobacco).

In North Carolina, tobacco on sandy soils showed no increase in yields as a result of cultivation if weeds were otherwise controlled. On loam soils and on clay-loam soils increased tobacco yields resulted with one to two cultivations. In no case were yields increased by more than two cultivations.

TREATMENT OF SERIOUS PERENNIAL WEEDS

Serious perennial weeds in row crops should be treated while confined to a small area, before they spread. Ideally they should be eradicated by using a soil sterilant. Methyl bromide, a temporary soil sterilant, can be used on many species that do not recover from depths greater than this gaseous chemical penetrates. It is a relatively expensive and laborious treatment. With methyl bromide, crops can be seeded into the treated soil within a few days after treatment without injury.

More persistent soil sterilants can also be used. Although they are less expensive than methyl bromide, treated soil may remain toxic to crops for several years after treatment. (See Chapter 29 for details on soil sterilants.)

Perennial grasses such as quackgrass, johnsongrass, and bermudagrass can be controlled by dalapon before planting beans, corn, cotton, potatoes, milo, soybeans, and sugar beets. It is applied as a preplant, foliar spray at 6 lb/acre when johnsongrass is 8–12 in. tall; quackgrass 4–6 in. tall; or comparable foliar development on other perennial grasses to be controlled. This is followed by plowing or deep discing 3 days or more after application. Additional delays are required before planting some crops.

Trifluralin applied at about double the normal rate for 2 years provides effective control of both seedling and rhizome johnsongrass in soybeans and cotton. The soil should be thoroughly tilled, to work the trifluralin in and around each rhizome. Later cultivation is needed to cut off or bury occasional surviving plants. This program permits the production of soybeans or cotton without loss of a cropping year. Similar control of bermudagrass has been observed.

Glyphosate, a new herbicide, is particularly effective on perennial grasses, but also controls many broadleaf weeds. It is applied to foliage of the plant to be controlled. Relatively nonselective, it is being used as of 1974 only on noncrop land. However, glyphosate may be registered later for use in crops to control these troublesome, perennial weeds.

LITERATURE CITED

1. *EPA Summary of Registered Agricultural Pesticide Chemical Uses*, 1973.
2. *Farm Tech. and Agri-fieldman, 29*, (February) 33 (1973).
3. Buchanan, G. A., and C. G. McWhorter, *FAO International Conference on Weed Control*, WSSA, Urbana, Illinois, 1970, p. 163.
4. Hauser, E. W., P. W. Santelman, and G. A. Buchanan, *FAO International Conference on Weed Control*, WSSA, Urbana, Illinois, 1970, p. 305.
5. Knake, E. L., *FAO International Conference on Weed Control*, WSSA, Urbana, Illinois, 1970, p. 284.
6. Meggett, W. F., *FAO International Conference on Weed Control*, WSSA, Urbana, Illinois, 1970, p. 86.
7. Nieto-Hatem, J., *FAO International Conference of Weed Control*, WSSA, Urbana, Illinois, 1970, p. 79.
8. Phillips, W. M., *FAO International Conference on Weed Control*, WSSA, Urbana, Illinois, 1970, p. 101.

9. Santelmann, P. W., E. W. Hauser, and E. Knake, *FAO International Conference on Weed Control*, WSSA, Urbana, Illinois, 1970, p. 260.

10. Schweizer, E. E., and J. H. Dawson, *FAO International Conference on Weed Control*, WSSA, Urbana, Illinois, 1970, p. 344.

For chemical, see the manufacturer's label for method and time of application, rates to be used, weeds controlled, and special precautions. Label recommendations must be followed—regardless of statements in this book. Also see the preface.

24 Vegetable Crops

Losses caused by weeds in vegetable crops can easily mean the difference between profit and loss. To ascertain more clearly the size of these losses, Wisconsin scientists carried out a special research program (4). They allowed a 15% weed stand to remain for the first $5\frac{1}{2}$ weeks in carrots before removing the weeds; this treatment reduced the yield of carrot roots by 78%. A 50% weed stand reduced carrot yield by 91%.

In onions, leaving a 15% weed stand for the first 6 weeks before removal reduced onion-bulb weight by 86%. A 50% weed stand reduced yields by 91%.

The conclusion—the first 4 weeks of crop growth are the most critical in affecting crop yields.

In vegetable crops it is especially important that the weed program approach 100% control. A few scattered weeds allowed to survive may grow so rapidly and so large that results may be nearly as serious as a thick weed infestation.

Past methods of weed control in vegetable crops have centered about the cultivator and hoe. But hoeing is expensive and labor is in short supply. With increased labor costs, more efficient methods of weed control are necessary if the grower is to earn a profit. "Cultivation of Crops" is discussed on pages 14-18 so we need not pursue the subject here. Experienced growers usually follow these practices.

There are several review papers on weed control in vegetable crops (4, 5, 6, 7, 8).

With more selective herbicides, growers can control weeds far more efficiently, with less human effort, than in the past. Savings in labor are important to the grower as well as to the consumer; both reap the benefits.

Before growers may use, or the manufacturer sell, any chemical, the chemical must be proven safe for use on food crops concerned. This is required before the U.S. Government will approve a label. Label clearance means data have been furnished showing that the chemical is useful and safe. The label gives recommendations for use. These recommendations should be followed carefully to assure that the crop will be free of herbicide residues.

Vegetable crop production is a very specialized business with crops grown intensively under varying conditions of soil and climate. A herbicide may produce excellent results under one set of conditions and either injure the crop or fail to control weeds under other conditions. In our discussion we do not have space to explain all variables. Check your local agricultural agencies for specific recommendations.

If you are not experienced in spraying herbicides on a given crop, proceed cautiously. With experience you can expand your control program.

MINOR CROP HERBICIDES

Development of an agricultural pesticide is expensive: currently, $6 to $10 million per product. Most of our available herbicides have been developed in the United States; however, many have come from Western Europe and a few are now coming from Japan. Preliminary development of herbicides is done almost exclusively in the private sector. Chemical companies do the synthesis, primary screening, and patent the products of their research. In most cases, the costs of synthesis, screening, development, registration, and service to the product must be paid ultimately through sales of the product. These costs in turn must be passed to the consumer via the cost of the product. In addition, Federal, State University, and county researchers make substantial contributions to the ultimate use of materials.

Acreage of any one vegetable crop is small when compared with that of corn, soybeans, or cotton. Therefore, vegetables are considered minor crops, and it is, generally, not economically feasible for a chemical company to develop a herbicide for a minor crop if it is not suitable for a major crop as well. As a result, most herbicides used in vegetable crops were previously developed for a major crop.

Minor crops may be somewhat less tolerant to herbicides than major crops; therefore, they must be used with greater care. Normally at least a twofold safety factor is desired (twice as much herbicide is required to induce crop injury as is needed for weed control). Also, many minor crops are expensive high-cost crops. The liability or risk of lawsuits further reduces a company's interest in adding such crops to the label.

An additional problem with minor crops is that of obtaining Federal registration for a particular use. Although a considerable amount of data used for registration of the major crop also applies to minor crops (e.g., toxicology of the compound, environmental impact, etc.), a considerable amount of specific information on the minor crop must also be developed.

The cost of this research may exceed potential profit to the chemical company, but this has been somewhat alleviated by the formation of an Interstate Research Project (IR-4) supported by the Federal government.

Because of space limitations we can discuss weed control in only a few of the more important vegetable crops. For more detailed information, see EPA Summary of Registered Agricultural Pesticide Uses (1), state publications, or other guides (2) for additional herbicide uses.

ASPARAGUS

Asparagus is a perennial crop which may remain in production 10–20 years before replanting or rotating to another crop; therefore, you must provide year-round weed control without injuring the crop.

Because asparagus is dormant in the winter and crowns are planted relatively deep (8–12 in.), growers can control weeds by overall discing during this period. During the rest of the year, herbicides are effectively used.

Diuron (0.8–3.2 lb/acre) or simazine (2–4 lb/acre) are applied after discing in the spring, but before the spears and weeds appear. These herbicides will control most annual weeds throughout the cutting season. They may be applied again after harvest and before fern development. Under most conditions these treatments give essentially year-round control of most annual weeds. Neither herbicide controls weeds on high organic peat soils.

Diuron or simazine may also be used to control annual weeds in artichokes at the same rates used on asparagus. They are applied after the last fall tillage as a directed spray.

Salts of 2,4-D are also used to control annual weeds in established asparagus. They are applied in the spring after discing and before spear or weed emergence, after the cutting season, or both. Some twisting of the spears may occur if 2,4-D is applied directly to emerged spears.

2,4-D may also be used to control certain perennial broadleaf weeds, especially field bindweed, in established asparagus. It is usually applied as a directed spray after harvest and after the fern is well developed. Scientists at Washington State University have used 2,4-D safely to control broadleaved weeds. They sprayed asparagus after a normal cutting or harvesting with 2 lb of amine 2,4-D/acre to control Canada thistle and morningglory. They repeated the experiment for 5 years. 2,4-D did not reduce the number, size, or total weight of spears produced.

If annual weeds are thick in the seedbed, varsol or Stoddard solvent can be used before the asparagus spears emerge. Apply 100 gal/acre broadcast or 25 gal directly over the row. Paraquat can also be used to remove emerged weeds at $\frac{1}{4}$–$\frac{1}{2}$ lb/acre. In some areas, chloramben is applied at 3 lb/acre to direct seeded asparagus immediately after seeding.

BEANS (DRY, SNAP, LIMA)

The several types of beans vary considerably in their tolerance to various herbicides; therefore, check the manufacturer's label before using.

Trifluralin and EPTC are the two most commonly used herbicides for annual weed control in beans. They may also be used in combination to control a wider range of weeds than either herbicide alone. Both herbicides require incorporation into the soil and are usually applied preplant; however, EPTC may be also applied postemergence. Trifluralin is usually applied at $\frac{1}{2}$–1 lb/ acre and EPTC at 3 lb/acre.

Other herbicides used for annual weed control in beans include chloramben (2–3 lb/acre), chlorpropham (4 lb/acre), DCPA (4–10 lb/acre), dinoseb (3–9 lb/acre), nitralin ($\frac{1}{2}$–$1\frac{1}{2}$ lb/acre), and fluorodifen (3–$4\frac{1}{2}$ lb/acre).

COLE CROPS (BROCCOLI, BRUSSEL SPROUTS, CABBAGE, CAULIFLOWER)

All of the cole crops appear to respond similarly to herbicides; however, it is advisable to consult the manufacturer's label for the registered use for each before using.

Cole crops may be directly seeded in the field or transplanted. Since germinating seeds are usually more sensitive to herbicides than older plants, the herbicides used in these two types of culture are somewhat different. However, once the direct-seeded seedling becomes established, it is as tolerant to herbicides as the transplant.

Direct-Seeded

Bensulide (5–6 lb/acre), CDEC (4–6 lb/acre), DCPA (4–10 lb/acre), nitralin ($\frac{1}{2}$–$1\frac{1}{2}$ lb/acre), nitrofen (3–6 lb/acre), and trifluralin ($\frac{1}{2}$–1 lb/acre) (see Figure 24-1) are used to control annual weeds in direct-seeded cole crops.

The first four of these herbicides may be incorporated into the soil or applied preemergence immediately after seeding. With preemergence treatment, sprinkle irrigation should be used if rain does not fall within a few days after application. Trifluralin must be incorporated into the soil to be effective. Nitrofen is applied as a preemergence herbicide and its activity is reduced if it is incorporated into the soil.

Figure 24-1. Annual-weed control in broccoli with trifluralin. Center: treated band. Left and right: untreated.

Transplants

The following treatments can often be applied to established, direct-seeded cole crops as well as transplants; however, if a preemergence treatment was applied, a second postemergence treatment may not be required.

Nitralin ($\frac{1}{2}$–$1\frac{1}{2}$ lb/acre) or trifluralin ($\frac{1}{2}$–1 lb/acre) may be applied as a soil-incorporation treatment before transplanting cole crops to control annual weeds.

CDEC (4–6 lb/acre), DCPA (4–10 lb/acre), or nitralin ($\frac{1}{2}$–$1\frac{1}{2}$ lb/acre) may be applied after transplanting cole crops, but before weeds emerge. Nitrofen (4–6 lb/acre) may be applied after transplants are established or 2 weeks after direct-seeded plants have emerged. These treatments control annual weeds.

CARROT FAMILY (CARROTS, DILL, PARSLEY, PARSNIPS)

One of our oldest selective weed-control methods in vegetables is the highly refined petroleum solvents used to control annual weeds in members of the carrot family. These solvents are like paint solvent or dry-cleaning fluid. It is sold under several names including Stoddard solvent, Varsol, 350° thinner, selective-weed oil, and carrot oil. Kerosene or No. 1 fuel oil has been used in the past, but they often cause a kerosenelike flavor and should be avoided. Even with more refined products, they must be applied before the carrot root has reached $\frac{1}{4}$ in. in diameter to avoid off flavors. The solvent used should be fresh from the refinery. These solvents control most emerged annual weeds and can be applied any time after the carrot has two true leaves and before the root is ¼ in. in diameter.

Oilings may be repeated 2–3 times if necessary and if done early in the season. Weeds are easiest to kill when less than 2 in. tall. Wetting sprays are applied, usually requiring 50–100 gal/acre.

High temperatures, but not over 85°F, usually provide best results. Carrots are more likely to be injured if temperature is high, humidity is high, or plants are wet when they are sprayed. Increased plant toxicity when plants are damp can be used to the grower's advantage. This will permit use of less oil, or provide more effective weed control. Therefore, some producers have a standard practice of spraying between dusk and dawn.

The following summary of herbicidal treatments for annual weed control is restricted to carrots; however, several herbicides may also be safe to use on other members of the carrot family. Check Federal registration and the manufacturer's label before using these on crops other than carrots.

Trifluralin (½–1 lb/acre) or bensulide (5–6 lb/acre) may be used as a preplant, soil-incorporation treatment. The latter is restricted to Texas. Chloroxuron (3–4 lb/acre), linuron (½–1½ lb/acre), or nitrofen (2–6 lb/acre) may be applied as a preemergence or postemergence treatment. The time of postemergence treatment for all three herbicides is somewhat different; chloroxuron—after true leaves form; linuron—when the crop is at least 3 in. tall; and nitrofen—2 weeks after crop emerges.

CELERY

Celery seed is usually planted in seed beds and then young plants transplanted into the field. It is also possible to grow celery by direct seeding into the field and thinning to stand.

Seed Beds

Celery seed beds are often fumigated with a temporary soil sterilant such as allyl alcohol or metham before planting the seed. When this fumigation is practiced, additional herbicides usually are not required before removing transplants from the seed beds.

Transplants

Basically two types of herbicides are used for annual weed control in transplanted celery. CDAA, CDEC, and trifluralin control germinating weed seeds but not emerged seedlings. However, chloroxuron, linuron, nitrofen, and prometryn control both small, emerged weed seedlings as well as germinating weed seeds. Therefore, CDAA, CDEC, and trifluralin must be applied before emergence of weed seedlings, while the others can be applied later.

Trifluralin ($\frac{1}{2}$–1 lb/acre) can be applied as a pretransplant, soil-incorporation treatment. CDAA (3–4 lb/acre) can be applied 2–3 days after transplanting and again after 3 weeks—in Florida only. CDEC (4–6 lb/acre) can be applied immediately after transplanting and again 3 weeks later. Combinations of CDAA and CDEC (3 lb/acre of each) can be used at similar times—again in Florida only.

The other four herbicides have somewhat different times of application. Those four are applied to small, emerged weeds and also control germinating weed seedlings. Nitrofen (3–6 lb/acre) must be applied within 2 weeks after transplanting. Chloroxuron (2 lb/acre) is applied between 2 and 6 weeks after transplanting and is most effective when 1–2% of a nonphytotoxic oil is added to the spray solution. Prometryn (0.8–3.2 lb/acre) is also applied 2–6 weeks after transplanting and before the weeds are 2 in. high. Linuron ($1\frac{1}{2}$ lb/acre) is applied before the crop is 8 in. tall.

Direct Seeded in the Field

Trifluralin ($\frac{1}{2}$–1 lb/acre) may be used as a preplant, soil-incorporation treatment to control annual weeds in direct-seeded celery. Nitrofen (3–6 lb/acre) is applied preemergence or postemergence within 2 weeks after crop emergence. Chloroxuron (2 lb/acre) and prometryn (0.6–0.8 lb/acre) may be applied after celery seedlings have two true leaves. Stoddard solvent or other refined petroleum oils (see carrots) can be used at the two–four-leaf stage, but before crown-leaf cups are formed.

CORN, SWEET

Weed control methods described for field corn in Chapter 23 also generally apply to sweet corn. Some varieties of sweet corn may be somewhat more sensitive to certain herbicides than field corn.

CUCURBIT FAMILY (CUCUMBERS, MELONS, PUMPKINS, AND SQUASH)

Members of the cucurbit family are primarily warm-weather crops; thus, their major weed problems concern summer-annual grasses and broadleaf weeds. Some of these are barnyardgrass, pigweed, and lambsquarter.

Table 24-1. Herbicides Used in Cucurbits

	Crop							
	Cucumber		Melon		Pumpkin		Squash	
Herbicide	Rate (lb/acre)	Appli-cation	Rate (lb/acre)	Appli-cation	Rate (lb/acre)	Appli-cation	Rate (lb/acre)	Appli-cation
Bensulide	6	PPI	6	PPI[1]	—	—	6	PPI
CDEC	4	PE	4	PE[2]	—	—	—	—
Chloramben (ammonium salt)	—	—	—	—	4	PE	4	PE
Chloramben (methylester)	2–3	PE	2–3	PE[3]	—	—	—	—
DCPA	10½	Post(1)	10½	Post(1)[4]	—	—	10½	Post(1)
Dinoseb (phenol)	1	PE	—	—	—	—	—	—
Dinoseb (ethanol and isopropylamines)	3	PE	—	—	6	PE	6	PE
Dinoseb (triethanolamine)	3	PE	—	—	6	PE	6	PE
Napthalam	8	PE	6–8	PE	—	—	—	—
Napthalam	6	Post(2)	6	Post(2)[5]	—	—	—	—
Trifluralin	1	Post(3)	1	Post(3)[2]	—	—	—	—

[1] Cantaloupe, crenshaw, muskmelon, watermelon.
[2] Cantaloupe, watermelon.
[3] Cantaloupe.
[4] Cantaloupe, honeydew, watermelon.
[5] Cantaloupe, muskmelon, watermelon.
Abbreviations: PPI—preplant soil incorporated.
PE—preemergence.
Post(1)—6–10 weeks after seeding.
Post(2)—immediately after transplanting or just before vining (1 month after planting).
Post(3)—directed spray at three–four-leaf stage, soil incorporated.

Herbicides to control these weeds are listed in Table 24-1. Note that none of these can be used on all four types of cucurbit crops; in fact, there are even further restrictions within the melons (see footnotes of Table 24-1). For example, within melons, the methylester of chloramben can be used only on cantaloupes, whereas, bensulide can be used on cantaloupes, crenshaws, muskmelons, and watermelons. Restrictions are usually based on the lack of selectivity in particular varieties to a given herbicide, but in some cases it is merely a lack of sufficient information for registration.

LETTUCE AND OTHER LEAF CROPS

Unlike cucurbit crops, lettuce and other leaf crops are cool-weather plants. Their different annual-weed complex may include annual bluegrass, crabgrass, chickweed, henbit, common groundsel and many others. Herbicides to

Table 24-2. Herbicides Used in Other Leafy Vegetable Crops

	Crop			
	Collards, kale, mustard greens, turnip greens		Spinach	
Herbicide	Rate (lb/acre)	Application	Rate (lb/acre)	Application
CDEC[1]	3–4	PE	3–4	PE or Post
Chlorpropham	—	—	3–4	PE
Cycloate	—	—	3–4	PPI
DCPA	5–10	PE	—	—
Trifluralin	$\frac{3}{4}$	PPI	—	—

[1] CDEC may also be used as a preemergence treatment in endive, escarole, and Hanover salad at 4 lb/acre.

Abbreviations: PE—preemergence.
PPI—preplant soil incorporation.
Post—postemergence.

control these weeds in commercial leaf crops, except lettuce, are given in Table 24-2.

Lettuce

Although lettuce is usually a cool-weather crop, it may be subjected to high temperatures. In Arizona and Southern California lettuce is planted in late August and September when air and soil daytime temperatures may exceed 100°F. Lettuce will not germinate above 90°F. However, by keeping the beds wet, they are cooled by the constant evaporation of water from the bed surface in this arid climate. This cooling allows shallow-planted lettuce seeds to germinate. Under these conditions, the annual-weed complex tends to be more summerlike than those weeds named above.

Many herbicides which control annual weeds in lettuce are applied as a preplant, soil-incorporation treatment. These include benefin (1–1½ lb/acre) (see Figure 24-2), bensulide (5–6 lb/acre), CDEC (2–4 lb/acre), propham (4–6 lb/acre), and a combination of benefin (1–1½ lb/acre) plus propham (3–4 lb/acre).

Preemergence applications are also used: CDEC (2–4 lb/acre), chlorpropham (1–3 lb/acre), and propham (4–6 lb/acre). Propham (4–6 lb/acre) can also be used postemergence when lettuce has at least four or more leaves.

Figure 24-2. Annual-weed control in lettuce with benefin. Center: treated band. Left and right: untreated.

Paraquat can be used before the crop emerges as a contact spray to control emerged, annual weeds at ½–1 lb/acre; crop plants emerged at time of application will be killed. Also, CDEC and chlorpropham may injure lettuce in some areas. Always check labels and with local extension agents.

Pronamide has been evaluated for several years in lettuce and with recent registration, it will probably come into general use.

Other Leaf Crops

Other leafy vegetable crops included here are collards, kale, mustard greens, turnip greens, and spinach. Tolerances of these first four crops to several herbicides are quite similar; however, spinach reacts differently. (See Table 24-2.)

ONIONS

Although the home gardener often grows onions from bulbs (sets) or transplants, most commercial growers use seed. Weeds are particularly serious in onions produced from seeds because onions germinate and emerge relatively slowly, and their cylindrical-upright leaves make them poor competitors. However, these same leaves are an advantage when using certain contact herbicides whose spray droplets bounce off onion leaves, but remain on leaves of many broadleaf weeds.

Several herbicides used as early postemergence treatments for annual weed control in onions require that they be applied only at certain stages of growth to avoid injury to the crop. These stages are classified as loop (crook), flag, one

Figure 24-3. Barnyardgrass control in onions with DCPA .Center: treated band. Left and right: untreated.

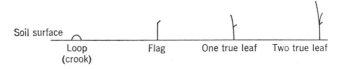

Figure 24-4. Stages of onion growth.

true leaf, two true leaf, and so on (Figure 24-4). The following discussion applies primarily to direct-seeded onions.

Highly refined petroleum solvents have been used to control annual weeds that emerge before the onions. It is applied at a rate of 30–50 gal/acre just before the onions emerge. It has no residual soil activity, hence a second flush of weeds may appear, particularly if the soil has been cultivated. If needed to control this "second round" of weeds, a special light aromatic oil, sometimes referred to as onion oil, has been used preemergence to onions at about 75 gal/ acre and postemergence to onions at 38 gal/acre.

DCPA (5–10 lb/acre), chlorpropham (4–8 lb/acre), bensulide (5–6 lb/acre), and nitrofen (3–4 lb/acre) control many broadleaf weeds and grasses in onions when applied preemergence. CDAA (6 lb/acre) can also be used preemergence to control annual weeds in onions, but application is delayed until just before onions emerge or in the loop stage.

DCPA, chlorpropham, and CDAA can also be used postemergence to onions at the above rates to control germinating weed seeds. They usually will not control emerged weeds. However, nitrofen (3–4 lb/acre) not only controls germinating weed seeds but also small emerged weeds of several species. It should not be applied after the two- to three-leaf stage of onions.

Chloroxuron (2–4 lb/acre) controls many broadleaf weeds when applied after onions reach the two- to three-leaf stage and weeds are 2 in. or shorter.

In some states only the lower rate is used and in some localities, 1 % emulsifiable nonphytotoxic oil is added to the spray solution.

Chlorpropham, chloroxuron, and CDAA should not be used on sandy soils because the crop may be injured. Bensulide can be used only in Texas and New Mexico.

PEAS, GREEN OR ENGLISH

Because peas are a cool-weather crop, the primary weed problem is annual, cool-weather weeds such as chickweed, henbit, shepherdspurse, wild oats, and annual bluegrass. The perennial weed, Canada thistle, is also a problem in some areas.

Several herbicides have been developed to control annual weeds in peas. Nitralin ($\frac{1}{2}$–1 lb/acre), trifluralin ($\frac{1}{2}$–$\frac{3}{4}$ lb/acre), and EPTC (2 lb/acre) are used as a preplant, soil-incorporation treatment. EPTC is used only in western Washington. Chlorpropham (4–6 lb/acre), propachlor (4–5 lb/acre), CDAA (4 lb/acre), and fluorodifen (3–4 lb/acre) can be used preemergence.

The several forms of dinoseb have been used for many years to control annual weeds in peas at various stages of crop growth. The triethanolamine salt (6 lb/acre) and a mixture of ethanol and isopropanolamine salts (9 lb/acre) are used as a preemergence treatment. The triethanolamine salt may also be used at the cracking stage (emerging pea seedling cracks the soil, but has not fully emerged) at 3 lb/acre.

Four forms of dinoseb may also be used as a postemergence treatment: (1) the phenol at the 4–6-in. stage—1.3 lb/acre; (2) the ammonium salt at the 2–8-in. stage—1.2 lb/acre; (3) the ethanol-isopropanolamine salt combination at the 2–8-in. stage—2.25 lb/acre; and (4) the triethanolamine salt at the 2–6-in. stage—1.13 lb/acre. These postemergence treatments should be applied before the flower buds appear. The area should not be grazed nor the "hay" fed to livestock within 40 days after treatment.

Dalapon ($\frac{3}{4}$ lb/acre) may be applied to control annual grasses at the 2–6-in. stage. MCPB may be used to control susceptible, annual broadleaf weeds and Canada thistle when the peas have six–twelve nodes, but do not apply later than three nodes before flowering.

POTATOES

The major weed problem in potatoes is annual, broadleaf weeds and grasses; in some areas quackgrass can also be serious. Dalapon (10 lb/acre) is sometimes used to control quackgrass as a preplant treatment to emerged quackgrass, followed by plowing 2–3 weeks later and before planting the potatoes.

Since potatoes are planted relatively deep and are often "hilled up" or "dragged off," cultivation is an effective method to control annual weeds. In fact it may be the only method used in some areas. This is particularly true on light, sandy soils with low organic matter where several of the following herbicides have been shown to injure the crop.

Space limits a detailed description of the great variety of times and methods of applying these herbicides. Check the manufacturer's label, local extension agents and with company representatives. The following herbicides are used to control annual weeds in potatoes by a soil application; some are applied before planting and others after planting: CDAA (3–4 lb/acre), CDEC (3 lb/acre), dalapon (4–10 lb/acre), DCPA (10 lb/acre), diallate (2 lb/acre), dinoseb (1–6 lb/acre), diuron (0.8 lb/acre), diphenamid (6 lb/acre), EPTC (4–6 lb/acre), linuron (2 lb/acre), oryzalin ($1\frac{1}{2}$–2 lb/acre), maloran (3 lb/acre), metabromuron (3 lb/acre), trifluralin (1 lb/acre), and vernolate (3–4 lb/acre). Some of these are restricted to specific states or areas of the country.

Dinoseb, paraquat, and petroleum solvents can be used as a contact spray to kill emerged, annual weeds before potatoes emerge. They are also used for preharvest vine killing.

TOMATOES

Tomatoes are grown commercially from transplants or direct-seeded in the field. Although annual weeds are the major problem by either planting practice, they are particularly serious in direct-seeded tomatoes. Weeds can be removed from direct-seeded tomatoes during thinning, but by this time they have already delayed the growth of young seedlings. Furthermore, the cost of labor for thinning may be doubled or tripled by heavy, annual-weed infestation.

Some common, annual weeds of tomatoes are barnyardgrass, crabgrass, foxtails, lambsquarters, purslane, mustard, smartweed, and nightshade. Fields containing perennial weeds such as quackgrass, johnsongrass, and field bindweed should be avoided, because there are no selective methods to control these weeds in tomatoes.

Blind tillage is sometimes practiced in direct-seeded tomatoes. In this method a small ridge about 1 in. high is placed over the seed row during planting. Just before tomato seedlings grow into the ridge, the ridge and emerging weeds are removed with a duck-foot shoe. This allows the tomato seedlings to emerge relatively free of weeds. Timing is critical for removing the ridge to obtain effective weed control and not injure the tomatoes. This technique has also been used effectively on peppers.

Stoddard solvent or similar selective weed oils may be used in direct-seeded tomatoes when annual weeds emerge before the tomatoes. About 20–40 gal/acre of oil are applied when 2–4% of the crop has emerged. Emerged tomato plants will be killed, but a considerable excess of tomato seeds is usually sown; and many seedlings are removed in the thinning operation. Uniform emergence of tomato plants is essential for this method to be effective. As with blind tillage, the technique of preemergence, selective oiling can be used on several other crops.

Paraquat ($\frac{1}{2}$–1 lb/acre) in 5–10 gal of water by aircraft or 20–60 gal of water by ground rig can replace Stoddard solvent in this procedure for tomatoes and certain other crops.

Bensulide (5 lb/acre), CDAA (3 lb/acre), CDEC (4–6 lb/acre), and diphenamid (4–6 lb/acre) can be applied preemergence to control annual weeds in direct-seeded tomatoes. In areas of limited rainfall these herbicides are usually incorporated into the soil before planting. Other preplant, soil-incorporated herbicides for tomatoes are pebulate (4 lb/acre), diphenamid (4 lb/acre) plus trifluralin ($\frac{1}{4}$ lb/acre) (see Figure 24-5), and diphenamid (4–6 lb/acre) plus pebulate (4 lb/acre). CDAA plus CDEC (3 lb/acre of each) as a preemergence treatment, and CDEC plus pebulate (3 lb/acre of each) as a preplant, soil-incorporation treatment have also been used in some areas.

Figure 24-5. Annual-weed control in direct-seeded tomatoes with diphenamid plus trifluralin; band treatment on top of beds. Foreground: not treated. (Elanco Products Company, a Division of Eli Lilly and Company.)

In general, transplants are more tolerant to herbicides than the direct-seeded crop. However, when the direct-seeded crop has four true leaves, it is generally tolerant to herbicides. Most of the following herbicidal treatments apply to transplants as well as the established, direct-seeded crop. The size of the plant, method of application, and other critical factors vary for each of these herbicides; therefore, check detailed information from the manufacturer's label, local extension staff and company fieldmen before use.

These herbicides can be used to control annual weeds in transplanted or established plants of tomatoes: bensulide (5 lb/acre), CDAA (6 lb/acre), CDEC (6 lb/acre), chloramben (4 lb/acre), chlorpropham (4 lb/acre), DCPA (5–10 lb/acre), diphenamid (4–6 lb/acre), nitralin ($\frac{1}{2}$–$1\frac{1}{2}$ lb/acre), pebulate (3–6 lb/acre), and trifluralin ($\frac{1}{2}$–1 lb/acre). Pebulate also suppresses the growth of yellow and purple nutsedge.

Except for trifluralin and nitralin, the preceding tomato herbicides have a relatively short period of soil activity; therefore, it is often necessary to apply both a preemergence or preplant treatment as well as a lay-by treatment to give seasonlong control.

LITERATURE CITED

1. *EPA Summary of Registered Agricultural Pesticide Chemical Uses*, 1973.
2. *Farm Tech. and Agri-fieldman, 29*, (February) 33 (1973).
3. Bruns, V. F., and W. J. Clore, *Weeds, 4(4)*, 393 (1956).
4. Dallyn, S., and R. Sweet, *FAO International Conference on Weed Control*, WSSA, Urbana, Illinois, 1970, p. 210.
5. Danielson, L. L., *FAO International Conference on Weed Control*, WSSA, Urbana, Illinois, 1970, p. 245.
6. Menges, R. M., and T. D. Longbrake, *FAO International Conference on Weed Control*, WSSA, Urbana, Illinois, 1970, p. 229.
7. Orsenigo, J. R., *FAO International Conference on Weed Control*, WSSA, Urbana, Illinois, 1970, p. 198.
8. Romanowski, R. R., *FAO International Conference on Weed Control*, WSSA, Urbana, Illinois, 1970, p. 184.
9. Shadbolt, C. A., and L. H. Holm, *Weeds, 4(2)*, 11 (1956).

For chemical, see the manufacturer's label for method and time of application, rates to be used, weeds controlled, and special precautions. Label recommendations must be followed—regardless of statements in this book. Also see the preface.

25 Fruit and Nut Crops

Weeds can damage fruit and nut crops seriously and in many ways. In newly planted crops, weeds compete directly with young trees for soil moisture, soil nutrients, carbon dioxide and perhaps light. In older plantings weed competition may be less serious, but still may reduce yields noticeably.

In addition, weeds may harbor plant diseases, insects, and rodents such as field mice that may girdle the trees. Also, weeds such as poison ivy may interfere with harvest. Where crops are picked up from the ground, such as some nut crops, weeds may seriously interfere with harvesting. The wasteful use of water by weeds is always important, especially in arid regions.

The crops discussed in this chapter are perennial plants. They are propagated in nurseries and transplanted to fields. Our discussion is limited to field-weed control. Although nurseries have serious weed problems, they are not covered here because tree tolerance to most herbicides varies according to age, soil type, and so on. See the herbicide label for more specific instructions.

Troublesome perennial weeds, such as quackgrass, johnsongrass, bermuda-grass, nutsedge, field bindweed, or Canada thistle, should be brought under control before setting a new orchard or making a small fruit planting. Perennial weeds can be controlled much easier and at less cost before planting than afterwards.

In most perennial crops where perennial weeds are absent or controlled before planting the crop, the primary weed problem is annual weeds. However, perennial weeds may invade the area later. This is especially true if perennial weeds are tolerant to the herbicides used. Also, absence of competition provides an essentially noncompetitive environment for tolerant weeds, including perennials, and they flourish.

A total weed-control system in fruit and nut crops may combine several methods including cultivation, mowing, mulching, and herbicides.

Weed control in deciduous fruit and nut crops has been reviewed by A. H. Lange (4). L. S. Jordan and B. E. Day (3) reviewed weed control in citrus crops.

CULTIVATION

When considered as a single weeding operation in tree crops, cultivation is just as effective and economical as with other tilled crops. Growers know well the advantages of cultivation. As for the disadvantages, shallow feeder roots are torn up, soil structure is changed, soil erosion increases especially on hill sides, weeds under trees are difficult to control by mechanized methods, and cultivation brings new weed seeds to the surface where they may germinate and grow. For one or more of these reasons, growers are depending less and less on cultivation by itself.

MOWING

Mowing is popular in borders and areas between trees of many fruit and nut crops, especially where soil erosion is serious. These areas can be maintained as short turf, effectively controlling erosion while keeping plant or weed competition to a minimum. Mowed turf gives a clean and neat appearance. The area under trees and vines is kept weed free by appropriate use of herbicides or mulches.

MULCHING

Mulching usually controls weeds by depriving them of light. Also, most mulches conserve soil moisture. Decomposable organic matter, such as straw, is used in many horticultural crops including strawberries and young fruit-tree plantings. Also, in recent years, both black plastic sheets and paper sheets impregnated with asphalt have been used. Plants are set through small holes made in these sheets.

HERBICIDES

Orchards

Twelve tree fruits are grouped under "Orchards." They are apples, apricots, avocados, cherries, peaches, pears, and plums—plus five citrus crops; grapefruit, lemons, limes, oranges, and tangerines. They are listed in Table 25-1 along with four other tree fruits and four nut crops. Seventeen commonly used herbicides are also given.

Weed oil, dinoseb, and paraquat are contact herbicides applied to foliage of annual weeds.

2,4-D is a foliar-translocated herbicide used primarily to control perennial broadleaf weeds, especially field bindweed and Canada thistle. It also controls annual broadleaf weeds. To avoid the volatility hazard, use a salt form.

Dalapon, also a foliar-translocated herbicide, mainly controls perennial

Table 25-1. Herbicides Used on Established Orchard Crops[1]

	Bromacil	2,4-D Amine	Dalapon	Dichlobenil	Dinoseb	Diphenamid	Diuron	DSMA, MSMA	EPTC	Napropamide[2]	Oryzalin	Paraquat	Simazine	Terbacil	Trifluralin	Weed Oil
Almond				X	X			X	X	X	X	X			X	X
Apple		X	X	X	X	X	X			X	X	X	X	X		X
Apricot			X		X					X	X	X	X		X	X
Avocado				X						X	X	X				X
Cherry			X	X						X	X	X	X			X
Date				X						X						X
Fig			X							X	X					X
Filbert			X	X						X	X	X				X
Grapefruit	X		X	X	X		X	X	X	X	X	X	X	X	X	X
Lemon	X		X	X	X		X	X	X	X	X	X	X	X	X	X
Lime			X	X	X	X		X		X						X
Nectarine			X							X	X	X			X	X
Olive						X				X	X					X
Orange	X	X	X	X	X	X	X	X	X	X	X	X	X	X	X	X
Peach			X	X	X	X				X	X	X	X	X	X	X
Pear	X	X	X	X	X		X			X	X	X				X
Pecan			X	X		X				X					X	X
Plum			X	X	X					X	X	X	X	X		X
Tangerine		X	X			X	X	X		X	X				X	X
Walnut, English			X	X		X		X		X	X	X			X	X

[1] Most of these control annual weeds. However, specific ones also control certain perennial weeds. Several also have restrictions as to soil type, geographical location, use on bearing or nonbearing crops, and minimum time between application and harvest. Read labels carefully and check with local agricultural authorities for exact usage.

[2] Use limited to California

grasses, especially johnsongrass, quackgrass, and bermudagrass. It also controls annual grass weeds.

DSMA and MSMA, also foliar-translocated, are primarily used to control johnsongrass and nutsedge as well as some annual weeds.

The remaining herbicides in Table 25-1 are applied to the soil primarily to control germinating annual broadleaf weeds and grasses. Certain of these soil-applied herbicides may also control specific perennial weeds. Most of these herbicides are applied to the soil surface and carried into the soil by rainfall or sprinkle irrigation. However, EPTC and trifluralin require soil incorporation or other techniques because of their relatively high volatility. EPTC may also be metered into irrigation water. Beyond controlling many annual weeds, EPTC also suppresses growth of nutsedge. Trifluralin controls certain perennial weeds, for example field bindweed, when applied as a subsurface layer in the soil. (See Figures 7-8 and 7-9.)

Some of these soil-applied herbicides are relatively persistent in soil and with one annual treatment may control annual weeds for a full year. (See Table 4-1.)

Orchards may be maintained with cultivation or noncultivation methods. Noncultivation methods are usually mowing or the use of herbicides. The specific method used depends somewhat on the crop, but perhaps more

Figure 25-1. Strip application of a herbicide in an almond orchard. The untreated areas between the treated strip will later be mowed, cultivated, or both.

important is the slope of the land and soil type. Some type of ground cover is desirable on hillsides to prevent erosion; mowing and use of a herbicide is important. On relatively flat terrain, cultivation, mowing, and herbicides are used.

Often a combination of these methods is used. Frequently a herbicide band is used down the tree row and cultivation or mowing is used between rows. This permits mowing or cultivation and minimizes the amount of herbicide used (see Figure 25-1). Cultivation close to the trees may cause mechanical damage to the trunks and to the root systems.

Resistant Species

The key to successful chemical weed control in orchards is closely related to the weed species present. Once a program of preemergence herbicides has been selected, it will usually have to be revised periodically to prevent an increase of resistant weed species; for example, if simazine or terbacil is being used, the annual grasses will soon be a problem. If diuron and nitralin are being used, weeds such as groundsel, burclover, wild oats and other resistant species will usually take over. Alternating herbicides on sequential years or the use of a combination of two or more herbicides usually prevents the development of a resistant-weed problem.

Grapes

Weeds can be serious competitors of the grapevine. This is especially true of deep-rooted perennial weeds in unirrigated vineyards of the West where little, if any, rain falls during the growing season. Here the grapevine depends totally on moisture stored in the soil from winter rains, and this is usually not sufficient for both the grapevine and perennial weeds. Annual weeds also compete especially in young plantings.

As with orchards, annual weeds in the crop row are often controlled with herbicides and the interrow area is cultivated or mowed. Cultivation close to the grapevine often injures the plant. In some soils, repeated cultivations for weed control can hasten the development of a plow sole which can impede water penetration. A mowed cover crop between rows with a herbicide-treated strip down the row is used in some areas. See Figure 25-2.

Annual weeds in vineyards can be controlled by about ten herbicides including contact sprays such as weed oil, dinoseb, or paraquat; and soil applications of diuron, EPTC, dichlobenil, simazine, oryzalin, or trifluralin. Some contact sprays should not touch the trunk and foliage. (See the label.)

Field bindweed, a serious perennial broadleaf weed, can be controlled by careful selective placement of 2,4-D amine in grapes, which are quite sensitive

Figure 25-2.   Strip application of a herbicide in a vineyard. The untreated areas between
the treated strip will be mowed or cultivated later.

to 2,4-D. To place the herbicide correctly, drag a wax bar impregnated with
2,4-D amine over the weeds, or use a hooded boom and low-pressure whirl-
chamber nozzles or flooding nozzles to deliver large droplets of a 2,4-D
formulation. Only acid or amine forms of 2,4-D should be used; they are
essentially nonvolatile. Avoid contact with the grapevines. Apply when field
bindweed is in the bloom stage and growing vigorously. This application is
safest on grapes after shatter following bloom, but before shoots reach the
ground.

Field bindweed can also be controlled in vineyards by injecting trifluralin
as a subsurface layer 4–6 in. deep. (See Figures 7-8 and 7-9.)

Perennial grasses in vineyards can be controlled with directed sprays of
dalapon. Repeat applications to regrowth are required about once a month,
or about five times during the growing season. Avoid spraying fruit or foliage.
Do not apply within 30 days before harvest.

Blueberries

Weed problems and general methods of control of weeds in blueberries are
similar to those for grapes. The herbicides usually used are weed oil, dinoseb,
2,4-D amine, chlorpropham, dichlobenil, diuron, oryzalin, and simazine.

Caneberries

Caneberries or brambles include mainly blackberries, boysenberries, dew-berries, loganberries, and raspberries. To control annual weeds in these crops, weed oil, 2,4-D amine, dichlobenil, dinoseb, diphenamid, chlorpropham, oryzalin, and simazine are used.

Weed oil and dinoseb are contact sprays used to control annual broadleaf weeds and grasses. 2,4-D amine is used to control sensitive, annual and peren-nial broadleaf weeds. All three herbicides are applied to the foliage of weeds to be controlled. Directed sprays with large spray droplets should be used to avoid drift to crop plants.

The remaining herbicides listed above are applied to the soil to control annual weeds as they germinate. These are not effective on emerged weeds.

Strawberries

Weed control in strawberries may cost up to several hundred dollars per acre per year. By using proper herbicides and other management practices, this cost can be reduced drastically. The lack of available labor for hand weeding has stimulated the use of herbicides in this crop.

Figure 25-3.   Black polyethylene plastic used as a mulch for weed control in strawberries. It also conserves moisture and keeps berries clean.

Strawberry beds are usually abandoned after 1–3 years as a result of either disease or serious weed infestation. Appropriate use of herbicides can considerably extend the productive life of many plantings, especially as more effective disease controls become available.

Varieties vary considerably in their tolerance to herbicides. Research work is needed to establish varietal tolerance to each herbicide treatment.

In some areas, fields intended for strawberries are fumigated with a mixture of methyl bromide and chloropicrin before planting. This treatment not only kills many weed seeds but also soil-borne disease organisms; for example, verticillium and nematodes. Methyl bromide is the primary weed killer of this mixture. Malva, burclover, and filaree seeds are often not killed by this treatment.

Methyl bromide and chloropicrin are gases under normal temperatures and pressures. They are sold in pressurized containers as a liquid. Both are injected into the soil about 6 in. deep through chisels spaced about 12 in. apart on a tool bar. Immediately after treatment the treated area is sealed with a gastight tarpaulin (polyethylene) for at least 48 hr. Injection of the fumigant and covering with the tarpaulin is completed in one operation with all equipment mounted on a single tractor. See Figure 25-3.

Strawberries may be planted three days after removing the tarpaulin. If weeds resistant to this treatment emerge later, they can usually be controlled with a postemergence treatment of chloroxuron at least 60 days before harvest.

Other herbicides which control annual weeds in strawberries are chloroxuron, dinoseb, DCPA, diphenamid, and sesone. Varietal the responses of strawberries to herbicides vary considerably and the degree of dormancy differs in various locations. Check the label and with your local Extension agent or other authoritives before using any of these herbicides.

LITERATURE CITED

1. *EPA Summary of Registered Agricultural Pesticide Chemical Uses*, 1973.
2. *Farm Tech. and Agri-fieldman*, *29* (February) 33 (1973).
3. Jordan, L. S., and B. E. Day, *FAO International Conference on Weed Control*, WSSA, Urbana, Illinois, 1970, p. 128.
4. Lange, A. H., *FAO International Conference on Weed Control*, WSSA, Urbana, Illinois, 1970, p. 143.

For chemical, see the manufacturer's label for method and time of application, rates to be used, weeds controlled, and special precautions. Label recommendations must be followed—regardless of statements in this book. Also see the preface.

26 Pastures and Ranges

Hundreds of kinds of weeds infest pastures and ranges. These include trees, brush, broadleaf herbaceous weeds, poisonous plants, and undesirable grasses. Control of trees and brush in pastures and range will be emphasized in the next chapter. We will limit this chapter primarily to control of weeds and grasses.

Almost half of the total land area of the United States is used for pasture and grazing (7, 18). Nearly all of this forage land is infested with weeds, some of it seriously so. These weeds interfere with grazing, lower yield and quality of forage, increase costs of managing and producing livestock, slow livestock gains, and reduce quality of meat, milk, wool, and hides. Some weeds are poisonous to livestock. The total cost of these losses is hard to estimate. (See Figure 1-2.) Losses from undesirable woody plants are given in Chapter 27. Controlling heavy infestations of some woody weeds has increased forage yields from 2 to 8 times.

Grass yields increased 4 times after the removal of sagebrush in Wyoming (1). Forage consumed by cattle increased 318% on native Nebraska pasture after perennial broadleaved weeds were controlled by improved agronomic practices and use of 2,4-D. This 318% increase was a result of better pasture species, deferred and rotational grazing, and effective weed control (8). Spraying with 2,4-D on high-level-fertility but weedy pastures increased consumption of forage by cattle 1000 lb/acre over the no-weed-control, high-fertility treatment (12).

These results emphasize the usual need for improved management practices along with a better weed-control program. Both must be used together.

Weed control programs for pastures and ranges often include a combination of mechanical, fire, and chemical methods. Biological control has been effective on certain species. (See Figure 1-10.) The cost of control programs must be critically evaluated in relation to potential return, especially on range land. See Figure 26-1.

Table 26-1. Chemical Composition of Grassland Weeds Compared to Timothy and Red Clover. Sampling Date June 5–10 (22)

Plant	Growth Stage	Number of Samples	Mean Percentage Composition (Air-Dry Basis)				
			N	P	K	Ca	Mg
Timothy	Early heading	19	1.55	0.26	2.17	0.34	0.10
Red Clover	In buds, before bloom	19	2.84	0.25	1.09	1.88	0.42
Tufted vetch	Early bloom	3	3.58	0.30	1.52	1.52	0.30
Yarrow	In buds, before bloom	8	1.56	0.31	2.35	0.82	0.18
Oxeye daisy	50% heads in bloom	7	1.63	0.34	2.48	0.94	0.21
Daisy fleabane	In buds, before bloom	11	1.47	0.38	2.12	1.12	0.20
Common dandelion	Mostly leaves	14	2.25	0.44	3.39	1.21	0.43
Yellow rocket	After bloom	7	1.44	0.24	1.55	1.23	0.17
Plantain	Mostly leaves	11	1.48	0.30	2.10	2.55	0.46
Narrowleaf plantain	Mostly leaves	5	1.85	0.37	1.90	1.90	0.33
Yellow dock	50% heads in bloom	13	1.84	0.30	2.29	1.11	0.42
Tall buttercup	In bloom	7	1.45	0.31	1.98	0.94	0.25
Wild carrot	Vegetative growth leaves	5	2.52	0.54	2.37	1.92	0.44
Mouseear chickweed	In bloom	11	1.73	0.41	3.14	0.70	0.26
Cinquefoil	Early bud stage	2	1.49	0.28	1.31	2.08	0.33
Common milkweed	Vegetative growth	2	3.02	0.47	3.08	0.80	0.45
Sensitive fern	Vegetative growth	6	2.27	0.48	2.50	0.65	0.39
Quackgrass	Before heading	11	1.82	0.28	2.14	0.36	0.10

CHEMICAL COMPOSITION OF GRASSLAND WEEDS

Nutrient or chemical composition of grassland weeds is important to livestock farmers for two reasons. First, how much do weeds contribute to the livestock ration? Second, weeds compete for nutrients and water needed by more palatable and more desirable species and thus cut down yields of desirable forage.

Scientists collected forage and weed samples before mowing in an intensive agricultural area of the Connecticut Valley in Massachusetts. Table 26-1 shows nutrient content of various plants as determined by chemical analysis (22). In addition to chemical composition, freedom from toxins, palatability, yield and persistence must be evaluated in determining the worth of a species for forage.

Choice of the most desirable forage species for the area is perhaps the first step in pasture improvement.

Proper management will control some weeds by itself. Proper management favors heavy plant growth, and many annual weeds are crowded out. For example, broomsedge disappears from pastures of the southeastern United States when the pastures are properly fertilized and seeded to high-yielding species such as Ladino clover, orchardgrass, or fescue.

Some weeds are favored by the recommended agronomic program. For example, dock responds to high soil fertility and favorable moisture. It is favored as much or more than the desired species.

In some cases weedy pastures will best be plowed, fertilized if needed, and reseeded. These are usual steps in areas favored with adequate moisture. In many other areas desirable species will flourish if grazing pressure is temporarily removed or time of use adjusted, and weedy species brought under control. This latter program is often most practical where rainfall is limited. Reseeding may also hasten establishment of desirable species in dryland areas. A weed-control program is usually necessary to allow new seedlings to become established.

MOWING

Mowing in the past was often recommended to control pasture weeds. Now, mowing is of little importance for rangeland weed control, and it is steadily becoming less important on more intensively grazed areas. Mowing has an effect on some kinds of weeds, but not on others; it is more effective on upright-growing annuals, but is ineffective on those with leaves and seed heads

Figure 26-1. Top: Before treatment with herbicides for weed control. Bottom: After treatment showing a nearly weed-free pasture. (Fisons Pest Control, Ltd., Chesterford Park, England.)

close to the ground. Mowing during early flowering usually slows down or stops seed production of the upright types. Repeated mowing will kill some tall-growing, perennial weeds. With most perennials mowing is best done during early flowering and repeated as needed.

Mowing is often disappointing. It tends to improve the appearance of the area, but kills few perennial weeds. In North Carolina, 5 years of monthly mowing was required to control horsenettle. Weekly mowing for 3 years (about 18 times during each growing season) reduced wild garlic plants in bermudagrass turf by only 52%.

In Nebraska, mowing a native grass pasture in either June or early July for 3 years left 65% of the perennial broadleaved weeds still living at the end of the experiment (8). After 20 years of mowing, 24–38% of the ironweed plants persisted (10).

Mowing may be used to good advantage in new grass-legume seedings to lessen weed competition. Clipping the tops off of broadleaved weeds may sufficiently reduce weed competition to permit survival of seedling grasses and legumes. However, mowing also clips the tops off the forage species, setting them back to some extent.

HERBICIDE CONTROL

Pasture and grazing areas are well suited to chemical weed control. As effective selective herbicides are developed, it is often possible to control a weed with little or no injury to desirable forage species. These forage species then respond with increased ground cover and larger yields. See Figure 26-2, 26-3.

Control of brush and woody weeds is covered in Chapter 27, while weed control in small-seeded legumes is discussed in Chapter 22. Control of these weed problems in pastures need not be repeated here. Many broadleaved weeds of pastures and grazing areas can be controlled by using 2,4-D;2,4,5-T, or dicamba. Picloram is used in some states for special problems.

Paraquat has also been used to renovate pastures where it is desirable to kill all emerged annual plants before land preparation and reseeding. It has no soil residual effect and therefore desirable species can be sown immediately after treatment.

LIVESTOCK POISONING AND ABORTION

Most herbicides in common use in grazing lands are relatively nonpoisonous. If recommendations on the label are followed, no poisoning should occur.

Some herbicides, such as 2,4-D and similar compounds, are thought to

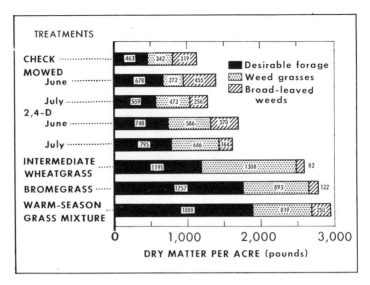

Figure 26-2. The effect of weed-control treatments on average pounds of dry matter per acre eaten by cattle. The seedbed for wheatgrass, bromegrass, and the warm-season grass mixture was plowed, killing many weeds. In addition, these plots were sprayed annually with 2,4-D for weed control. (D. L. Klingman and M. K. McCarty, U. S. Department of Agriculture and University of Nebraska (7).)

Figure 26-3. Left: In research plots, tebuthiuron removed weeds, leaving desirable grass forage. Right: No herbicide treatment, with heavy weed infestation. (Dr. J. F. Stritzke, Oklahoma State University.)

temporarily increase the palatability of some poisonous plants. Livestock may eat these plants after treatment, whereas before spraying animals would have avoided them. Under such conditions livestock may be poisoned after plants have been sprayed, even though the herbicide is nonpoisonous.

Hydrocyanic acid (HCN) or prussic acid may be produced from glucosides which are found in various species of the sorghum family and in wild cherry.

If cattle, sheep, and other animals with ruminant stomachs eat these plants, HCN is produced and animals may be poisoned. Two studies of wild cherry (5, 9) indicated no increase in HCN content following treatment with 2,4-D or 2,4,5-T.

"Tift" Sudan grass was treated when 8 in. tall with 1 lb/acre of 2,4-D, 2,4,5-T, and MCPA. Samples were taken 32 days after treatment. Chemical analysis showed that HCN increased after all three chemical treatments (17). The use of phenoxy herbicides evidently does not change the usual precautions necessary to prevent HCN poisoning. Ruminant animals should not be permitted access to plants containing HCN, whether treated or not.

Livestock may also be poisoned by eating plants having a high nitrate content. Nitrate is reduced to nitrite by microorganisms in the animal's intestinal tract. Nitrite in the bloodstream interferes with effective transport and use of oxygen and the animal dies of asphyxia (suffocation). The lethal dose of potassium nitrate is 25 g/100 lb of animal weight (2).

Lambsquarters, pigweed, and smartweed treated with 2,4-D had a very high nitrate content as compared with the controls (23).

A more extensive study of the effect of herbicides on nitrate content of 14 weed species was conducted in Michigan. Herbicides used included 2,4-D, 2,4,5-T, MCPA, dinoseb, Chlorpropham, and MH. Before treatment, 10 of the 14 weeds contained enough nitrate to cause poisoning if consumed in considerable quantities. Following herbicide treatment four species showed no change in nitrate content, five species showed increases, and one a definite decrease. Variation in the other species prevented drawing definite conclusions. Scientists concluded that many weeds contain enough nitrate to cause poisoning if eaten by livestock, whether sprayed by herbicides or not (4).

A high nitrate content in plants has been associated with abortion in cattle in Wisconsin (16). In Portage County, 400 abortions in cattle were reported in 1954. Reproductive diseases and pathogens accounted for only a very small number of abortions. "Poisonous weeds" were considered as a possible explanation. Further study revealed that the muck soils were high in nitrogen but lacking in phosphorus and potassium; this condition is conducive to nitrate storage in plants. Weed species were analyzed for nitrate nitrogen and classified according to nitrate content (Table 26-2).

Next, pastures were treated with 2,4-D to eliminate weeds thought to contribute to the high abortion rate. Pasture areas were divided for experimental purposes. On one pasture 2,4-D was applied in both 1956 and 1957. The area was weed free in 1957. Ten heifers that grazed on this area calved normally in 1957, but all 11 heifers that grazed on nontreated and weedy pastures aborted in the same year.

A feeding trial was conducted to test the effectiveness of dosing pregnant cattle with nitrate to induce abortion. Three 700-lb heifers that were given

Table 26-2.   Nitrate Nitrogen Content of Plants (16)

High NO$_3$ Content (above 1000 ppm)	Medium NO$_3$ Content (300–1000 ppm)	Little, or No NO$_3$ Content (below 300 ppm)
Elderberry	Goldenrod	Linaria
Canada thistle	Cinquefoil	Meadow rue
Stinging nettle	Boneset	Yarrow
Lambsquarters	Mints	Vervain
Redroot pigweed	Foxtail	Dandelion
White cockle	Aster	Milkweed
Burdock	Groundcherry	Willow
Smartweed	Toadflax	Dogwood
		Spiraea

3.56 oz of potassium nitrate each day aborted after 3–5 days. The aborted fetuses and placentas were similar to those aborted on the weedy pastures (13).

In summary, herbicides are generally nonpoisonous to livestock if used as directed on the label. Poisoning may occur if palatability is increased following spraying so that livestock consume larger-than-usual quantities of poisonous weeds. Killing poisonous weeds with herbicides may reduce the poisoning hazard.

PREVENTING LIVESTOCK POISONING BY WEEDS

Poisonous plants should be immediately isolated from livestock to prevent livestock poisoning. You may need to fence the infested area or remove livestock from the area. A small number of poisonous plants may be cut and removed from the pasture or killed by herbicide treatment.

Hundreds of plants cause livestock poisoning. Usually eradication, or at least very effective control is needed. Cutting and removal of plants, or treatment with an effective herbicide may be most desirable. If the area is small, a soil sterilant may be the best answer, as discussed in Chapter 29. Small, isolated areas can probably be treated best with hand equipment or with granular materials. Large areas may be treated best by broadcast-type equipment, either ground operated or aerial.

LITERATURE CITED

1. Alley, H. P., and D. W. Bohmont, *Wyoming Agricultural Experiment Station Bulletin*, B-354, 1958.
2. Bradley, W. B., H. F. Eppson, and O. A. Beath, *Wyoming Agricultural Experiment Station Bulletin, 241*, 1 (1940).

3. Elder, W. C., *Controlling Perennial Ragweed, Oklahoma A & M Bulletin*, B-369, 1951, p. 1.

4. Frank, P. A., and B. H. Grigsby, *Weeds, 5(3)*, 206 (1957).

5. Grigsby, B. H., and C. D. Ball, *North East Weed Control Conference Proceedings, 6*, 327 (1952).

6. Kingsbury, J. M., *Poisonous Plants of the United States and Canada*, Prentice Hall, New York, 1964.

7. Klingman, D. L., *FAO International Conference on Weed Control*, WSSA, Urbana, Illinois, 1970, p. 401.

8. Klingman, D. L., and M. K. McCarty, *U.S.D.A. Bull., 1180*, 1 (1958).

9. Lynn, G. E., and K. C. Barrows, *North East Weed Control Conference Proceedings, 6*, 331 (1952).

10. McCarty, M. K., D. L. Klingman, and L. A. Morrow, *U.S.D.A. Technological Bulletin, 1473*, 1974.

11. Muenscher, W. C., *Poisonous Plants of the United States*, Macmillan, New York, 1949.

12. Peters, E. J., and J. F. Stritzke, *U.S.D.A. Technological Bulletin 1430*, 1971.

13. Simon, J., J. M. Sund, M. J. Wright, and A. Winter, *J. A. Vet. Med. Assoc., 132*, 164 (1958).

14. *Southern Weed Conference Proceedings Research Report, 11*, 35 (1958).

15. *Southern Weed Conference Proceedings Research Report, 13*, 360 (1960).

16. Sund, J. M., and M. J. Weight, *Down to Earth*, (Dow Chemical Co.) *15(1)*, 10 (1959).

17. Swanson, C. R., and W. C. Shaw, *Agron. J., 46(9)*, 418 (1954).

18. U.S.D.A., *A.R.S. Agricultural Information Bulletin 168*, (1957).

19. U.S.D.A., *A.R.S. Special Report 22-46*, May, 1958.

20. U.S.D.A., *Agricultural Information Bulletin No. 327*, Washington D.C., 1968.

21. U.S.D.A. and U.S.D.I., *Chemical Control of Range Weeds*, Washington D.C., 1966.

22. Vengris, J., M. Drake, W. G. Colby, and J. Bart, *Agron. J., 45(5)*, 213 (1953).

23. Willard, C. J., *North Central Weed Control Conference Proceedings, 7*, 110 (1950).

For chemical, see the manufacturer's label for method and time of application, rates to be used, weeds controlled, and special precautions. Label recommendations must be followed— regardless of statements in this book. Also see the preface.

27 Brush and Undesirable Trees

Control of woody plant growth is a problem affecting most types of property. This includes grounds surrounding homes and industrial plants; telephone, telegraph, highway, and railroad rights-of-way; and recreation and grazing areas.

There are about 1 billion acres of pasture, pasturelands, and grazing lands in the United States. On parts of nearly all of this area, woody plants present some problem. In range and pasture areas it is often desirable to eliminate all or most of the woody plants, leaving only grasses and legumes for livestock grazing.

On many Western dryland ranges, native grasses increase rapidly where brush is controlled. In Wyoming, range forage yields doubled the first year after sagebrush was treated and increased fourfold during a 5-year control program. In Oklahoma, control of brush with chemicals plus proper management (principally keeping livestock off during the first summer) increased the growth of grass 4–8 times during the first 2 years after treatment.

Figure 27-1 shows the relationship between mesquite control on Southwestern rangeland and increase of perennial grass forage. With 150 mesquite trees/acre, production of grass forage was reduced by about 85% and total production including that produced by the mesquite was reduced by nearly 60%.

In the Southeastern United States over 100 million acres of land well suited to loblolly and short-leaf pine are being invaded by less desirable hardwoods and heavy brush undergrowth. The same is true in Northwestern United States and Canada where dominant, but inferior, hardwoods may invade stands of Douglas fir, balsam fir, and spruce. See Figure 27-2.

Aside from ranges, pastures, and forests, woody plants must be controlled along electric power and telephone lines, firebreaks, highways, railroads, farm

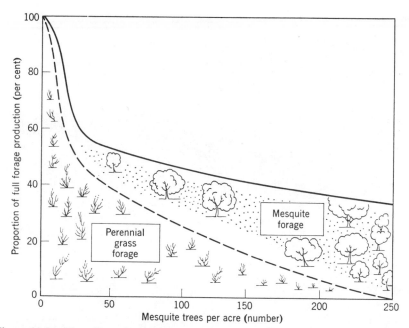

Figure 27-1. The effect of mesquite on forage production. (Adapted from *U.S.D.A. Leaflet No. 421*, 1957.)

Figure 27-2. Red pine released from oak overstory by aerial spraying with 1 lb/acre of 2,4,5-T. Photograph was taken 1 year after treatment. (J. L. Arend, U. S. Forest Service.)

Figure 27-3. Woody plants controlled by 2,4-D and 2,4,5-T sprays. Forage grasses have increased. (O. A. Leonard, University of California, Davis.)

fences, field borders, irrigation and drainage ditches, and in recreation areas. See Figure 27-3.

Some brush species act as alternate hosts for disease organisms affecting other plants. Common barberry harbors the organism causing stem rust in wheat and some other grasses. Other brush plants are poisonous to man, such as poison ivy, poison sumac; and some, including wild cherry and locoweed, are poisonous to ruminant animals.

Control methods for woody plants vary with nature of the plant and size of the job. Some plants can be easily killed by one cutting. For example, many conifer trees have no adventitious buds on the lower parts from which new shoots or sprouts may develop. However, on large acreages even these species are controlled more efficiently by herbicidal sprays than by mechanical means.

Plants with underground buds from which new shoots or sprouts may develop are hardest to control. Some woody plants such as velvet and honey mesquite have these buds on the lower part of the trunk; other plants have their buds on underground, horizontal roots, or stems. (See Figure 27-4.) Most of the serious woody weeds are in one of these categories. As long as such plants have enough stored food and are not inhibited in some other way, they will send up new shoots. Some plants continue to send up new shoots for as long as 3 years or more, even when all top growth is repeatedly removed.

As discussed in Chapter 1, higher plants produce their own food. This food-building process starts with photosynthesis. If areas where photosynthesis takes place (mainly the leaves) are repeatedly destroyed, the plant will eventually die of starvation.

Control methods for woody species are similar to those for other weeds, except that they must be adapted to woody and heavy type of growth. Methods include repeated cutting or defoliation, digging or grubbing, cabling and crushing, burning, girdling, and herbicide treatments.

REPEATED CUTTING OR DEFOLIATION

The object of repeated cutting or repeated defoliation is to reduce food reserves within the plant until it dies. Plants commonly die during winter months if

Figure 27-4. Section of mesquite stump. The wartlike structures are buds that produce sprouts. (Texas A & M University, Department of Range and Forestry.)

reserve foods have been reduced enough to make them more susceptible to winter injury. (See Figure 1-9.)

Most woody species bear their leaves well above ground, making repeated cutting or defoliation effective. Most species must be defoliated several times each season if they are to be controlled. Plants may be cut by hand, by mowers, by special roller-type cutters, or by saws. They may be defoliated with special equipment or by chemicals. Species having palatable leaves may be sufficiently defoliated by livestock. Sheep and goats have eradicated undesirable shrubs and tree sprouts, and insects may also defoliate a plant, gaining control as a biological predator.

Plants have well-established, annual patterns of food usage and food storage. This knowledge of root reserves as they vary during the year is valuable in reducing the number of treatments needed. With most deciduous plants, root reserves are highest during the fall. These reserves gradually diminish through the winter months as the plant uses its energy to survive. In the spring the plant needs considerable energy to send up new stems and leaves. Thus food reserves in roots are usually lowest just before full-leaf development.

Many plants reach this stage at about spring flowering or slightly before; therefore, mowing or defoliation is best started at that time. (See Figure 1-9). Repeat treatments are ordinarily needed for 1 or more seasons. Usually treatments are repeated when leaves reach full development. For more precise timing, root-reserve studies must be made on each plant species.

As leaves become full grown, they start synthesizing more food than they need. As the excess accumulates, at least part of it is moved back to replenish the depleted root reserves.

Apical Dominance

The terminal bud of a twig or a shoot may hold back development of buds lower on the stem or roots. When the apical bud is cut off, lateral buds are no longer inhibited and they may develop shoots. This mechanism is controlled by hormones or auxins within the plant.

When apical dominance is destroyed, more stems may develop than were present originally. Following cutting, the stand may appear thickened rather than thinned. However, if cuttings are persistently continued so as to keep root reserves low, the stand will start to thin as its food supply is depleted. Under such conditions, the increase in number of shoots may actually speed reduction of root reserves (see "Mowing", page 18).

DIGGING OR GRUBBING

Plant species that sprout from an individual stem or root and remain as a "bunch" or "clump" can be effectively controlled by digging or grubbing. Whether this method is practical depends on the ease of removing the "clump," and density of stand. Mesquite is an example of this type of plant. Young plants can easily be removed by grubbing.

Giant bulldozers have been equipped with heavy steel blades or fingers (root plows) that run under the mesquite, lifting out the bud-forming crown along with the rest of the tree. The operation is often expensive and hard on desirable forage species. Other weeds may quickly invade the disturbed area. Reseeding the disturbed area with desirable grasses will aid in reducing invasion by weeds. The disadvantages of this method include: (1) remaining roots may resprout, (2) soil surface is severely disturbed, (3) its difficulty in rocky soils, and (4) dormant weed seeds are brought to the surface to germinate and invade the area.

CABLING AND CRUSHING

Cabling is a technique of pulling out trees without digging. A heavy chain is dragged between two tractors; this action tends to uproot most of the trees if the soil is moist. Then the chain may be pulled in the opposite direction to finish tearing trees loose from the soil. The method cannot be used on large, solidly rooted trees, and it is not effective on small, flexible brush, or on any species that sprouts from underground roots or rhizomes.

Crushing is used to mash solid stands of brush to the ground by driving over them with a bulldozer. This treatment breaks off most of the branches and main stems. Crushing has one disadvantage—it disturbs the soil, releasing buried, dormant weed seed to germinate.

BURNING

Prior to modern weed science (about 1950), many of our grasslands and woodlands were burned regularly. These fires were started by lightning, Indians, and white men. Some fires were accidental, others intentional.

Although effective chemical methods have been developed to control many brush species on range land, burning is still used in many places. This is because of the low cost and the fact that certain species are resistant to most herbicides but controlled by burning (e.g., red cedar). In Kansas experiments, late-spring burning controlled brush best and injured desirable forage species least (1).

We commonly consider forest fires only as destructive. This is usually true of large, uncontrolled fires that occasionally ravage our forests.

But fire in forest areas can produce benefits too. Since about 1950 "pre-scribed burning" or "controlled burning" has come to mean periodic burning with clear-cut purposes. When done properly, this method leaves desirable forest trees uninjured. Such burning removes some undesirable trees, brush, leaves, branches, and other debris, prepares a seedbed, and serves other useful purposes. By removing undesirable trash (1) there is less competition from "weed" species, (2) the seedbed is improved for desirable tree seedlings and made less desirable for others, (3) less fuel remains for wild fires, and (4) some disease organisms may be held in check.

Thickness of bark is important in determining susceptibility to fire (2). If the phloem and cambium of the trunk are killed by heat, the tree may have an effect similar to girdling.

Most pine seedlings develop best with less than $1\frac{1}{2}$ in. of debris on the forest floor; oak seedlings need more debris. One reason for prescribed burning is to prepare seedbeds favorable to pine and unfavorable to oak (4, 12).

All tree seedlings, when small, are susceptible to fire. Most pine seedlings become resistant after the first year and will withstand considerable fire after the trunk has reached a 2-in. diameter at breast height (DBH) (3, 11, 12). Oak and most other hardwoods the same age are still susceptible. Therefore, where pine forests are desired, controlled fire can effectively and selectively kill the deciduous hardwoods. Frequency of prescribed burning will depend on local conditions; usually every 2–3 years is best (11, 12, 14).

Before setting a fire, consult local authorities for regulations and precau-tions, consider firebreaks, fire-fighting equipment, wind direction and pos-sibility of a shift, and other fire hazards. A heavy share of the cost of burning is for men and equipment on standby for emergency use. Obviously an escaped fire can be extremely costly.

During several years of protection from fire, considerable debris and under-growth may accumulate, If so, the first burning may be hot enough to damage even large trees. Burning when woods are only partially dry will reduce in-tensity of the fire.

GIRDLING

Girdling consists of completely removing a band of bark around the woody stem. The method is effective on most woody, dicotyledonous plants, but is too laborious to be efficient on thick stands of smaller-stemmed species. In most species, sprouting near the base of the tree is eliminated or at least consider-ably reduced on large, mature trees. Young trees may sprout profusely follow-ing girdling.

Plant foods move down to the roots through the phloem, and water and nutrients move upward through the xylem (see Chapter 3). When bark is removed phloem tissue is destroyed, and plant foods no longer can move to the roots. Since roots depend on tops for food, the roots eventually die of starvation.

This process usually takes from 1–3 years. The length of time depends upon the rate of food usage and amount of food stored in the roots at the time of girdling. If girdling is properly done, the top of the tree shows little injury until the root is near death. As soon as the root dies, the top also succumbs.

The girdle must be wide enough to prevent healing and deep enough to prevent the cambium from developing a new phloem. The girdled area should dry in a short time. Chemicals similar to those used for frill treatment or heat may help kill an active cambium. The xylem should not be injured. If the top is killed too rapidly through injury to the xylem (sapwood), root reserves may not be depleted enough to prevent shoot or sprout growth.

If sprouts or lateral branches develop below the girdle, these will soon provide enough food for the roots to keep the tree alive. Thus any sprouts that develop may be cut, girdled, or treated with suitable chemical sprays to prevent food from returning to the roots.

HERBICIDES

Use of herbicides to control brush and undesirable trees has expanded rapidly since 1950. Proper use of chemicals has generally proven to be more effective and less costly than most other methods.

Herbicides can be applied in many different ways to accomplish different purposes. These can be classified as foliage sprays, bark treatment, trunk injections, stump treatment, and soil treatment.

Do not evaluate the effectiveness of herbicides too quickly; wait at least until the next growing season. In some cases, trees may die 2–3 years after herbicide treatment, after repeatedly leafing out followed by defoliation.

Foliage Sprays

Foliage sprays are applied with the intent of getting uniform coverage. The need for thorough wetting will depend on the chemical used and the species treated. In most cases, uniform coverage is more important than thorough wetting; the latter is more often simply a means of obtaining uniform coverage. See Figure 27-5.

Figure 27-5. Applying a foliage spray. (F. A. Peevy, Agricultural Research Service, U. S. Department of Agriculture.)

Effectiveness of foliage sprays varies considerably among species and with size of woody plants. Several chemicals kill above-ground parts, but sprays have been less effective in killing roots. A resurgence of sprouts is common with some species. Repeat treatments are often needed as new sprouts develop.

Absorption by leaves is the first major problem of herbicide effectiveness. Many woody plants have coarse, thick leaves with a heavy cuticle. On such plants a nonpolar substance should be, and apparently is, absorbed more effectively than a polar substance. Thus, esters of 2,4-D, 2,4,5-T, silvex, and other oil-soluble herbicides are absorbed better than salt formulations. For the same reasons, an effective wetting agent added to phenoxy salts should and does increase absorption.

Once absorbed, water-soluble or polar forms (especially of 2,4-D) are translocated more readily through the phloem than the esters (5, 7, 8). Esters appear to be hydrolyzed to their respective acids upon entering the plant.

A second important problem is acute toxicity versus chronic toxicity. Sometimes for promotional reasons a "quick show" of herbicide effects is needed. Therefore, chemicals and rates may be used to "burn" the plant's leaves quickly. This may be contrary to certain physiological principles of effective translocation. For example, the excessively high application rate may inhibit translocation and the chemical is not translocated to the roots. Thus root kill is not attained.

For example, 2,4,5-T was dissolved in a nontoxic oil and in diesel oil, which is toxic, and applied to mesquite in greenhouse experiments. As a solvent the nontoxic oil gave superior absorption and translocation of 2,4,5-T as compared to diesel oil (8).

This result suggests three practical steps: (1) low-toxicity oils and wetting agents should be used for maximum absorption and translocation of growth substances through *phloem tissues*, (2) phenoxy compounds should be applied at rates which cause chronic toxicity, but not acute toxicity, and (3) low rates of treatment and repeated treatment often give superior final results.

Foliage sprays of growth substances (such as 2,4-D, 2,4,5-T, and silvex) have usually given best root kills when large quantities of food are being translocated to the roots (6). With species that become somewhat dormant in summer, this translocation period usually occurs about the time of full-leaf development in the spring; a second peak period may occur in the fall. With plants that continue rapid growth and translocation through the summer, treatment is possible all summer long.

Rates of treatment vary considerably according to species, age of plants, and method of application. Adequate recommendations are usually found on the label. Chemicals are applied by hand sprayers with mist nozzles; power-mist applicators; truck-, trailer-, or tractor-mounted, power-spray equipment; and airplanes. With ground equipment, water is normally used as the carrier.

Late-season plants develop a thickened, waxy cuticle. On such plants oil-water mixes are more effective. Airplane application may use water, oil, or oil-water emulsions depending on the type of foliage to be treated and the relative humidity of the atmosphere.

Herbicides such as 2,4-D, 2,4,5-T, and silvex are applied at rates of 1–4 lb/acre. Amitrole, 2 lb/100 gal of water, is effective on poison ivy and poison oak.

Picloram and dicamba are often effective as a foliar or soil application for certain woody species. They are used on species resistant to phenoxy-type herbicides mentioned above. They are often combined with 2,4-D, 2,4,5-T, or both, to control a very broad spectrum of woody species. Dicamba is applied at the rate of $\frac{1}{4}$–$\frac{1}{2}$ lb/acre as foliar sprays. Picloram is not registered for use on food crops, pasture, or range.

Ammonium sulfamate is another effective foliage treatment. It is usually suggested where phenoxy compounds are too hazardous to use because of spray drift. From 100 to 400 lb/acre in a wetting spray are usually required.

Bark Treatment

In bark treatments the chemical is applied to the stem near the ground and absorbed through the bark. Two methods are used: broadcast bark treatment and basal bark treatment.

Broadcast Bark Treatment

Broadcast bark treatment involves application of the herbicide to the entire stem area of the plant. This treatment is especially adapted to thick stands of small-stemmed, woody plants. Spray is usually applied between fall leaf drop and midwinter. Phenoxy herbicides are effective on several species; from 4 to 12 lb/acre are applied, mixed in oil. Apply the spray so that the greatest amount of it will cover the stems evenly. For example, treating upright stems with a horizontally directed spray exposes the greatest surface to the spray.

Basal Bark Treatment

Here the herbicide is applied so as to wet the stem base, 10–15 in. above ground, until rundown drenches the stem at groundline. The chemical can be applied from a sprayer, or from a container that simply lets the chemical "trickle" on the stem base. A small number of trees can easily be treated with

Figure 27-6.   Basal bark application is effective on trees up to 6 in. in diameter. (Amchem Products, Inc., Ambler, Penn.)

Figure 27-7. Basal bark application of 4% 2,4,5-T solution was dribbled to saturate the lower 15 in. of this red oak tree trunk in November. Eleven months later, the tree is dead. (D. B. Cook, Cooxrox Forest, Stephenson Center, N.Y.)

a tin can to pour the chemical on the base of the tree trunk. Brush or large trees are left standing. This method is especially effective on brush and trees less than 6 in. in diameter (see Figure 27-6). Also, it controls some plants that sprout from horizontal, underground roots and stems, especially if applied during early summer.

This method has the advantage that the operator can selectively treat only those trees he wishes to kill, without injury to other trees.

Herbicides used most often are phenoxy compounds such as 2,4-D, 2,4,5-T, or silvex. From 8 to 16 lb active ingredient of herbicide are mixed in enough diesel oil to make 100 gal of mixture. The lower rate is suggested when trees are growing actively and the higher rate for dormant work. The herbicide mixture is usually applied at the rate of 1 gal/100 in. of tree diameter; for example, fifty 2-in.-diameter or thirty-three 3-in.-diameter trees. See Figure 27-7.

The herbicide usually is translocated upward rapidly from the treated area; scientists think that there is little or no translocation to roots and stems below.

The tops of the plant may quickly develop symptoms of herbicide treatment, even though roots may live 1 year or more before they die.

Scientists do not fully understand the way basal bark treatment kills plants. It is possible that treatment either kills the phloem or immobilizes it to the point that the tree is chemically girdled. In addition, the herbicide may inhibit bud formation and sprout development. If the tree is chemically girdled, the tree root dies from starvation, as it does with girdling.

A variation of basal bark treatment is the "oil pour" treatment used for mesquite growing in sandy soils. Diesel oil is poured around the base of the tree, saturating the soil and covering the basal buds. This results in a very effective root kill. It is effective at any time of the year, but late winter or spring is best. No herbicide, in addition to the oil, is needed.

Injections

Trunk injections often help the herbicide penetrate through the bark. In some cases, injections may serve as a girdle, with the herbicide acting as a chemical girdle. Techniques used are: frilling or notching with an ax or tree injector.

Frill Treatment

Frill treatment consists of a single line of overlapping, downward ax cuts around the base of the tree. The herbicide is then sprayed or squirted into the cut entirely around the tree. The method is effective on trees too large in diameter for basal bark treatment. See Figure 27-8.

Phenoxy compounds, such as 2,4-D and 2,4,5-T, are the principal herbicides; they are used at rates from 8 to 16 lb, diluted in enough diesel oil to make 100 gal of spray mixture.

Notching

The tree trunk may be notched with an axlike frill treatment, except that one notch is cut for every 6 in. of trunk circumference. This method is slightly less effective than frill treatment, which uses a continuous notch around the tree.

Research work in California on live oak (*Quercus wislizenii*), black oak, and Madrone (*Arbutus menziesii*) indicated that best kills are obtained by making cuts close to the ground line and deep enough to place the chemical in the xylem. 2,4-D was the best herbicide; however, 2,4,5-T and silvex also gave good results. MCPA was less effective, and monuron and amitrole applied in the notches were not effective (9).

Amine salts of phenoxy compounds were more effective than esters. Using

Figure 27-8. Frill application is effective on all trees. It is used primarily on trees that are too large for basal bark applications. (Dow Chemical Company, Midland, Mich.)

chemicals containing 4 lb of active ingredient/gal, the most practical dosage appeared to be 4 ml/cut. This would be enough to treat 946 cuts/gal. Late fall and winter were the most effective seasons for treatment. Trees died at various times for as long as 5 years after treatment (9).

The tree-injector tool speeds notch treatment, and when properly used does a satisfactory job. The oil-soluble amine or ester form of the phenoxy compounds plus diesel oil (2:9 ratio) should be used; one cut/every 2 in. in trunk diameter. This tool is especially effective against elm, postoak, white oak, live oak, and willow. More resistant trees are treated by spacing injections closer together. Trees which are more resistant include ash, cedar, hackberry, hickory, blackjack oak, red oak, persimmon, and sycamore.

Cups or notches made at 6-inch intervals with an ax in the base of the tree, and filled with 1 teaspoonful of ammonium sulfamate crystals per cup, will kill many kinds of broadleaved trees. See Figures 27-9 and 27-10.

Stump Treatment

Stumps of many species may quickly sprout after trees or brush are cut. Most trees can be prevented from sprouting by proper stump treatment. However, weedy trees which can develop sprouts from underground roots or stems are

Figure 27-9.   Application of ammonium sulfamate in notches. (F. A. Peevy, Agricultural Research Service, U. S. Department of Agriculture.)

Figure 27-10.   The tree injector speeds notch application of herbicides, and permits selection of the tree to be killed. (W. C. Elder, Oklahoma State University.)

Figure 27-11. The stump will not sprout if the top and sides are saturated with the proper herbicide. (Dow Chemical Company, Midland, Mich.)

Figure 27-12. Right: In research plots, tebuthiuron was applied two years prior to taking this picture. The brush and trees are dead. Reseeding the area with grass has converted the area to a highly productive grazing area. Left: Not treated. (Elanco Products Company, a Division of Eli Lilly and Company.)

difficult to control by stump treatment alone. With most such species, stump treatment is effective if followed by a midsummer basal bark treatment of the sprouts when they develop. See Figure 27-11.

Phenoxy herbicides (such as 2,4-D, 2,4,5-T, and silvex) are usually used in stump treatment at the rate of 8–16 lb of acid in enough diesel oil to make 100 gal of spray solution. Apply enough spray to "wet" the tops and sides of the stump so that rundown drenches the stem at groundline.

Table 27-1.   Herbicides Used to Control Some Common Woody Plants[1]

Plant	2,4-D	2,4,5-T	2,4-D plus 2,4,5-T	Silvex[2]	Dicamba[2]	Picloram[2,3]	Amitrole[3]	Tebuthiuron[3]
Alder	F,B	F,B,C	F,B,C		C			S
Ash		F,B,C	F,B,C		C	S	F	S
Blackberry		F,B	F,B		F		F	S
Buckbrush	F,B	F,B	F,B					S
Elderberry	F,B	F,B	F,B					
Elm	C	F,B,C	F,B,C			F		S
Locust	F,B	F,B,C	F,B,C			S	F	S
Manzanita	F,B	F,B	F,B			F		
Maple		F,B,C	F,B,C		C	S	F	
Mesquite		F,B,C			F	F		S
Oak	F,C	F,B,C	F,B,C	F,B,C	C	S		S
Poison Ivy or Oak	F,B	F,B	F,B		F	F	F	S
Rabbitbrush	F,B		F,B					S
Rose		F,B	F,B			S		S
Sagebrush	F,B	F,B	F,B					S
Saltcedar	F,B,C		B,C	F,B,C	S	S		S
Sassafrass		F,B	F,B					
Sumac	F,B	F,B	F,B		F		F	S
Willow	F,B	F,B,C	F,B,C		F			S

[1] All formulations of these herbicides are not suitable for all the uses indicated. Check manufacturer's label for specific species controlled with various formulations available.

[2] Dicamba, picloram, and silvex are often used in combination with 2,4-D, 2,4,5-T, or both.

[3] Noncrop land only.

Abbreviations: F—foliar, B—stems, C—cutsurface (frill injection or stump treatment), S—soil.

Ammonium sulfamate crystals are also highly effective on many species. Crystals are usually applied to the cut top of the stump at $\frac{1}{4}$ oz (about $1\frac{1}{2}$ tea-spoonfuls)/in. of diameter.

Stump treatments are most effective if applied immediately after the tree is cut. Sprouting from older cuts can usually be controlled by increasing the amount of chemical.

Soil Treatments

Some herbicides control woody plants when applied to the soil. Most of these herbicides are used as dry pellets. They require rainfall to leach them into the soil as deep as the feeder roots. Therefore, they are usually applied just before or early in the rainy season. Herbicides used this way usually persist in the soil for more than 1 year for maximum efficacy. Effects may develop slowly and not be apparent for 1–2 years after treatment. See Figure 27-12.

Research has shown that dicamba, fenuron, karbutilate, monuron, pi-cloram, TBA, and tebuthiuron can be used effectively by this method. The species to be controlled determines which of these herbicides is used. Some-times combinations of these have been more effective than either one alone (see Table 27-1).

LITERATURE CITED

1. Aldous, A. E., *Kansas Agricultural Experiment Station Technological Bulletin*, 38, 1934.
2. Brend, J. L., *J. Forestry*, *48*, 129–130 (1950).
3. Bruce, D., *J. Forestry*, *45*, 809–814 (1947).
4. Bruce, D., and C. A. Bickford, *J. Forestry*, *48*, 114–117 (1950).
5. Crafts, A. S., *J. Agr. Food Chem.*, *1*, 51–55 (1953).
6. Elder, W. C., and J. E. Webster, *Oklahoma Technological Bulletin*, *T-80*, 1–11 (1959).
7. Emrick, W. E., and O. A. Leonard, *J. Range Mgt.*, *7*, 75–76 (1954).
8. Hull, H. M., *Weeds*, *4(1)*, 22–42 (1956).
9. Leonard, O. A., *Weeds*, *5(4)*, 291–303 (1957).
10. Leonard, O. A., and W. A. Harvey, *California Agricultural Experiment Station Bulletin 812*, 1965.
11. Little, S., J. P. Allen, and E. B. Moore, *J. Forestry*, *46*, 810–819 (1948).
12. Little, S., and E. B. Moore, *Ecology*, *30*, 223–233 (1949).
13. Reynolds, H. G., and F. H. Tschirley, *U.S.D.A. Leaflet No. 421*, Washington D.C., 1957.
14. Squires, J. W., *J. Forestry*, *45*, 815–819 (1947).
15. U.S.D.A. and U.S.D.I., *Chemical Control of Range Weeds*, Washington D.C., 1966

For chemical, see the manufacturer's label for method and time of application, rates to be used, weeds controlled, and special precautions. Label recommendations must be followed—regardless of statements in this book. Also see the preface.

28   Aquatic-Weed Control

Aquatic plants, as used here, include those plants which normally start in water and complete at least part of their life cycle in water.

Excessive growth of aquatic weeds causes many serious problems for people who use ponds, lakes, streams, and irrigation and drainage systems. Excess weeds (1) obstruct water flow and increase water losses; (2) interfere with navigation, fishing and other recreational activities; (3) destroy wild life habitats; (4) cause undesirable odors and flavors; (5) lower real-estate values; (6) create health hazards, and (7) speed up the rate of silting by increasing the accumulation of silt and debris.

On the plus side, aquatic weeds may reduce erosion along shore lines, and some plant species provide food and protection for aquatic invertebrates, fish, fowl, and game. Algae are the original source of food for nearly all fish and marine animals; and swamp smartweed, wildrice, wild millet, and bulrush provide food and protection for waterfowl, especially ducks.

Controlling aquatic weeds sometimes causes problems other than those of the chemical itself. For example, the rapid killing of dense, weedy growth may kill fish, which happens even though the chemical is nontoxic to the fish. During photosynthesis, living plants release oxygen and fish depend on this oxygen for respiration. When plants are killed, they produce no more oxygen. Worse yet, dead plants are decomposed by microorganisms which require oxygen for respiration. These two actions may reduce oxygen content in the water, causing the fish to suffocate. The answer is to treat only a part of very heavily infested areas at one time; fish will move to the untreated part.

Suppose you owned a recreation area suitable for both swimming and fishing. The right fertilization favors microscopic plants, used as food by fish. This heavy growth of microscopic plants makes the water appear cloudy or dirty and may give it an undesirable odor; hence it is less desirable for swimming. Here you can choose swimming or fishing, but not both—at least not at their best.

The use made of water will largely determine treatments that can be made for weed control.

METHODS OF CONTROLLING AQUATIC WEEDS

1. Proper construction of pond
2. Competition for light
3. Pasturing
4. Drying
5. Mowing
6. Hand cleaning
7. Chaining
8. Dredging
9. Burning
10. Biological control
11. Chemical control

Proper Construction of Pond

Proper pond construction is highly important in controlling pond weeds. Many rooted aquatic plants are not easily established in deep water. Build the pond so that as much water as possible is at least 3 ft deep. You can have water 3 ft deep only 9 ft from shoreline if all the edges of the pond have a slope of 3 to 1. Such a slope greatly reduces the area where cattails, rushes, and sedges first start growing. However, steep banks are hazardous for swimming. Gentle slopes should be provided for swimming areas.

Competition for Light and Fertilization

Ponds adequately fertilized develop millions of tiny plants and animals which give the water a cloudy appearance (bloom). If the water has a bloom and is at least 3 ft deep, submerged aquatic weeds have almost no chance to grow due to inadequate light. The benefits of fertilization include:

1. Increased growth of beneficial microscopic life, including phytoplankton and zooplankton,
2. Increased food supply for fish from the food chain that develops from the above, and
3. Effective weed control by shading. Plants that do reach the surface should be cut off; otherwise they will be stimulated by the fertilizer.

A 16-20-4 or similar fertilizer is suggested at about 50 lb/acre. Put on the first application in early spring and repeat as needed to maintain the cloudy appearance. A light-colored object should not be visible 1.5 ft below the surface. Fertilization of ponds is practical only where there is little loss of water from the pond, because fertility is lost with the overflow.

Pasturing

Pasturing is economical and effective in controlling marginal aquatic grasses, weeds, and some woody species. A good legume-grass pasture mixture, if properly managed and grazed, will give the banks and dam a lawnlike appearance. A good sod also protects the banks against erosion and helps to control undesirable species.

Excessive trampling may destroy the banks and muddy the water. Also, leaches in the water may attack animals and some diseases are spread in the water to livestock, especially to dairy animals.

Drying and/or Freezing

Drying is a simple way to control many submerged aquatics. If the water can be drawn from the pond or ditch, leaf and stem growth of submerged weeds will be killed after several days exposure to sun and air. Drying usually must be repeated to control regrowth from roots or propagules in the bottom mud or sand.

Especially in cold climates, if the water is drawn down in late fall and the lake not allowed to refill until early spring, many aquatic weeds will be killed.

Mowing

Mowing effectively controls some ditchbank weeds. Power equipment can be most easily used where the banks are relatively smooth and not too steep. Also, underwater power-driven weed saws and weed cutters are available. The effects usually are short lived. Mowing is usually required at rather frequent intervals and disposal of mowed weeds is often difficult.

Hand Cleaning

In lightly infested areas, hand cleaning may be the most practical method of control. A few hours spent in pulling out an early infestation may prevent the weed from spreading. The method is particularly effective on new infestations of emergent weeds such as cattail, arrowhead, and willow.

Chaining

Chaining aquatic weeds resembles cabling woody plants (see Chapter 27). A heavy chain, attached between two tractors, is dragged in the ditch. The chain tears loose the rooted weeds from the bottom. The method is effective against both submerged and emergent aquatics.

Chaining should be started whenever new shoots of emersed weeds rise about 1 ft above the water or when submersed weeds reach to the water surface. It should be repeated at regular intervals. Dragging the chain both ways may be effective in tearing loose most of the weeds.

The method is limited primarily to ditches of uniform width, accessible from both sides with tractors, and free of trees and other obstructions. After chaining it is usually necessary to remove plant debris from the ditch to keep it from collecting and stopping the flow of water.

Dredging

Dredging is a common method of cleaning ditches that are accessible from at least one side. The dredge may be equipped with the usual bucket, or a special weed fork may be used. Dredging may solve two problems; the removal of weeds, and the removal of silt and debris.

Dredging has been tried in ponds from specially built pontoons, but in general the pontoon dredge has not proven practical. Dredging is an expensive operation, both because of high equipment costs and the large amount of labor involved.

Burning

Burning may be used to control ditchbank weeds such as cottonwoods, willows, perennial grasses, and many of the annual weeds. Green plants are usually given a preliminary searing. After 10–14 days, vegetation may be dry enough to burn from its own heat.

Large trucks have been used in irrigation districts in the West. Each truck is equipped with a 30 –35-ft maneuverable boom with oilspraying nozzles at the end; lighter-weight equipment is also available (see Figure 1-11.)

Burning can also be combined with chemical- or mechanical-control programs. Burning the previous year's debris allows better spray coverage of regrowth. It may be desirable to burn the dead debris after chemical treatment. Mowing followed by burning the dried weeds may increase the effectiveness of the mowing.

Biological Control

Aquatic weeds have been controlled, according to several reports, by fish, snails, insects, microorganisms, and higher plants. Ducks often effectively control duckweed in small ponds. Biological control has appeal because of the continuing control potential and the nonuse of chemicals in the water.

However, as with other biological-control methods, care must be taken not to introduce a biological-control organism that will have undesirable side effects; for example, the use of a fish that reduces the population of game fish.

Some freshwater fish will eat aquatic vegetation (2). The white amur (Chinese grass carp), tilapia, and silver dollar fish are used to control aquatic weeds in certain areas of the world (12, 23). The white amur has been used in the People's Republic of China, Czechoslovakia, Poland, and the Soviet Union.

A large freshwater snail (*Marisa cornuarietis*) has been used to control aquatic weeds in the southern United States and Puerto Rico (19). It also eats the eggs of disease-carrying snails, an additional advantage. However, it may also feed on certain desirable plants such as rice, watercress, and water chestnuts. Another freshwater snail (*Pomacea australis*) also shows promise as an aquatic-weed-control organism.

For alligator-weed control, three insects are available in the Florida area. These are flea beetle (*Agasicles hygrophila*), a species of thrips (*Amynothrips andersoni*), and a moth (*Vogtia malloi*) (7).

Low-growing species of aquatic plants which only grow a few centimeters tall have been used to compete with undesirable species, (e.g., pondweeds) perhaps several meters long. Slender spikerush (*Eleocharis acicularis*) has shown considerable promise for control of submersed weeds in California (24).

Although biological control holds great potential, its actual use has been limited. The federal government now supports considerable research on this method. More information on biological control of aquatic weeds may be found in review papers (19, 24).

Chemical Control

Chemicals effectively control many aquatic and ditchbank weeds. To control these you need to know: (1) the name or names of the weed species, (2) the appropriate chemical, recommended rate, and time of treatment, and (3) the amount of water or size of area to be treated.

Water-surface areas are usually measured in acres, like field areas. One acre is 43,560 ft^2, or an area 208.7 ft^2. One acre-ft of water means 1 acre of water, 1 ft deep—that is, 43,560 ft^3 of water; 325,828 gal, or 2,719,450 lb. Thus, to produce a chemical concentration of 1 ppm, add 2.7 lb of the chemical (active ingredient)/acre-ft of water; 5 ppm would require 13.5 lb, and 5 ppm to an average depth of 3 ft would require 40.5 lbs. A closely related technique involves treating the "bottom acre-foot" (6). See page 414.

Running water is measured by several methods. Rates are usually given as *cubic feet per second*; 1 ft^3/sec is equal to 450 gal/min. Usually the rate of water flow is determined by the use of a weir and gauge.

CHEMICALS USED IN AQUATIC WEED CONTROL

The more important chemicals used in aquatic weed control are discussed below. The use of specific chemicals for the control of certain aquatic weeds is given in Table 28-1. Those factors of particular importance to aquatic weed control are included here. For more complete discussions of chemicals see Chapters 8 through 20.

Restrictions on the use of water that has come in contact with an aquatic herbicide are extremely important. In domestic water supplies, only ionic copper has a legal residue tolerance for *finished potable water*. Herbicides with residue tolerance for *raw potable water* are the dimethylamine salt of 2,4-D (0.1 ppm), endothall (0.1 ppm) and diquat (0.001 ppm). (See later sections for their respective restrictions.) The use of water treated with endothall must be delayed from 7 to 25 days depending on concentration, formulation, and location. Other aquatic herbicides have various restrictions such as forbidding application to water used for human consumption, irrigation, agricultural sprays, watering dairy animals, and so on. Detailed instructions and restrictions are printed on the label.

Acrolein

Acrolein controls most submersed water weeds and many snails. It was effective against *Potamogeton crispus*, *Elodea densa*, and other submersed species (21). When tested in small irrigation channels in Western states at rates of 2 gal per ft^3 per sec flow of water, this chemical controlled weeds from 1 to 8 miles downstream, with an average of $3\frac{1}{2}$ miles (18). However, the principal use of acrolein is in large canals where the usual treatment is 0.6 parts per million on a volume basis (ppmv) for 8 hr, or 0.1 ppmv for 48 hr. Weed control with these treatments may extend 20–50 miles downstream.

Treated water did not harm crops when used for irrigation at low concentrations. Higher concentrations may cause injury to susceptible crops, such as cotton.

Acrolein is a potent irritant and lacrimator, but with proper equipment and precautions it can be safely applied.

Amitrole

Amitrole is also safe on fish at normal rates. It is especially effective on cattails. A special form of amitrole and sodium thiocyanate known as amitrole-T is very effective on waterhyacinth and certain other emergent and floating species. Amitrole is applied at 5–10 lb/acre active ingredient as a foliage spray.

Table 28-1. Herbicides Used to Control Several Common Aquatic Weeds[1]

	Class[2]	Acrolein	Amitrole	Aromatic Solvents	Copper Ion	2,4-D	Dalapon	Dichlobenil	Diquat	Endothall	Fenac	Silvex
Algae,												
Unicellular	S				x					x		
Filamentous	S				x					x		
Chara	S				x			x				
Alligatorweed	E					x						x
Arrowhead	E					x						x
Bladderwort	S	x		x		x	—	x	x	x	x	
Bulrush	E					x	x					
Cattail	E		x			x	x					
Coontail	S	x		x		x		x	x	x	x	
Duckweed	F					x			x			
Elodea	S	x		x		x		x	x		x	x
Naiad	S	x		x		x		x	x	x	x	
Pickerelweed	E					x						x
Pondweed	S	x		x		x		x	x	x	x	x
Rush	E					x						
Spikerush	S,E					x					x	
Waterhyacinth	F					x			x			
Waterlily	E					x		x				x
Watermilfoil	S,E	x		x		x		x	x	x	x	x
Waterprimrose	S,F					x		x		x	x	x

[1] These uses are suggested in one or more of the following sources, but not necessarily in each source: References 1, 3, 7, or 20.

[2] Classes: F—Floating (unattached, tops above water).
               E—Emersed (rooted under water, top above water or growing on wet soil).
               S—Submersed (tops mostly under water, usually rooted in soil).

Like dalapon, it should be applied on cattails between the time of flowering and seed maturity in the eastern United States. In the West, fall application is best for cattail control.

Aromatic Solvents

Aromatic oils have been particularly effective in controlling weeds in ditches with running water. These chemicals are deadly to fish but the movement of water in irrigation canals usually has higher priority than saving fish in the canals. In the West, hand-weeding and dredging costs of $80–$400/mile have been cut to $10–$40/mile with the aromatic solvents.

Aromatic solvents are sold as varsol, Stoddard's solvent, paint thinners, general solvents, and for dry-cleaning clothes. Commercial forms specially developed for aquatic-weed control are available. The most effective ones had a flash point of not less than 80°F, distillation ranging between 278 and 420°F, and an aromatic content of not less than 85% (6).

Velocity, temperature, exposure time, and rate all affect results in treating ditches. But do not wait until the ditch is clogged seriously. Water velocities of $\frac{3}{4}$ to $1\frac{1}{4}$ ft/sec produced the best results. Water temperatures should be 70°F or above; in fact, the warmer the water, the lower the concentration required. The length of exposure time is also important. A 30-min. exposure in running water is considered minimal. A longer exposure in stagnant water (ponds) permits lower concentrations.

In the western states, rates of 5.4–5.9 gal of solvent per minute per cubic feet per second of water flow over a period of 30 min (400–440 ppm) controlled waterweed (*Anacharis canadensis* Michx.), horned pondweed (*Zannichellia palustris* L.), and leafy pondweed (*Potamogeton foliosus* Raf.) for periods up to 6 weeks, $\frac{3}{4}$–$1\frac{1}{2}$ miles down the channel below the point of application (6).

A rate of 8.1 gal of the aromatic solvent per cubic feet per second of flow (600 ppm) applied over a period of 30 min, gave effective control of these species for as long as 8 weeks; this treatment also controlled the more resistant sago pondweed (*Potamogeton pectinatus* L.) (6).

Ten gallons of aromatic solvent per min per cubic feet per second of flow (740 ppm), and applied for 30 minutes, were needed to control Richardson pondweed [*Potamogeton richardsonii* (Ar. Benn.) Rydb.], gigantic sago pondweed [*Potamogeton pectinatus* L. var. *interruptus* (Kit.) Aschers.], American pondweed (*Potamogeton nodosus* Poir.), and waterstargrass [*Heteranthera dubia* (Jacq.) MacM.]. This higher rate gave effective control for distances up to 5 miles without "booster" applications (6).

Aromatic solvents strong enough to kill most submersed aquatic weeds caused no important injury when used to irrigate alfalfa, beans, lima beans, carrots, cotton, sorghum, white clover, lettuce, oats, sweet corn, potatoes, sugar beets, or potatoes (6).

Aromatics are very toxic to fish, crayfish, snails, plankton, and mosquito larvae. These chemicals often kill fish at concentrations as low as 5–10 ppm. If given the opportunity, fish will usually move from treated areas. Aromatics cause strong and persistent off-flavor in fish. Naphtha fumes have been reported as toxic to waterfowl.

Livestock tend to avoid treated water so there is no apparent danger to them. Guinea pigs forced to drink treated water showed no ill effects (6).

Copper Sulfate

Copper sulfate (bluestone, blue copperas, blue vitriol) or copper-triethanol-amine complex are very effective against most kinds of algae, including chara. Soon after treatment the algae's color changes to a grayish white. Several days after treatment, the water should be nearly free of algal growth. See Figure 28-1.

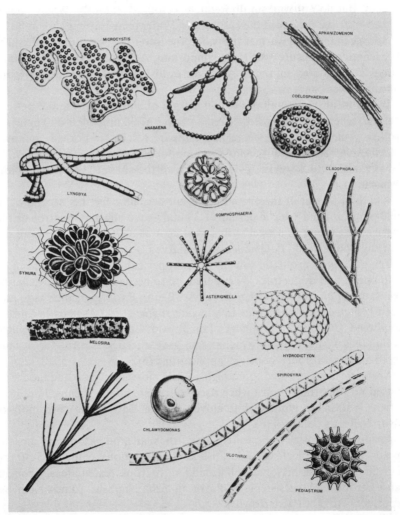

Figure 28-1. Algae often responsible for creating nuisance blooms. (C. M. Palmer and R. A. Taft, U.S. Public Health Service.)

Rates of 0.1–1 ppm copper ion (0.4–4.0 ppm copper sulfate) are usually adequate. With some resistant species, treatments of 0.5 ppm, repeated each day for 3–5 days, have given better control than a single treatment. To provide a concentration of 1 ppm copper ion, 10.8 lb of copper sulfate/acre-ft of water are required. If the lake has an average depth of 2 ft, then 21.6 lb of copper sulfate/acre of surface would provide the 1-ppm concentration of copper.

Copper sulfate can be applied as a spray, or the crystals can be placed in a cloth bag and towed behind a boat until the chemical is dissolved. Copper sulfate is easily dissolved in water at ordinary room temperatures, being soluble to about 25% by weight.

Copper sulfate applied continuously in sufficient quantity to maintain a 0.6–1.0-ppm concentration in moving water throughout the growing season has given effective control of algae and rooted submerged weeds in large canals and reservoirs carrying water for industrial and potable uses.

Concentration should be maintained at 1 ppm early in the season. It may be reduced gradually after midsummer as low as 0.6 ppm later in the growing season (22).

The U.S. Public Health Service has ruled that water used for drinking shall not have over 4 ppm of copper sulfate or 1 ppm of the copper ion. Concentrations up to 100 ppm have been safe as drinking water for 2-yr-old cattle. The concentration of copper sulfate considered "safe" for fish is about 1 ppm depending on species and age of fish, and hardness of the water. Water with 1 ppm copper sulfate can be used for irrigation water for most crops.

2,4-D and Silvex

Spraying plant parts above water is an effective way of killing many weeds. However, in most aquatic plants 2,4-D and silvex do not translocate readily to submerged parts. Slow killing of the plants favors translocation downward. This can best be obtained by repeated application of low rates of the chemical.

If surface sprays are used, they will be most effective if the entire plant can be treated. Therefore, lowering the pond water level will usually increase herbicide effectiveness. For effective absorption, the chemical must be on the plant for 2–10 hr. Oil-soluble forms resist washing from the plant and non-polar (oil-soluble) forms are more quickly absorbed than polar formulations. (See Chapters 6 and 7).

In addition to surface sprays, many plants are susceptible to 2,4-D and silvex dissolved in the water at rates of 1–5 ppm by weight. To provide a concentration of 10 ppm in 1 acre-ft of water, 27 lb of 2,4-D (acid equivalent) are required. Water-soluble forms of 2,4-D or granular forms may be used. The granular form falls to the pond bottom where a relatively high concentration may develop at the soil-water interface. Applying a heavy granular ester form of 2,4-D at the rate of 20–70 lb/acre gave effective control of duckweed, coontail, watermilfoil, eelgrass, and several species of Potamogeton (9). The use of granular formulations holds considerable promise for aquatic weed control in still water, but not in flowing water.

Pure 2,4-D acid, 100-ppm concentration, caused slight mortality to fingerling bream and large-mouth bass. The sodium salt was only slightly more toxic. Scientists concluded that these forms would not be dangerous to these two species under practical application for aquatic-weed control (11). Small rainbow trout have tolerated rates of 112 ppm of the sodium salt of 2,4-D (7).

In another experiment, fish were kept in the 2,4-D sodium salt solution for 1 week. The LD_{50} concentration of 2,4-D for minnows 1.5–2.5 in. long was approximately 2000 ppm; for sunfish 2.5–4 in. long, 1000 ppm; and for catfish 5–6 in. long, 2000 ppm (10).

Other work has shown ester forms of 2,4-D to be toxic to fish. Fingerling bluegills suffered losses of 40–100% from the butyl ester of 2,4-D at concentrations of 1–5 ppm (16). Esters are oillike and oil-soluble materials, and oils are known to be toxic to most fish. Esters may be acting similar to oils, or it may be the oil solvents, emulsifying agents, or other additives that are killing the fish. When pure 2,4-D ester is used in granular form, without surface-active agents, there is considerably less hazard to fish.

Dalapon

Dalapon is harmless to fish at normal rates. Lake emerald shiners apparently suffered no ill effects from 3 days in 3000 ppm of dalapon, but 5000 ppm was fatal (8).

Dalapon is especially effective against grasses and cattails. It is best applied as a foliage spray. Most species are susceptible if not rooted in standing water. Draining is advisable, if possible, several weeks before treatment with dalapon. Dalapon (5 lb/acre) plus amitrole (2 lb of amitrole/acre) applied as a spray between flowering and seed formation has provided excellent cattail control in the eastern United States. In the West, 15–20 lb/acre of dalapon alone, or 10 lb/acre plus 2 lb/acre of amitrole is required.

Dichlobenil

Granular dichlobenil is used to control many aquatic weeds in static water such as lakes and ponds. It is particularly effective on elodea, watermilfoil, coontail, chara, and pondweeds (*Potamogeton* spp.). It is applied in early spring at rates of 7–10 lb/acre before weeds start to grow. Although it does not harm fish at these rates, humans may not eat these fish until 90 days after treatment. It cannot be used in commercial fish or shellfish waters. Do not apply to water that will be used for irrigation, livestock, or human consumption.

Diquat

Diquat controls many submersed aquatic weeds and algae in static water. Bladderwort, coontail, elodea, naiad, pondweeds, watermilfoil, spirogyra, and pithophora are among the species controlled. It is applied at rates that give a concentration of $\frac{1}{2}$ ppm by weight ($1\frac{1}{2}$ lb cation/acre-ft of water). The liquid form is injected beneath the surface. For best results, apply before weeds reach the surface of the water. Because diquat is a contact spray, repeated applications may be necessary to give seasonlong control.

Do not use in muddy water because diquat is rendered ineffective by its tight adsorption on soil particles. Do not use treated water for irrigation, agricultural sprays, swimming, or domestic purposes within 10 days after treatment. It is not harmful to most fish at the recommended rate.

Endothall

Endothall is available as a number of derivatives and in several liquid and granular formulations. The form chosen depends on properties of the water and method of application. It controls most algae and submersed aquatic weeds in static and in some flowing waters. However, elodea is not controlled. It is applied at a broad range of rates—from 0.3 to 10 lb/acre-ft of water. The rate depends primarily on the derivative selected, but also on the species to be controlled, properties of the water, and formulation. Treated weeds may take 2–4 weeks to die when treated with the K or Na salts. The considerably more active cocoamine salts kill plants in a few days but are also extremely toxic to fish. Do not use treated water for irrigation, agricultural sprays, animal consumption, or domestic purposes within 7–14 days. Fish may be used as food 3 days after treatment.

Fenac

Fenac can be used to control most submersed aquatic weeds where water can be removed from the area. Unlike other aquatic herbicides, fenac is applied to the soil surface, not to the water. Draw down water to expose area where weeds root. Treat only exposed soil. Keep water down for three weeks after treatment. It is particularly useful in drainage ditches, ponds, and margins of lakes. However, it is not permitted where these waters are used for irrigation, livestock consumption, or domestic needs. Applications can be made any time, but a treatment in the fall before freezing weather gives the best results. The rate is 15–20 lb/acre.

AQUATIC WEEDS OF SPECIAL IMPORTANCE

Aquatic weeds are commonly classified by their growth habits: (1) *floating*, (2) *emersed*, and (3) *submersed*. Floating aquatic weeds are not attached to soil, and their tops are above water. Emersed aquatic weeds are rooted in soil with their tops above water. Submersed aquatic weeds are usually rooted in soil with their tops entirely or mostly under water. Table 28-1 gives the classification of 20 common aquatic weeds and herbicides which effectively control them. See Figures 28-2 and 28-3.

The following are descriptions of some of the more important aquatic weeds.

Algae

Algae are so tremendously important as aquatic weeds that you might read more about them in an encyclopedia or botany book. Algae may annoy bathers, causing a type of dermatitis and symptoms of hay fever. Blue-green algae have been known to cause poisoning of horses, cattle, sheep, hogs, dogs, and poultry. Odors and fishy tastes often result from decaying algae in water reservoirs. Extremely heavy algae growth may suffocate fish by depleting the supply of oxygen in the water at night.

Figure 28-2. Several submersed aquatic weeds. Left to right: water-milfoil, coontail, eelweed, pondweed, and large-leaf pondweed (11).

Figure 28-3. Emersed and floating aquatic weeds. Left to right: burreed, duckweed, white waterlily, floating-leafed pondweed (11).

Algae are thallus-like plants with chlorophyll (green color). They may consist of a single cell, a single filament, branched filaments, or they may grow 200 ft long like the giant algae of the Pacific (known as kelp). They are often called pond scums or seaweeds. (See Figure 28-1.)

Plankton algae are microscopic, free-floating (unattached) organisms. They frequently cause "water bloom" of mid- to late summer. They may become so numerous that they give the water a thickened, green, pea-soup appearance.

Algae are extremely important synthesizers (producers) of food. They are the "original" sources of food and energy for most fish and aquatic animals. Algae live under moist conditions on land, in the soil, or in water—both salt and fresh.

Different temperatures of the water favor different kinds of algae. Some prefer warm water, others cold, so we find spring annual algae, summer annuals, winter annuals, and perennials.

Constant high turbidity in the water prevents algae from receiving adequate light, thereby reducing algae growth. Carp and young crayfish may keep a pond turbid; or boat propellers may stir up the muddy bottom enough to inhibit algae.

Chara (Chara spp.)

Chara is a large, green algae characterized by branched, erect stems that have cylindrical whorls of branches along the stem. It is anchored to the bottom. The general appearance resembles a higher plant. There are many different species (see Figure 28-1). These plants reproduce both vegetatively and by sexual fruiting bodies. After chara becomes encrusted with calcareous deposits, the plant takes on a coarse, gritty feeling and a gray-green shade. Chara, a submersed aquatic weed, thrives best in clear, hard water, and is found in water up to 20 ft deep.

Alligatorweed

Alligatorweed (*Alternanthera philoxeroides*) grows mostly along the coast from Texas to Virginia, reaching 25–150 miles inland, and it has been reported on the West Coast. It is widely distributed in the tropical areas of Central and South America.

It is a coarse, fleshy plant, forming dense mats in shallow water and on mud flats. It has narrow, opposite leaves and short spikes of whitish flowers (see Figure 28-4).

Alligator weed depends on vegetative reproduction (stems) for its spread. Broken-off branches root easily and spread rapidly.

Arrowhead, Duck-Potato (*Sagittaria* spp.)

Leaves of this group of plants give it its general name of arrowhead (see Figure 28-5). These plants are usually perennial, reproducing by seed and

Figure 28-4.   Alligatorweed.

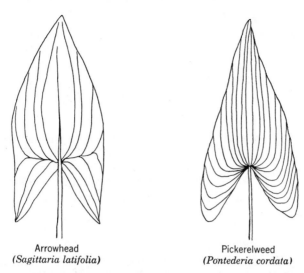

Arrowhead
(*Sagittaria latifolia*)

Pickerelweed
(*Pontederia cordata*)

Figure 28-5. Left: Arrowhead (Sagittaria latifolia). Right: Pickerelweed (Pontederia cordata).

tuber-bearing rootstocks. Indians used the tubers as food. The stems usually have a milky juice. The plant is best adapted to shallow water or muddy shores. It can usually be found near the edges of the pond as a marginal weed.

Bladderwort (*Utricularia* spp.)

The bladderwort family includes 250 different species with many of them adapted to aquatic conditions (15). The plant family is widely distributed in tropical areas. It has finely divided, simple leaves. The rootless plant either floats or is submersed, and the stem base may become anchored to the bottom of the pond. Some species have bladderlike leaf segments which act as a trap door to catch small aquatic animals. The plant reproduces from seed or by vegetative growth.

Bulrush (*Scirpus* spp.)

The Bulrush genus has about 150 species, widely distributed. The genus *Scirpus* is a member of the sedge family and is not a true rush. Most grow in marshes, ponds, and lakes. The stems are either triangular or rounded and sheathed at the base.

Cattails (*Typha* spp.)

Cattails dominate extensive marshy areas of the world. They are tall, grass-like plants with fleshy leaves having no midvein. They spread rapidly by rhizomes and by small, airborne seeds. Seeds remain viable for 5 years or more (14).

Small areas of cattails can be eliminated by pulling. Repeated underwater mowing is also effective. Cut when the first spikes reach two-thirds full size; then repeat when the cattail is again 2 ft tall. Most kinds of cattail are killed if submerged in 4 ft of water; *T. domingensis* requires deeper flooding.

Coontail (*Ceratophyllum demersum*)

Coontail is a submersed aquatic herb with whorls of finely dissected leaves (see Figure 28-2). In the spring the lower stem may anchor in the mud with plant growing upright. Later, the plant may become a tangled mass with filamentous algae. Leaves are crowded toward the tip of the stem, giving it the coontail appearance. The plant spreads by both seed and vegetative growth.

Coontail is common in shallow lakes, ponds, and sluggish streams. It is favored by water rich in organic material.

Duckweed Family (*Lemnaceae*)

Duckweeds are small, stemless, floating plants. They lack true leaves, but have thickened or rounded fronds. The family includes the smallest of the flowering plants. Duckweeds propagate mainly from vegetative growth which develops at the edge or base of the parent frond. Also, during the fall, small bulblets may develop at the edge of the leaf. These sink to the pond bottom, live through the winter, and germinate the next spring. Seed is usually produced sparingly, or not at all (see Figure 28-3).

Elodea (*Elodea canadensis*)

Elodea, also known as waterweed, is a common plant of home aquariums. The leaf is veinless; leaf margins have microscopic teeth. Its leaves are attached to the stem in a whorled arrangement. Stems are floating and usually rooted to the bottom.

Figure 28-6. Cattails treated with dalapon. Top: Soon after treatment. Bottom: Mid-summer of the next season shows effective control. (Dow Chemical Company, Midland, Mich.).

The plant is dioecious. Male flowers break loose or grow to the surface to shed their pollen. Female flowers grow to the surface after fertilization and produce up to five spindle-shaped seeds.

Hydrilla (*Hydrilla verticillata*)

Hydrilla was first found in the United States near Miami, Florida, in 1960 (5). It is now widely spread and it has become a troublesome aquatic weed in the United States. Since its introduction, it has spread widely in Florida as well as into Georgia and Alabama (4).

Hydrilla is a submersed vascular aquatic weed, rooted to the bottom with long branching stems. The lower leaves are opposite and small, whereas the medium and upper leaves are in whorls of fours and eights. Leaves are verticillate and narrow lanceolate. Hydrilla is easily confused with elodea unless flowers with the threadlike pedicel are present. In addition, hydrilla differs from elodea in that hydrilla has sharply toothed leaf margins, a "harsh" feel, and prominent red veins. Vegetative reproduction from broken shoots, subterranean shoots, and turions appear to be more important in its spread than seeds (4).

Current recommendation for chemical control of hydrilla is 1 ppm of copper sulfate plus 1 ppm of diquat.

Naiad (Najadaceae Family; Genus Najas)

The genus *Najas* has eight species found in the United States (15). Common names include naiad and bushy pondweed. It grows as a submersed aquatic. The leaves are slender and usually arranged in an opposite manner. Prominent bracts (stipules) may occur on the stem at the point of leaf attachment. The stem is slender, much branched, and rooted at the lower nodes. The plant reproduces by seed and by rooted stems.

Appropriate pond fertilization helps control naiad.

Pickerelweed (*Pontederia cordata*)

Pickerelweed has fleshy, heart-shaped to lance-shaped leaf blades which are easily confused with arrowhead (see Figure 28-5). It has thick, creeping rootstocks, strongly rooted to the bottom. The flowers are blue spikes.

Pickerelweed grows primarily as a marginal weed in shallow water, and it reproduces by creeping rootstocks and from seed.

Pondweed (*Potamogeton* spp.)

Potamogeton, with 60 species, is the largest genus of true aquatic seed plants of temperate regions (15).

Pondweed plants grow in both fresh and salt water. They grow best in water 1–2 ft deep, but some species grow in water 6 ft deep or more. Some species have two different kinds of leaves. Those remaining submerged are usually thick, transparent, long, and slender. Those emerging are firm and leathery and are usually attached to the stem by a petiole, with leaves alternately arranged. The stems are jointed, with roots forming at the lower nodes (see Figure 28-2).

Pondweeds reproduce by seed, by turions and tubers that develop in the leaf axils of some species, and from rootstock which may produce tubers.

Deep water with appropriate fertilization will help to control pondweed.

Rush (*Juncus* spp.)

Rush plants resemble grasses, except the leaves are rounded. The stems are mostly hollow, but may be filled with pith. Flower clusters are broad and more or less flat topped.

Rush plants are well adapted to marshes and bog areas. Some grow in water. Needle rush (*J. roemerianus*) is an important feed for water fowl.

Rush, Spike (*Eleocharis* spp.)

The spikerush genus has about 150 species, and as bulrush, is a member of the sedge family. It is especially well adapted to marshy areas with shallow water. It spreads by means of seed and creeping rootstocks.

Waterhyacinth (*Eichornia crassipes*)

Waterhyacinth includes five species, but only one is of major importance in the United States. It is a tropical plant, confined primarily to the Southern Coastal States and to California. It is found principally in ponds and quiet streams.

Waterhyacinth has a blue flower that is somewhat funnel shaped and two lipped. The plant is normally free floating, buoyed by bladder like, inflated leaf petioles. It forms a matlike growth. The leaf blade may be kidney shaped to somewhat rounded.

The plant reproduces mainly by vegetative offsets. Several thousand new plants may develop during one season from one original plant. Some reproduction results from seed.

Waterlily (*Nymphaeaceae*)

Species of the waterlily family are primarily perennial aquatic herbs. The leaves are generally shaped like a heart or shield and usually emerge above the water or float on the surface; the flowers are usually large and showy.

White Waterlily (*Nymphaea tuberosa*)

The white waterlily flower is normally green on the outside and the remainder may be white, yellow, pink, and occasionally blue. The leaf blade is nearly circular with veins radiating from the point of petiole attachment. Leaves vary in size from 2 to 25 in. in diameter. Reproduction is mainly from tuberous offshoots and to a lesser extent from seed.

Spatterdock (*Nuphar advena*)

Spatterdock is a perennial waterlily with creeping rhizomes. The leaf blade is rounded with a deep cleft at the base, and its veins come mostly from the midrib and often extend to the margin without forking. Most of the leaves float on the surface, some submerged leaves being very thin and delicate.

Yellow flowers, possibly tinged with purple, are large and borne individually on a thick stem. Its flowers develop during the entire summer. The plant reproduces by seed and creeping rhizomes.

Spatterdock is difficult to control because it lives in water 5 ft deep. In 4-ft water, mowing during heavy flowering and again about 1 month later has given up to 90% control.

Watermilfoil (*Myriophyllum* spp.)

Watermilfoil includes 20 different species with parrotfeather (*Myriophyllum brasiliense*) and eurasian watermilfoil (M. spicatum) being troublesome species (15). The lower leaves are finely divided and threadlike; each leaf resembling a small, green feather, while the upper leaves may resemble bracts (see Figure 28-2). The plant usually grows submersed in water, but stems often emerge above water during the last half of the summer. Flowers and seeds are produced sessile in the leaf axils.

Reproduction is from seed, creeping rhizomes, and broken stem parts which take root.

Pond fertilization will help to control watermilfoil species. Light infestations and small areas can be removed by hand.

Waterprimrose (*Jussiaea michauxiana*)

Waterprimose, a perennial aquatic weed, usually begins to grow at the bank of bodies of water. Once a flotant has formed it will begin to grow as a floating plant. The floating leaves in the spring are ovate, but later the erect branches bear elongated to elliptical leaves. Flowers are quite conspicious, 3–5 cm in diameter, with five yellow petals. Water primrose is found in ponds, slow-moving canals, and stagnant waters.

LITERATURE CITED

1. *Agri-Fieldman*, *29*, 118 (1973).

2. Avault, J. W., R. O. Smitherman, and E. W. Shell, *FAO Fisheries Bulletin*, *44*, 109 (1966).

3. Bayer, D. E., C. L. Elmore, W. A. Harvey, L. S. Jordan, A. H. Lange, W. B. McHenry, and R. Yeo, *Non-Crop Farm*, *Industrial and Aquatic Weed Control Recommendations*, Univ. of California, Davis. 1972, p. 21.

4. Blackburn, R. D., and L. W. Weldon, *Hyacinth Control J.*, *8*, 4 (1970).

5. Blackburn, R. D., L. W. Weldon, R. R. Yeo, and T. M. Taylor, *Hyacinth Control J.*, *8*, 17 (1969).

6. Bruns, V. F., J. M. Hodgson, H. F. Arle, and F. L. Timmons, *U.S.D.A. Circular No. 971*, 1955, p. 1.

7. Burkhalter, A. P., L. M. Curtis, R. L. Lazor, M. L. Beach, and J. C. Hudson, *Aquatic Weed Identification and Control Manual*, Florida Department of Natural Resources, Tallahassee, Florida, 1974.

8. Cope, O. B., *U.S.D.I.*, *Fish and Wildlife Service Memoria*, 1946, p. 1.

9. Grigsby, B. H., and R. H. Hamilton, *Weed Society of America*, *Abstracts*, 1958, p. 56.

10. Harrison, J. W. W., and E. W. Rees, *Am. J. Phar.*, *118(12)*, 4 (1946).

11. King, J. E., and W. T. Penfound, *Ecology*, *27(4)*, 372 (1946).

12. Lahser, C. W., *Progressive Fish-Culturist*, *29*, 48 (1967).

13. MacKenthun, K. M., *Committee on Water Pollution*, Madison, Wisconsin, 1958.

14. Martin, A. C., R. C. Erickson, and J. H. Steenis, *U.S.D.I. Circular No. 19*, 1957.

15. Muenscher, W. C., *Aquatic Plants of the United States*, Comstock, Ithaca, New York, 1944.

16. Snow, J. R., Thesis, Alabama Polytechnic Institute, 1948, p. 1.

17. Stephens, J. C., A. L. Craig, and D. S. Harrison, *Florida Agricultural Experiment Station Circular S-97*, 1957, p. 1.

18. Timmons, F. L., L. Weldon, R. R. Yeo, H. F. Arl'e, V. F. Bruns, G. M. Hodgson, and D. D. Suggs, *Weed Science Society of America, Abstracts*, 1960, p. 64.

19. Timmons, F. L., *FAO International Conference on Weed Control*, WSSA, Urbana, Illinois, 1970, p. 357.

20. U.S.D.A., *Agricultural Handbook 332*, (1969).

21. van Overbeek, J., W. J. Hughes, and R. Blondeau, *Science, 129(3345)*, 335 (1959).

22. Weight, W. K., *Weed Science Society of America, Abstracts*, 1956, p. 65.

23. Yeo, R. R., *Weeds, 15*, 27 (1967).

24. Yeo, R. R., and T. W. Fisher, *FAO International Conference on Weed Control*, WSSA, Urbana, Illinois, 1970, p. 450.

For chemical, see the manufacturer's label for method and time of application, rates to be used, weeds controlled, and special precautions. Label recommendations must be followed— regardless of statements in this book. Also see the preface.

29 Total Vegetation Control

Total control of vegetation is the removal of all higher green plants and maintenance of these areas vegetation free. Complete absence of vegetation is desirable on many sites such as railway road beds, industrial areas, highway brims, fence rows, and irrigation and drainage-ditch banks. Industrial sites include storage and work areas, lumber yards, utility transmission stations and railroad right-of-ways. See Figures 29-1 and 29-2.

Vegetation can be controlled totally by mechanical or chemical methods. Sometimes both methods are used. Brush and trees may be removed by mechanical methods at first clearance, although this method adds considerable cost. These areas are then maintained weed free with herbicides. Herbaceous species can be eliminated by either mechanical or chemical methods. Discing or other mechanical means may need to be repeated several times a season to keep the area weed free. But persistent herbicides at high rates are needed only annually or less often.

FOLIAR HERBICIDES

To obtain vegetation-free areas, certain nonselective foliar-applied herbicides are used to control emerged, herbaceous weeds before or with persistent, nonselective, soil-applied herbicides. These foliar herbicides include contact herbicides such as paraquat, cacodylic acid, dinoseb, or weed oil; or the translocated herbicide amitrole. Also, to broaden the spectrum of weed species controlled, other herbicides are added. For example, 2,4-D, 2,4,5-T, silvex, picloram, dicamba, and fenac are used in this way. Foliar-applied herbicides are often applied to areas treated with the persistent nonselective herbicides discussed below.

Figure 29-1. Pipeline and storage-tank areas kept weed-free by an annual application of monuron. (E. I. du Pont de Nemours and Company.)

SOIL HERBICIDES

Persistent, nonselective herbicides used for total vegetation control are given in Table 29-1. Chemical and physical properties of these compounds, as well as other uses, have been discussed in previous chapters. Although some of these herbicides are also used as selective herbicides, the rate of application is usually lower for selective uses than for nonselective uses.

The soil-applied herbicides may remain toxic to higher green plants for more than 1 year. Factors affecting length of time that a herbicide remains toxic in the soil were discussed in Chapter 4. In general, dry weather with little or no leaching, cool or cold temperatures, and heavy soils tend to lengthen the time that a herbicide will remain toxic.

Under any given condition the length of time that a herbicide will remain toxic can be predicted with reasonable accuracy (see Table 4-1).

Annual applications of most of these herbicides at rates that persist somewhat longer than 1 year are usually more economical than massive rates that will persist for 2 years or more.

These persistent, nonselective soil-applied herbicides must be leached into the rooting or seed-germination zone of the weeds to be effective. Therefore, they are usually applied just before or during the rainy season.

Figure 29-2. Bermudagrass control with three annual treatments of karbutilate on noncrop land. Foreground: Untreated. Background: Treated. (W. B. McHenry, University of California, Davis.)

Figure 29-3. Tebuthiuron, at increased rates, provides effective total vegetation control for industrial areas and railroads. (Elanco Products Company, a Division of Eli Lilly and Co.)

Table 29-1. Herbicides Commonly Used for Total Vegetation Control of Annual Weeds on Noncrop Land[1]

Herbicide Soil-Applied	Rate (lb/acre)	Herbicide Foliar-Applied	Rate (lb/acre)
Atrazine	2.5–5	Amitrole	2.5–5
Borate	200–400	Cacodylic acid	6–8
Bromacil	2.5–5	Dinoseb	1.5–3
Chlorate	200–800	Paraquat	0.5–1
Diuron	2.5–5	Weed oil	50–100 (gal/acre)
Erbon	100–150		
Karbutilate	2.5–5		
Monuron	2.5–5		
Prometon	2.5–5[2]		
Simazine	2.5–5		
Tebuthiuron	1.5–5		

[1] Various combinations of these herbicides are often used to give greater persistence and/or broader spectrum of weeds controlled.

[2] Prometon has considerable foliar activity as well as soil activity.

PREVENT INJURY TO TREES, SHRUBS, ORNAMENTALS, AND LAWNS

Nearly all nonselective herbicides will kill all types of plant growth. As you would expect, some plants are more susceptible to some herbicides than to others. At high rates they should all be considered as very effective against trees, shrubs, ornamental flowers, and lawns. Never apply soil sterilants to the rooting zone of such plants or so that the chemical is washed into their rooting zone.

For chemical, see the manufacturer's label for method and time of application, rates to be used, weeds controlled, and special precautions. Label recommendations must be followed—regardless of statements in this book. Also see the preface.

30 Lawn, Turf, and Ornamentals

Lawn- and turf-weed problems have been largely underestimated. Nearly every home and apartment house has a weed problem. Other turf areas which have weed problems are golf courses, public and private parks, and other recreation areas, grounds surrounding many commercial and governmental buildings, and roadsides.

No other type of weed control directly affects so many people. In the United States over 5 million acres are in home lawns, and an additional 10 million acres are in other types of turf (1).

Because there are millions of consumers, numerous turf species, and a multitude of ornamental plants, the job of educating users is more complex than in other areas of weed control.

The importance of management practices that produce a strong and vigorous turf cannot be overemphasized. In general these practices are well understood for each area of the United States. They are specific for each geographic area. They include choice of an adapted lawn grass, proper grading and seedbed preparation, fertilization, mowing, and watering, as well as the control of insects, diseases, and weeds.

LAWN

Weed control alone cannot guarantee a beautiful lawn. Other recommended practices must also be followed. For example, ridding a lawn of crab grass may leave a bare yard unless plans are made to encourage desirable turf grasses to become established. However, desirable turf grasses are usually present in areas where crabgrass predominates. With elimination of crabgrass and proper fertilization, mowing, and watering, grasses such as Kentucky bluegrass or bermudagrass soon cover the area.

Topsoil Added to New Lawns

After new homes or buildings are finished, only subsoil may remain for starting a lawn. Often topsoil is hauled in to cover the area 2–4 in. deep. It usually contains weed seeds and weedy plants that soon infest the area.

The owner has three choices. He can: (1) use the topsoil and hope to control the weeds later. Also, he should attempt to get top soil from fields known to be free of serious perennial lawn weeds; (2) use a soil fumigant such as methyl bromide to rid the soil of weeds; or (3) not use the topsoil. *Proper fertilization*, including thorough mixing into the upper 3–4 in. of soil, makes it possible to grow many turf plants on subsoil. Turf plants are favored by adding peat moss at the rate of 1 bale (7 cubic feet) per 200 square feet of surface plus liberal fertilization, with both worked into the surface 3–4 in. of subsoil. *Well-rotted* sawdust is as good as peat; however, the fertilization program depends upon the degree of sawdust decomposition. Well-decomposed sawdust will require less fertilizer than fresh sawdust. Also, *well-rotted* manure (fresh manure will have weed seeds) can be used in place of the peat moss. A surface mulch of peat moss at seeding time may discourage weeds, reduce soil erosion, and keep the surface from drying rapidly.

Before Planting

Serious weeds that cannot be controlled after turf is established should be eliminated before planting. Such methods may delay planting, but will reduce the work required after the turf is planted.

Weed control before seeding may include shallow cultivation after emergence of most annual weeds, or the use of herbicides. A contact spray such as paraquat or cacodylic acid after weed emergence is usually superior to shallow cultivation because ungerminated weed seeds are not brought to the surface where they will germinate. It is also possible to control emerged weeds with translocating herbicides; 2,4-D kills broadleaf weeds and dalapon controls grasses. Herbicides must be specific for the weeds involved.

Soil fumigants such as calcium cyanamide, dazomet, or metham not only control weeds but also many soil insects, diseases, and nematodes. Another soil fumigant, methyl bromide, can be used, but usually requires a professional applicator because it is highly toxic to man. Methyl bromide is the most effective for the control of nutsedge and bermudagrass.

The length of time that the chemical remains toxic in the soil will determine when the turf can be seeded, sprigged, or sodded. Trial seeding of a small area with a quick-germinating grass and legume will help determine freedom from herbicide toxicity (see Table 4-1).

At Planting

Siduron can be applied at seeding of cool-season turf (bluegrass, ryegrass, fescue, and some bent grasses) to control annual grasses. Diphenamid can be used at planting of dichondra for control of annual grasses and some broadleaf weeds.

After Planting

Postplanting control is needed after turfgrass species have emerged, but before they become well-established. This can be done by hand weeding, mowing, use of herbicides, or a combination. Mowing of newly planted turf will control some erect broadleaf weeds, but is not effective on grasses or prostrate broadleaf weeds. While the lawn is young, mowing height should be kept high so that a minimum of foliage of the turf species is removed.

Herbicides must be used with care on young turfgrass species to avoid injury. Uniform distribution and rate of application are essential. Many emerged broadleaf weeds can be controlled by low rates of bromoxynil,

Figure 30-1. Crabgrass control in dichondra with bensulide. Left: untreated. (C. L. Elmore, University of California, Davis.)

2,4-D, silvex, or dicamba once the grass seedlings have reached the three-to four-leaf stage. At this stage, dosage rates should be cut to about one-fourth to one-half of those recommended for established turf (2). As the grass becomes older these rates can be gradually increased.

Siduron selectively controls germinating, annual-weed grasses for about 1 month; for example, crabgrass, foxtail, and barnyardgrass in both newly seeded and established bluegrass, bentgrass, fescue, redtop, and ryegrass turf.

In fall plantings of turf species an early spring application of several herbicides has controlled several annual weedy grasses (2). Wait at least until the turf species reach the three- to four-leaf stage before applying any herbicides. They include bensulide, DCPA, siduron, and terbutol. See Figure 30-1.

ESTABLISHED TURF

With good management practices, weeds in established turf often can be controlled by hand pulling or cutting the occasional weed out of a home lawn.

Since an established turf tolerates herbicides much better than a new planting, many other herbicides may be used. A turf is established when the grasses have developed an extensive root system and are well tillered or when the rhizome (runner) system is well grown.

Herbicides used on established turf are either soil-active or foliar-active. Generally, soil-active compounds are applied about 2–3 weeks before germination of weed seeds to be controlled. They are preemergence-type herbicides. Some require immediate sprinkle irrigation to minimize foliar injury to desirable turf species. All must be leached into the soil within a few days after application by sprinkle irrigation or rainfall to be effective.

Foliar-active herbicides are applied to emerged weeds. Because it usually takes several hours or sometimes a few days for maximum absorption, they should not be applied if rain is expected or if sprinkle irrigation is to be applied during that time.

Table 30-1 shows herbicides available to control several common turf weeds in four grass species and dichondra. Because formulations of these herbicides available on the market vary considerably, and formulations have considerable influence on their activity, application rates are not given. The manufacturer's label gives the recommended rate. Many formulations contain two or more herbicides to increase the number of weed species controlled.

Nonselective general-contact sprays such as Stoddard solvent, cacodylic acid, and paraquat are especially effective for killing *all vegetation* in brick walks, along borders, or under woody ornamentals. For best results, weeds should be treated when 1–2 in. tall. Be careful not to saturate the trunks of woody ornamentals, because saturation may girdle the plant. Wet the plants to be killed. In some cases, it is desirable to add to each gallon of varsol 1 liquid oz of concentrated pentachlorophenol (4 lb active ingredient/gal). This mixture is extremely toxic to green-plant material. With normal rates there is no hazard to shrubbery or tree roots which extend under the treated area.

ORNAMENTALS

Ornamentals include shade trees, shrubs, herbaceous annuals and perennials, bulbs, and nonturf groundcovers. Weeds in ornamentals present serious problems for nurserymen as well as professional and home gardeners. Weed control methods include *hand weeding, mulches, cultivation,* and *herbicides.* Selecting the most suitable method or methods of controlling weeds in ornamentals is a difficult task.

Hand weeding, although tedious, is often very useful for the home gardener; however, for the nurseryman or professional gardener it is usually too expensive. *Mulches* are commonly used quite effectively. To control most annual weeds, a mulch 2 or 3 in. thick is sufficient (2). Good mulching material includes wood or bark chips, sawdust, peatmoss, small grain straw free of weed seeds, pine needles and gravel or stones. An ideal mulch allows free passage of moisture and air, but smothers growth of young weeds and prevents germination of seeds that require light. *Cultivation* is used widely by the nurseryman, somewhat by the professional gardener, but little by the home gardener.

Herbicides can be used for weed control in ornamentals. However, because of limited information on selectivity of specific herbicides to the hundreds of species grown, we are prevented from making specific suggested uses here. Another obstacle—landscaping usually involves a mixture of species to obtain the desired affect and selectivity of a specific herbicide to these various species may be different.

The nurseryman often can use a variety of herbicides to control weeds in his plantings because he has relatively large areas planted to a single species. However, the professional gardener must use a herbicide that is selective to the several species that may be present within a given area. The home gardener's selection of herbicides for weed control in ornamentals is even more restricted because he deals with small areas and a great variety of species.

In spite of these limitations, there are several herbicides that are fairly selective to most *woody* ornamentals and are widely used. These include trifluralin, DCPA, oryzalin, nitralin, diphenamid, dichlobenil, simazine, chlorpropham, chloramben, and nitrofen.

To help select proper herbicides for weed control in ornamentals, ask your local county extension agent or other public agencies, local landscape architects, and read the herbicide manufacturer's literature and labels. However,

Table 30-1. Herbicide Selection Guide for Turf[1,2]

Weeds to be Controlled in the given Turf Species	Bentgrass	Bermudagrass	Bluegrass, Kentucky	Fescue, Fine-Leaf	Dichondra
Barnyardgrass *Echinochloa crus-galli*	Bensulide	Benefin Bensulide DCPA	Benefin Bensulide DCPA	Benefin Bensulide DCPA	Bensulide Diphenamid
Bermudagrass *Cynodon dactylon*	—	—	—	—	Dalapon
Bindweed, field *Convolvulus arvensis*	2,4-D amine Dicamba	2,4-D amine Dicamba	2,4-D amine Dicamba	2,4-D amine Dicamba	—
Bluegrass, annual *Poa annua*	Bensulide	Benefin Bensulide Terbutol	Benefin Bensulide Terbutol	Benefin Bensulide	Bensulide Diphenamid
Burclover, California *Medicago polymorpha* var. *vulgaris*	Dicamba MCPP[3]	Dicamba MCPP	Dicamba MCPP Silvex	Dicamba MCPP	Monuron
Catsear, spotted *Hypochaeris radicata*	2,4-D amine	2,4-D amine	2,4-D amine	2,4-D amine	—
Chickweed, mouseear *Cerastium vulgatum*	Dicamba MCPP[3]	Dicamba MCPP	Dicamba MCPP Silvex	Dicamba MCPP	—
Chickweed, Common *Stellaria media*	Dicamba	Dicamba	Dicamba	Dicamba	Monuron Diphenamid

Appendix

Table A-1. Alphabetical Listing of Herbicides by Common Name and Corresponding Trade Name or Names Used in the United States

Common Name	Trade Name	Common Name	Trade Name
Acrolein	Aqualin	Chloroxuron	Norex, Tenoran
Alachlor	Lasso	Chlorpropham	Furloe
Ametryn	Evik	Cyanazine	Bladex
Amitrole	Amino triazole Weedazol	Cycloate	Ro-Neet
		Cyprazine	Outfox
AMS	Ammate	2,4-D	Numerous names
Asulam	Asulox	Dalapon	Dowpon
Atrazine	AAtrex	Dazomet	Mylone, Micro-fume
Barban	Carbyne	2,4-DB	Butoxone, Butyrac
Benefin	Balan, Quilan	DCPA	Dacthal
Bensulide	Betasan, Prefar	2,4-DEP	Falone
Bentazon	Basagran	Desmedipham	Bethanol-475
Benthiocarb	Saturn	Diallate	Avadex
Benzadox	Topcide	Dicamba	Banvel
Bifenox	Modown	Dichlobenil	Casoron
Bromacil	Hyvar	Dichlorprop	Weedone 2,4-DP
Bromoxynil	Buctril	Dinitramine	Cobex
Butachlor	Machete	Dinoseb	Numerous names
Butralin	Amex-820	Diphenamid	Dymid, Enide
Butylate	Sutan	Dipropetryn	Sancap
Cacodylic acid	Phytar, Rad-E-Cate	Diquat	Ortho Diquat
Carbetamide	Legurame	Diuron	Karmex
CDAA	Randox	DSMA	Ansar, Weed-E-Rad
CDEC	Vegadex	Endothall	Numerous names
Chloramben	Amiben, Vegiben	EPTC	Eptam
Chlorbromuron	Bromex, Maloran	Erbon	Erbon

409

Table A-1 (continued)

Common Name	Trade Name	Common Name	Trade Name
Fenac	Fenac	Penoxalin	Prowl
Fenuron	Dybar	Phenmedipham	Betanal
Fluchloralin	Basalin	Picloram	Tordon
Fluometuron	Cotoran, Lanex	Potassium azide	Kazoe
Fluorodifen	Preforan, Soyex	Profluralin	Tolban
Glyphosate	Roundup	Prometon	Pramitol
Hexaflurate	Nopalmate	Prometryn	Caparol
		Pronamide	Kerb
Isopropalin	Paarlan	Propachlor	Ramrod
Karbutilate	Tandex	Propanil	Stam
Linuron	Lorox, Lexone	Propazine	Milograd
MAA	Numerous names	Propham	Chem Hoe
MCPA	Numerous names	Prynachlor	Basamaize
MCPB	Can-Trol, Thistrol	Pyrazon	Pyramin
Metham	Vapam, VPM	Siduron	Tupersan
Methazole	Probe	Silvex	Kuron,
Metribuzin	Sencor		Weedone 2,4,5-TP
MH	Numerous names	Simazine	Princep
Molinate	Ordram	Sodium azide	Smite
Monuron	Telvar	Sodium boroate	Numerous names
Monuron TCA	Urox	Sodium chlorate	Sodium Chlorate
MSMA	Numerous names	2,4,5-T	Numerous names
Napropamide	Devrinol	2,3,6-TBA	Numerous names
Naptalam	Alanap	TCA	Sodium TCA
Nitralin	Planavin	Tebuthiuron	Spike, Perflan
Nitrofen	TOK	Terbacil	Sinbar
Norea	Herban	Terbutol	Azak
Norflurazon	Zorial	Terbutryn	Igran
Oryzalin	Surflan, Dirimal	Triallate	Avadex BW,
Oxadiazon	Ronstar		Far-Go
Paraquat	Ortho Paraquat	Trifluralin	Treflan, Elancolan
Pebulate	Tillam	Vernolate	Vernam

410

Table A-2. Alphabetical Listing of Herbicides by Trade Name Used in the United States and Corresponding Common Name

Trade Name	Common Name	Trade Name	Common Name
AAtrex	Atrazine	Falone	2,4-DEP
Amex-820	Butralin	Fenac	Fenac
Alanap	Naptalam	Furloe	Chlorpropham
Amiben	Chloramben	Herban	Norea
Amino Triazole	Amitrole	Hyvar	Bromacil
Ammate	AMS	Igran	Terbutryn
Ansar	DSMA	Karmex	Diuron
Aqualin	Acrolein	Kazoe	Potassium Azide
Asulox	Asulam	Kerb	Pronamide
Avadex	Diallate	Kuron	Silvex
Avadex BW	Triallate	Lanex	Fluometuron
Azak	Terbutol	Lasso	Alachlor
Balan	Benefin	Legurame	Carbetamide
Banvel	Dicamba	Lorox	Linuron
Basagran	Bentazon	Maloran	Chlorbromuron
Basalin	Fluchloralin	Machete	Butachlor
Basamaize	Prynachlor	Milograd	Propazine
Betanal	Phenmedipham	Modown	Bifenox
Betasan	Bensulide	Nopalmate	Hexaflurate
Bethanol 475	Desmedipham	Norex	Chloroxuron
Bladex	Cyanazine	Ordram	Molinate
Bromex	Chlorbromuron	Ortho Diquat	Diquat
Buctril	Bromoxynil	Ortho Paraquat	Paraquat
Butoxone	2,4-DB	Outfox	Cyprazine
Butyrac	2,4-DB	Paarlan	Isopropalin
Can-Trol	MCPB	Perflan	Tebuthiuron
Caparol	Prometryn	Phytar	Cacodylic acid
Carbyne	Barban	Planavin	Nitralin
Casoron	Dichlobenil	Prefar	Bensulide
Chem Hoe	Propham	Preforan	Fluorodifen
Cobex	Dinitramine	Primitol	Prometon
Cotoran	Fluometuron	Princep	Simazine
Dacthal	DCPA	Probe	Methazole
Devrinol	Napropamide	Prowl	Penoxalin
Dirimal	Oryzalin	Pyramin	Pyrazon
Dowpon	Dalapon	Rad-E-Cate	Cacodylic acid
Dymid	Diphenamid	Ramrod	Propachlor
Enide	Diphenamid	Randox	CDAA
Eptam	EPTC	Ro-Neet	Cycloate
Erbon	Erbon	Ronstar	Oxadazon
Evik	Ametryn		

411

Table A-2 (continued)

Trade Name	Common Name	Trade Name	Common Name
Roundup	Glyphosate	TOK	Nitrofen
Sancap	Dipropetryn	Tolban	Profluralin
Saturn	Benthiocarb	Topcide	Benzadox
Sencor	Metribuzin	Tordon	Picloram
Sinbar	Terbacil	Treflan	Trifluralin
Smite	Sodium azide	Tupersan	Siduron
Sodium TCA	TCA	Urab	Fenuron TCA
Soyex	Fluorodifen	Urox	Monuron TCA
Spike	Tebuthiuron	Vapam	Metham
Stam	Propanil	Vegadex	CDEC
Surflan	Oryzalin	Vernam	Vernolate
Sutan	Butylate	VPM	Metham
Tandex	Karbutilate	Weedazol	Amitrole
Telvar	Monuron	Weed-E-Rad	DSMA
Tenoran	Chloroxuron	Weedone 2,4-DP	Dichlorprop
Thistrol	MCPB	Weedone 2,4,5-TP	Silvex
Tillam	Pebulate	Zorial	Norflurazon

Table A-3. Susceptibility of Some Common Annual Weeds to Several Selective Herbicides at Usual Selective Rate[1]

	Alachlor	Atrazine	Chloramben	Chlorpropham	2,4-D	Diuron	EPTC	Linuron	Trifluralin
Barnyardgrass	S	I	S	I	T	S	S	S	S
Chickweed		S		S	I	S	S	S	S
Cocklebur	T	S	I	T	S	I		S	T
Crabgrasses	S	T	S	I	T	S	S	S	S
Fall panicum	S	I	I	I	T	S	S	I	S
Foxtails	S	S	S	I	T	S	S	S	S
Jimsonweed	I	S	I	I	S			S	T
Johnsongrass seedlings	I	T	I	T	T	I	S	I	S
Knotweed	T	S		S	I	S	S	I	S
Lambsquarters	I	S	S	T	S	S	S	S	S
Morningglory annual	T	S	T	I	S	I	S	S	I
Mustards	I	S	I		S	S	T	S	T
Nightshades	I	S	S		I		I	S	I
Nutsedges	I	I	T	T	I		I	T	T
Pigweeds	S	S	S	I	S	S	I	S	S
Purslane	S	S	S	S	I	S	S	S	S
Ragweed	I	S	S	T	S	S	I	S	I
Smartweed	I	S	S	S	I			S	I
Velvetleaf	T	S	I	T	I	I	I	S	T
Wild cucumber	T	S	I	T	I			S	I

[1] S—susceptible; I—intermediate; and T—tolerant.

413

Liquid Measure

1 gallon (U.S.) = 3785.4 milliliters (ml); 256 tablespoons; 231 cubic inches; 128 fluid ounces; 16 cups; 8 pints; 4 quarts; 0.8333 imperial gallon; 0.1337 cubic foot
    1 liter = 1000 milliliters; 1.0567 liquid quarts (U.S.)
    1 gill = 118.29 milliliters
    1 fluid ounce = 29.57 milliliters; 2 tablespoons
    3 teaspoons = 1 tablespoon; 14.79 milliliters
    1 gallon of water = 8.355 pounds; 1 cubic foot of water = 62.43 pounds; 7.48 gallons

Weight

    1 gamma = 0.001 milligram (mg)
    1 grain (gr) = 64.799 milligrams
    1 gram (g) = 1000 milligrams; 15.432 grains; 0.0353 ounce
    1 pound = 16 ounces; 7000 grains; 453.59 grams
    1 short ton = 2000 pounds
    1 long ton = 2240 pounds

Linear Measure

    12 inches = 1 foot
    36 inches = 3 feet; 1 yard
    1 rod = 16.5 feet
    1 mile = 5280 feet; 1760 yards; 160 rods; 80 chains; 1.6094 kilometers (km)
    1 chain = 66 feet; 22 yards; 4 rods; 100 links
    1 inch = 2.54 centimeters (cm)
    1 meter = 39.37 inches; 10 decimeters (dm)
    1 micron (μ) = $\frac{1}{1000}$ millimeter (mm)

Area

    1 township = 36 sections; 23,040 acres
    1 square mile = 1 section; 640 acres
    1 acre = 43,560 square feet; 160 square rods; 4840 square yards; 208.7 feet square; an area $16\frac{1}{2}$ feet wide and $\frac{1}{2}$ mile long
    1 hectare = 2.471 acres

Capacity (Dry Measure)

1 bushel (U.S.) = 4 pecks; 32 quarts; 35.24 liters; 1.244 cubic feet; 2150.42 cubic inches

Pressure

1 foot lift of water = 0.433 pound pressure per square inch (psi)
1 pound pressure per square inch will lift water 2.31 feet
1 atmosphere = 760 millimeters of mercury; 14.7 pounds; 33.9 feet of water

Geometric Factors ($\pi = 3.1416$; r = radius; d = diameter; h = height)

Circumference of a circle = $2\pi r$ or πd
Diameter of a circle = $2r$
Area of a circle = πr^2 or $\frac{1}{4}\pi d^2$ or $0.7854d^2$
Volume of a cylinder = $\pi r^2 h$

Weight of Dry Soil

Type	Per Cubic Foot (lb)	Per acre, 7″ deep (lb)
Sand	100	2,500,000
Loam	80–95	2,000,000
Clay or silt	65–80	1,500,000
Muck	40	1,000,000
Peat	20	500,000

Temperature, degrees

$$F° = C° + 17.78 \times 1.8 \qquad 1°C = 1.8°F$$
$$C° = F° - 32.00 \times \tfrac{5}{9} \qquad 1°F = \tfrac{5}{9}°C$$

°C	°F	°C	°F
100	212	30	86
90	194	20	68
80	176	10	50
70	158	0	32
60	140	−10	14
50	122	−20	−4
40	104	−30	−22

Length of Row Required for One Acre

Row Spacing	Length or Distance
24 in.	7260 yards = 21,780 ft
30 in.	5808 yards = 17,424 ft
36 in.	4840 yards = 14,520 ft
42 in.	4149 yards = 12,445 ft
48 in.	3630 yards = 10,890 ft

Nozzle Capacities in Gallons Per Hour Required for Various Rates of Application

Rate of Application

Nozzle Spacing (In.)	7½ Gal/Acre			10 Gal/Acre			12½ Gal/Acre			35 Gal/Acre		
	3 mph	4 mph	5 mph	3 mph	4 mph	5 mph	3 mph	4 mph	5 mph	3 mph	4 mph	5 mph
6	1.4	1.8	2.3	1.8	2.9	3.0	2.3	3.0	3.8	6.3	8.5	10.6
8	1.8	2.4	3.0	2.4	3.2	4.0	3.0	4.0	5.0	8.5	10.3	14.2
10	2.3	3.0	3.8	3.0	4.0	5.0	3.8	5.0	6.3	10.6	14.1	17.6
12	2.7	3.6	4.5	3.6	4.8	6.0	4.5	6.0	7.6	12.7	16.9	21.2
14	3.2	4.2	5.3	4.2	5.7	7.0	5.3	7.0	8.8	14.8	19.8	24.7
16	3.6	4.8	6.0	4.8	6.5	8.1	6.0	8.1	10.1	16.9	22.6	28.3
18	4.1	5.4	6.8	5.4	7.3	9.1	6.8	9.1	11.3	19.0	25.4	31.7
20[1]	4.5	6.0	7.5	6.0	8.1[1]	10.0	7.5	10.0	12.6	21.2	28.2	35.3
21	4.8	6.3	7.9	6.3	8.5	10.6	7.9	10.6	13.2	22.2	29.8	37.0
22	5.0	6.6	8.3	6.6	8.9	11.1	8.3	11.1	13.9	23.3	31.1	38.8
24	5.4	7.2	9.0	7.2	9.7	12.1	9.0	12.1	15.1	25.4	33.9	42.4
30	6.8	9.1	11.3	9.1	12.1	15.1	11.3	15.1	18.8	31.7	42.3	52.8
36	8.1	10.9	13.6	10.8	14.5	18.1	13.6	18.1	22.6	38.0	50.7	63.5
42	9.5	12.7	15.8	12.7	16.9	21.2	15.8	21.2	26.4	44.3	59.3	74.1
48	10.9	14.5	18.1	14.5	19.3	24.2	18.1	24.2	30.1	50.7	67.8	84.6

[1] Example: If a spray application rate of 10 gal/acre is desired, at 4 mph, and the sprayer has a boom with nozzles 20 in. apart: From the table above, a nozzle with a capacity of 8.1 gal/hr at the desired pressure would be required.

Width of Spray Pattern at Various Distances from Nozzle—Inches

Spray Angle Degrees	DISTANCE FROM NOZZLE ORIFICE													
	2"	4"	6"	8"	10"	12"	15"	18"	24"	30"	36"	42"	48"	60"
15°	.5	1.1	1.6	2.1	2.6	3.2	3.9	4.7	6.3	7.9	9.5	11.1	12.6	15.8
25°	.9	1.7	2.7	3.5	4.4	5.3	6.6	8.0	10.6	13.3	15.9	18.6	21.2	26.6
30°	1.1	2.1	3.2	4.3	5.4	6.4	8.0	9.7	12.8	16.0	19.3	22.4	25.9	32.0
40°	1.4	2.9	4.3	5.8	7.2	8.7	10.9	13.0	17.4	21.6	26.2	30.6	34.9	42.8
45°	1.7	3.3	4.9	6.6	8.2	9.9	12.4	14.9	19.8	24.8	29.8	34.8	39.7	49.6
50°	1.9	3.6	5.6	7.4	9.3	11.2	14.0	16.8	22.4	28.0	33.6	39.1	44.8	56.0
60°	2.3	4.6	6.9	9.2	11.4	13.9	17.3	20.8	27.6	34.6	41.6	48.4	55.4	69.2
65°	2.5	5.1	7.6	10.2	12.7	15.2	19.1	22.9	30.5	38.1	45.8	53.2	61.0	76.4
70°	2.8	5.6	8.2	11.2	14.0	16.8	21.0	25.2	33.6	42.0	50.4	59.8	67.2	84.0
73°	2.9	5.9	8.8	11.8	14.8	17.6	22.2	26.6	36.4	44.4	53.2	62.0	71.0	88.5
75°	3.1	6.1	9.2	12.3	15.3	18.4	23.0	27.6	36.8	46.0	55.2	64.2	73.5	92.0
80°	3.4	6.7	10.1	13.4	16.8	20.1	25.2	30.2	42.0	50.2	60.4	72.5	80.8	100.0
90°	4.0	8.0	12.0	16.0	20.0	24.0	30.0	36.0	48.0	60.0	72.0	84.0	96.0	120.0
100°	4.8	9.5	14.3	19.1	23.8	28.6	35.8	42.4	57.2	71.4	86.0	100.0	114.6	143.0
120°	6.9	13.9	20.8	27.8	34.7	41.6	52.0	62.4	83.0	104.0	125.0	145.8	166.2	208.0

Pressure Drop in Hose Due to Friction. These Figures Are for 25 Feet of Clean Hose with No Couplings

¼" ID HOSE		⅜" ID HOSE		½" ID HOSE		⅝" ID HOSE		¾" ID HOSE		1" ID HOSE		1¼" ID HOSE	
Flow GPM	Pr. Drop PSI per 25 Ft.	Flow GPM	Pr. Drop PSI per 25 Ft.	Flow GPM	Pr. Drop PSI per 25 Ft.	Flow GPM	Pr. Drop PSI per 25 Ft.	Flow GPM	Pr. Drop PSI per 25 Ft.	Flow GPM	Pr. Drop PSI per 25 Ft.	Flow GPM	Pr. Drop PSI per 25 Ft.
.2	.8												
.3	1.5												
.4	2.5	.5	.5										
.5	4.0	.6	.8										
.6	5.0	.8	1.3										
.8	9.0	1.0	1.8	1.0	.5								
		2.0	6.0	2.0	1.5								
		3.0	13.0	3.0	3.1	3.0	1.0						
				4.0	6.0	4.0	1.8						
				5.0	8.5	5.0	2.5	5.0	1.0				
				6.0	12.0	6.0	3.7	6.0	1.5				
						8.0	6.5	8.0	2.5	8.0	.6		
						10.0	9.5	10.0	3.7	10.0	1.0		
								15.0	8.0	15.0	2.0	15.0	.7
								20.0	14.0	20.0	3.4	20.0	1.2
										25.0	5.0	25.0	1.8
										30.0	6.5	30.0	2.5
										40.0	12.0	40.0	4.4
												50.0	6.0
												60.0	9.0
												70.0	13.0

Table of Available Commercial Materials in Pounds Active Ingredients per Gallon Necessary to Make Various Percentage Concentration Solutions[1]

Pounds of active ingredient in 1 gal of commercial product	Pounds of active ingredient/pint[1]	Liquid ounces of commercial product/1 gal of solution[1] to make:				
		$\frac{1}{2}\%$	1%	2%	5%	10%
		liquid oz	liquid oz	liquid oz	liquid oz	liquid oz
2.00	0.25	2.68	5.36	10.72	26.80	53.60
2.64	0.33	2.02	4.05	8.10	20.25	40.50
3.00	0.375	1.78	3.56	7.12	17.80	35.60
3.34	0.42	1.59	3.18	6.36	15.90	31.80
4.00	0.50	1.34	2.68	5.36	13.40	26.80
6.00	0.75	0.89	1.78	3.56	8.90	17.80

[1] Based on 8.4 lb/gal (weight of water) and 128 liquid oz = 1 gal, 16 liquid oz = 1 pint.

Equivalent Quantities of Liquid Materials When Mixed by Parts

Water	1–400	1–800[1]	1–1600
100 gal	1 qt	1 pt	1 cup
50 gal	1 pt	1 cup	$\frac{1}{2}$ cup
5 gal	3 tbs	5 tsp[1]	$2\frac{1}{2}$ tsp
1 gal	2 tsp	1 tsp	$\frac{1}{2}$ tsp

[1] Example: If a recommendation calls for 1 part of the chemical to 800 parts of water, it would take 5 tsp in 5 gal of water to give 5 gal of a mixture of 1–800.

Equivalent Quantities of Dry Materials (Wettable Powders) for Various Quantities of Water

Water	Quantity of Material					
100 gals[1]	1 lb	2 lb	3 lb	4 lb[1]	5 lb	6 lb
50 gals	8 oz	1 lb	24 oz	2 lb	$2\frac{1}{2}$ lb	3 lb
5 gals[1]	3 tbs	$1\frac{1}{2}$ oz	$2\frac{1}{2}$ oz	$3\frac{1}{4}$ oz[1]	4 oz	5 oz
1 gal	2 tsp	3 tsp	$1\frac{1}{2}$ tbs	2 tbs	3 tbs	3 tbs

[1] Example: If a recommendation calls for a mixture of 4 lb of a wettable powder to 100 gals of water, it would take $3\frac{1}{4}$ oz (approximately 12 tsp) to 5 gals of water to give 5 gals of spray mixture of the same strength.

Note: Wettable materials vary considerably in density. Therefore the teaspoonful (tsp) and tablespoonful (tbs) measurements in this table are not exact dosages by weight but are within the bounds of safety and efficiency for mixing small amounts of spray.

Index